Digital Control
of Dynamic Systems

GENE F. FRANKLIN
STANFORD UNIVERSITY

J. DAVID POWELL
STANFORD UNIVERSITY

ADDISON-WESLEY PUBLISHING COMPANY

READING, MASSACHUSETTS
MENLO PARK, CALIFORNIA
LONDON · AMSTERDAM
DON MILLS, ONTARIO
SYDNEY

Richard C. Dorf, Consulting Editor

Library of Congress Cataloging in Publication Data

Franklin, Gene F
    Digital control of dynamic systems.

    Includes bibliographical references and index.
    1. Digital control systems. 2. Dynamics.
I. Powell, J. David, 1938–   joint author.
II. Title.
TJ216.F72   1980   629.8'95   79-16377
ISBN 0-201-02891-3

Second printing, June 1981

ISBN 0-201-02891-3
ABCDEFGHIJ-MA-8987654321

To
Gertrude, David, Carole
and
Lorna, Daisy, Annika

Preface

This is a book about the use of digital computers in the real-time control of dynamic systems such as servomechanisms, chemical processes, and vehicles which move over water, land, air, or space. The material requires some understanding of the Laplace transform and assumes the reader has studied a first course in linear feedback controls. The special topics of discrete and sampled-data system analysis are introduced, and considerable emphasis is given to the z-transform and the close connections between the z-transform and the Laplace transform.

The emphasis of the book is on the design of digital controls to achieve good dynamic response and small errors while using signals that are sampled in time and quantized in amplitude. Both transform (classical control) and state-space (modern control) methods are described and applied to illustrative examples. The transform methods emphasized are the root-locus method of Evans and the log-magnitude and phase–versus–log-frequency method of Bode; to aid in the use of Bode's method, the w-plane is introduced. The state-space methods developed are the technique of pole assignment augmented by an estimator (observer) with feed-forward or zero assignment included, and optimal quadratic-loss control. On the latter topic, the emphasis is on the steady-state constant-gain solution; the results of the separation theorem in the presence of noise are stated but not proved. The topic of model making is treated via statistical identification of parameters by least squares and maximum likelihood.

The material in the book which is new to the student is the treatment of signals which are discrete in time and amplitude and which must coexist with those that are continuous in both dimensions. The philosophy of presentation is that the new material should be closely related to material already familiar, and yet, by the end, a direction to wider horizons should be indicated. This approach leads us, for example, to relate the z-transform to the Laplace transform and to describe the implications of poles and zeros in the z-plane to those known meanings attached to poles and zeros in the s-plane. Also, in developing the design methods we relate the digital control design methods to those of continuous systems. And yet, pointing to more sophisticated methods, we present the elementary parts of quadratic-loss gaussian design with minimal proofs to give some idea of the use of this powerful method and to motivate the study of its theory more thoroughly later. The subject matter is particularly suitable for treatment in a laboratory setting and algorithms suitable for programming on a laboratory computer are frequently given.

To review the chapters briefly, the methods of linear analysis are presented in Chapters 1 to 4. Here are introduced the z-transform in Chapter 2, methods to generate discrete equations which will approximate continuous dynamics in Chapter 3, and combined discrete and continuous systems in Chapter 4. This last chapter introduces the sampling theorem and the phenomenon of aliasing. The basic deterministic design methods are presented in Chapters 5 and 6, the root-locus and Bode methods in Chapter 5, and pole placement and estimators in Chapter 6. The state-space material assumes no previous acquaintance with the phase plane or state space, and the necessary analysis is developed from the beginning. Some familiarity with simultaneous linear equations and matrix notation is expected, and a few unusual or more advanced topics such as eigenvalues, eigenvectors, and the Cayley-Hamilton theorem are presented in Appendix C. In Chapter 7 the nonlinear phenomenon of amplitude quantization and its effects on system error and system dynamic response are studied. These first seven chapters comprise the syllabus of a ten-week first course on digital control.

Chapter 8 introduces parametric identification by starting with deterministic least squares, introducing random errors, and completing with an algorithm for maximum likelihood. In Chapter 9 is introduced optimal quadratic loss control; first the control by state feedback is presented and then the estimation of the state in the presence of system and measurement noise is developed, based on the recursive least-squares estimation derived in Chapter 8. The final chapter, Chapter 10, presents methods of analysis and design guidelines for the selection of the sampling period in a digital control system. In such a work, the selection of notation is always critical to be sure our symbols aid rather than interfere with learning. We list at the front of the book a glossary of terms which we use and which we commend to those teachers who use this book.

At Stanford University, two courses based on this material are given. The first course covers Chapters 1 through 7 and follows a course in linear control which may have used Dorf (1980) or Ogata (1970). The second course covers Chapters 8 through 10. Both courses are heavily dependent on laboratory work as a supplement to the lectures for learning. A very satisfactory complement of laboratory equipment is a digital computer capable of running BASIC and having an A/D and a D/A converter, an analog computer with ten operational amplifiers, and a strip recorder.

As do all authors of technical works, we must acknowledge the vast array of contributors on whose work our own presentation is based. The list of references gives some small measure of those to whom we are in debt.   On a more personal level, we wish to express our appreciation to those responsible for making Stanford an exciting place to work and to those students of E.207 and E.208 for whom this book was written.   We hope its publication will contribute to the education of their successors at Stanford and elsewhere.

We would like especially to express our appreciation to Judy Clark, who aided us in so many ways in the preparation of the notes which became the manuscript which became this book.

Stanford, California   G. F. F.
January 1980   J. D. P.

Contents

Glossaries

CONTROL GLOSSARY

Plant

Continuous case:

$$\dot{\mathbf{x}} = \mathbf{F}\mathbf{x} + \mathbf{G}\mathbf{u}(t - \lambda) + \mathbf{G}_1\mathbf{w}$$

$$\lambda = \text{pure time delay}$$

Discrete:

$$\mathbf{x}_{k+1} = \mathbf{\Phi}\mathbf{x}_k + \mathbf{\Gamma}\mathbf{u}_k + \mathbf{\Gamma}_1\mathbf{w}_k$$

$$\mathbf{x} = \text{state} = N_s \times 1 \text{ or } n \times 1$$

$$\mathbf{u} = \text{control} = N_c \times 1 \text{ or } m \times 1$$

$$\mathbf{w} = \text{input disturbance or plant noise } N_w \times 1$$

$$\mathbf{F} = \text{continuous system matrix}$$

$$\mathbf{\Phi} = \text{discrete system matrix}$$

$$\mathbf{G} = \text{continuous-control input matrix}$$

$$\mathbf{\Gamma} = \text{discrete-control input matrix}$$

$$\mathbf{G}_1 = \text{plant-noise input matrix}$$

$$\mathbf{\Gamma}_1 = \text{discrete plant-noise matrix}$$

$$\overline{\mathbf{w}} = \mathscr{E}\mathbf{w} = \text{average value of plant noise}$$

$$\mathbf{R_w} = \text{plant-noise spectral density matrix}$$

$$\mathscr{E}(\mathbf{w} - \overline{\mathbf{w}})(\mathbf{w} - \overline{\mathbf{w}})^T = \mathbf{R_w}\delta(t) \quad \text{continuous}$$
$$= \mathbf{R_w} \quad \text{discrete}$$

$$\lambda_i(\mathbf{F}) = p_i = \text{open-loop poles} = \text{``}x\text{'' on root locus}$$

Plant output or sensor equations

Continuous system:
$$\mathbf{y} = \mathbf{Hx} + \mathbf{Ju} + \mathbf{v}$$

Discrete system:
$$\mathbf{y} = \mathbf{H}_d\mathbf{x} + \mathbf{J}_d\mathbf{u} + \mathbf{v}$$

$\mathbf{y}$ = output measurements = $N_0 \times 1$ or $p \times 1$

$\mathbf{v}$ = output noise or disturbance = $N_0 \times 1$ or $p \times 1$

$\mathbf{H}$ = continuous output matrix

$\mathbf{J}$ = continuous-plant direct transmission matrix

$\mathbf{H}_d$ = discrete-system output matrix

$\mathbf{J}_d$ = discrete-plant direct-transmission matrix

$\mathbf{v}_{\bar{v}} = \mathscr{E}\mathbf{v}$ = average value of output noise = sensor bias

$\mathbf{R}_v$ = measurement-noise spectral density matrix

$\mathscr{E}(\mathbf{v} - \mathbf{v}_{\bar{v}})(\mathbf{v} - \mathbf{v}_{\bar{v}})^T = \mathbf{R}_v\delta(t)$ continuous
$$= \mathbf{R}_v \qquad \text{discrete}$$

Control law
$$\mathbf{u} = -\mathbf{Kx} \qquad \text{or} \qquad \mathbf{u} = -\mathbf{K\hat{x}}$$

Control characteristic polynomial
$$\alpha_c(s) \qquad \text{or} \qquad \alpha_c(z)$$

$\lambda_i(\mathbf{F} - \mathbf{GK}) = r_i$ = roots of closed-loop characteristic equation

("Δ" on root locus)

Controllability matrix $(n = N_s)$
$$\mathscr{C} = [\mathbf{G} \quad \mathbf{FG} \cdots \mathbf{F}^{n-1}\mathbf{G}] \qquad \text{or} \qquad [\mathbf{\Gamma} \quad \mathbf{\Phi\Gamma} \cdots \mathbf{\Phi}^{n-1}\mathbf{\Gamma}]$$

Estimator/observer

Continuous:
$$\mathbf{\dot{\hat{x}}} = \mathbf{F\hat{x}} + \mathbf{Gu} + \mathbf{L}(y - \hat{y})$$
$$\mathbf{\hat{y}} = \mathbf{H\hat{x}} + \mathbf{Ju}$$

Discrete:

One-step prediction
$$\mathbf{\hat{x}}_{k+1} = \mathbf{\Phi\hat{x}}_k + \mathbf{\Gamma u}_k + \mathbf{L}(\mathbf{y}_k - \mathbf{\hat{y}}_k)$$
$$\hat{y}_k = \mathbf{H}_d\mathbf{\hat{x}}_k + \mathbf{J}_d\mathbf{u}_k$$

Current estimator
$$\mathbf{\bar{x}}_{k+1} = \mathbf{\Phi\hat{x}}_k + \mathbf{\Gamma u}_k \qquad \text{Time update}$$
$$\mathbf{\hat{x}}_{k+1} = \mathbf{\bar{x}}_{k+1} + \mathbf{L}(\mathbf{y}_{k+1} - \mathbf{\bar{y}}_{k+1}) \quad \text{Observation update}$$

$$\bar{\mathbf{y}}_{k+1} = \mathbf{H}_d\bar{\mathbf{x}}_{k+1} \qquad (\mathbf{J}_d = 0)$$

$$\mathbf{L} = \text{estimator gain matrix}$$

$$\mathbf{P} = \mathscr{E}(\mathbf{x} - \hat{\mathbf{x}})(\mathbf{x} - \hat{\mathbf{x}})^T = \mathscr{E}\tilde{\mathbf{x}}\tilde{\mathbf{x}}^T = \text{state error covariance matrix}$$

$$\mathbf{R}_x = \mathscr{E}(\mathbf{x}\mathbf{x}^T)$$

$$\mathbf{R}_u = \mathscr{E}\mathbf{u}\mathbf{u}^T$$

Controller

Continuous:

$$\dot{\mathbf{x}}_c = \mathbf{A}\mathbf{x}_c + \mathbf{B}\mathbf{y} + \mathbf{M}\mathbf{r}$$
$$\mathbf{u} = \mathbf{C}\mathbf{x}_c + \mathbf{D}\mathbf{y} + \mathbf{N}\mathbf{r}$$

Discrete:

$$\mathbf{x}_c(k + 1) = \mathbf{A}\mathbf{x}_c(k) + \mathbf{B}\mathbf{y}(k) + \mathbf{M}\mathbf{r}(k)$$
$$\mathbf{u}(k) = \mathbf{C}\mathbf{x}_c(k) + \mathbf{D}\mathbf{y}(k) + \mathbf{N}\mathbf{r}(k)$$

$$\mathbf{x}_c = \text{controller state}$$

$$\mathbf{r} = \text{reference input} = N_0 \times 1$$

$$\mathbf{A} = \text{Controller system matrix}$$

$$\mathbf{B} = \text{Controller input distribution matrix}$$

$$\mathbf{C} = \text{Controller output matrix}$$

$$\mathbf{D} = \text{Controller direct-transmission matrix}$$

$$\mathbf{M} = \text{Controller reference-input distribution matrix}$$

$$\mathbf{N} = \text{Controller reference-input direct transmission matrix}$$

Optimal control

$$\mathscr{J} = \int_{t_0}^{t_f} l(\mathbf{x}, \mathbf{u}, t)\, dt + \psi(\mathbf{x}_f, t_f)$$

Quadratic loss

$$\mathscr{J} = \mathscr{E}\left\{ \int_{t_0}^{t_f} (\mathbf{x}^T\mathbf{Q}_1\mathbf{x} + \mathbf{u}^T\mathbf{Q}_2\mathbf{u})\, dt + \mathbf{x}_f^T\mathbf{Q}_0\mathbf{x}_f \right\}$$

Discrete quadratic loss

$$\mathscr{J} = \mathscr{E}\left\{ \sum_{k=j}^{N} (\mathbf{x}^T\mathbf{Q}_1\mathbf{x} + \mathbf{u}^T\mathbf{Q}_2\mathbf{u}) + \mathbf{x}_N^T\mathbf{Q}_0\mathbf{x}_N \right\}$$

ALPHABETICAL GLOSSARY

A	Controller system matrix
B	Controller input matrix

C	Controller output matrix
D	Controller direct matrix
F	Plant system matrix
G	Plant input matrix
$\mathbf{G}_1$	Plant disturbance input matrix
H	Plant output matrix
$\mathbf{H}_d$	Discrete plant output matrix
J	Plant direct matrix
$\mathbf{J}_d$	Discrete plant direct matrix
K	Control gain
L	Estimator gain
M	Controller-reference-input distribution matrix
N	Controller-reference-input direct matrix
P	$E\tilde{\mathbf{x}}\tilde{\mathbf{x}}^T$
$\mathbf{Q}_1, \mathbf{Q}_2, \mathbf{Q}_0$	Loss matrices, state, control, terminal, respectively
$\mathbf{R}_\mathbf{w}, \mathbf{R}_\mathbf{v}, \mathbf{R}_\mathbf{x}, \mathbf{R}_\mathbf{u}$	Spectral density matrices of $\mathbf{w}, \mathbf{v}$; covariance matrices of $\mathbf{x}, \mathbf{u}$
T	Sampling period
u	Control
v	Measurement noise
w	Plant or input noise
x	Plant state
$\mathbf{x}_c$	Controller state
y	Output
α_c, α_e	Control and estimator characteristic polynomials
Γ	Discrete plant control input matrix
$\mathbf{\Gamma}_1$	Discrete plant noise input matrix
λ	Plant delay time or transportation lag
Φ	Discrete plant system matrix

1 / Introduction

1.1 PROBLEM DEFINITION

The control of physical systems with a digital computer is becoming more and more common. Aircraft autopilots, mass transit vehicles, oil refineries, paper-making machines, and countless electromechanical servomechanisms are among the many existing examples. Furthermore, many new digital control applications are being stimulated by microprocessor technology including control of various aspects of automobiles and household appliances. Among the advantages of digital logic for control are the increased flexibility of the control programs and the decision-making or logic capability of digital systems which can be shared with the control function to meet other system requirements.

The digital controls studied in this book are for closed-loop (feedback) systems in which the dynamic response of the process being controlled is a major consideration in the design. A typical topology of the elementary type of system which will occupy most of our attention is sketched schematically in Fig. 1.1. This figure will help to define our basic notation and to introduce several features which distinguish digital controls from those implemented with analog devices. The process to be controlled is called the plant and may be any of the physical processes mentioned above whose satisfactory response requires control action.

By "satisfactory response" we mean that the plant output, $y(t)$, is to be forced to follow or track the reference input, $r(t)$, despite the presence of disturbance inputs to the plant [$w(t)$ in Fig. 1.1] and despite errors in the sensor [represented by $v(t)$ in Fig. 1.1]. It is also essential that the tracking succeed even if the dynamics of the plant should change somewhat during the operation. The process of holding $y(t)$ close to $r(t)$, including the case where $r \equiv 0$, is referred to generally as the process of *regulation*. A system which has good regulation in the presence of disturbance signals is said to have good *disturbance rejection*. A system which has good regulation in the face of changes in the plant parameters is said to have low *sensitivity* to these parameters. A system which has both good disturbance rejection and low sensitivity we call *robust*.

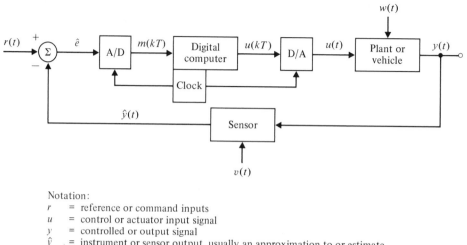

Notation:

r	=	reference or command inputs
u	=	control or actuator input signal
y	=	controlled or output signal
$\hat{y}$	=	instrument or sensor output, usually an approximation to or estimate of y. (For any variable, say θ, the notation $\hat{\theta}$ is now commonly taken from statistics to mean an estimate of θ).
$\hat{e}$	=	$r-\hat{y}$ = indicated error
e	=	$r-y$ = system error
w	=	disturbance input to the plant
v	=	disturbance or noise in the sensor
A/D	=	analog-to-digital converter
D/A	=	digital-to-analog converter

Fig. 1.1 Basic control system block diagram.

The means by which robust regulation is to be accomplished is through the control inputs to the plant [$u(t)$ in Fig. 1.1]. It was discovered long ago[1] that a scheme of feedback wherein the plant output is measured (or sensed) and compared directly with the reference input has many advantages in the effort to design robust controls over systems which do not use such feedback. Much of our effort in later parts of this book will be devoted to illustrating this discovery and demonstrating how to exploit the advantages of feedback. However, the problem of control as discussed thus far is in no way restricted to digital control. For that we must consider the unique features of Fig. 1.1 introduced by the use of a digital device to generate the control action.

We consider first the action of the analog-to-digital (A/D) converter on a signal. This device acts on a physical variable, most commonly an electrical voltage, and converts it into a stream of numbers. In Fig. 1.1, the A/D converter acts on the indicated error signal, $\hat{e}$, and supplies numbers to the digital computer. It is also common for the sensor output, $\hat{y}$, to be sampled and have the error formed in the computer. We need to know the times at which these numbers arrive if we are to analyze the dynamics of this system.

---

[1] See especially the book by Bode (1945).

In this book we will make the assumption that all the numbers arrive with the same fixed period T, called the *sampling period*. In practice, digital control systems sometimes have varying sample periods and/or different periods in different feedback paths. Typically there is a clock as part of the computer logic which supplies a pulse every T seconds, and the A/D converter sends a number to the computer each time the pulse arrives. Thus in Fig. 1.1 we identify the sequence of numbers into the computer as $m(kT)$. We conclude from the periodic sampling action of the A/D converter that some of the signals in the digital control system, like $m(kT)$, are variable only at discrete times. We call these variables *discrete signals* to distinguish them from variables like $\hat{e}$ and y which change continuously in time. A system containing only discrete variables is called a *discrete-time* system. A system having both discrete and continuous signals is called a *sampled-data* system.

In addition to generating a discrete signal, however, the A/D converter also provides a *quantized* signal. By this we mean that the output of the A/D converter must be stored in digital logic composed of a finite number of digits. Most commonly, of course, the logic is based on binary digits (i.e., bits) composed of 0's and 1's, but the essential feature is that the representation has a finite number of digits. A common situation is that the conversion of $\hat{e}$ to m is done so that m may be thought of as a number with a fixed number of places of accuracy. If we plot the values of $\hat{e}$ versus the resulting values of m we may obtain a plot like that shown in Fig. 1.2. We would say that m has been truncated to one decimal place, or that m is *quantized* with a q of 0.1 since m changes only in fixed quanta of, in this case, 0.1 units. (We will use q for quantum size, in general.) Note that quantization is a nonlinear function. A signal which is both discrete and quantized is called a *digital signal*. Not surprisingly, digital computers in this book process digital signals.

In a real sense the problems of analysis and design of *digital controls* are concerned with taking account of the effects of the sampling period T and the quanti-

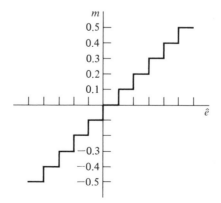

Fig. 1.2 Plot of output versus input characteristics of the A/D converter.

zation size q. If both T and q are small, digital signals are nearly continuous, and continuous methods of analysis and design can be used. We will be interested here in those cases which have significant values for both T and q. We will, however, treat the problems separately and first consider q to be zero and study discrete and sampled-data (combined discrete and continuous) systems which are linear. Later we will analyze in more detail the source and the effects of quantization[2] and discuss specific techniques for sample-rate selection.[3]

Our approach to the design of digital controls is to assume a background in continuous systems and to relate the comparable digital problem to its continuous counterpart. We will develop the essential results, from the beginning, in the domain of discrete systems, but we will call upon previous experience in continuous-system analysis and in design to give alternative viewpoints and deeper understanding of the results. In order to make these references to a background in continuous-system design meaningful, we will review the concepts and define our notation as required.

1.2 EXAMPLE SYSTEMS FOR STUDY

In order to guide the discussion in the following chapters we have developed models for six example control problems in Appendix A. The first of these is a satellite attitude control problem in which the plant transfer function is the double integrator

$$G_1(s) = 1/s^2. \tag{1.1}$$

This example is simple, but with two poles on the stability boundary, this transfer function must be controlled with care. The second example is a servomechanism, motivated by a radar tracking antenna. The basic control dynamics of the amplifier-motor-load system is second order and has a transfer function with one pole at the origin and one pole on the negative real axis. The transfer function can be normalized to be

$$G_2(s) = 1/s[(s + 1)]. \tag{1.2}$$

The third example to be used to illustrate digital control comes from the process industries—a mixing process. The normalized transfer function which represents this system is

$$G_3(s) = e^{-1.5s}/(s + 1). \tag{1.3}$$

The obvious feature of this model is the delay term, $e^{-1.5s}$, which is called a *transportation delay*. Such delays are very difficult to control, and the example affords us an opportunity to study these difficulties in a digital control context.

---

[2] In Chapter 7.
[3] In Chapter 10.

Also, since there is no internal integrator in this model, good steady-state accuracy requires integral control, which presents an interesting but common design problem.

In many electromechanical systems, there is a flexible mechanical member between the actuator and the position sensor. The effect of this flexibility is to introduce a complex pole into the transfer function with very small real part and thus very small damping. When normalized in amplitude and time and written to include the rigid body dynamics of the satellite attitude motion, a typical form is

$$G_4(s) = \frac{1}{s^2[s^2/25 + 0.02(s/5) + 1]}.$$ (1.4)

A variation of this problem occurs if the sensor is on the same mass as the actuator and the two masses which are coupled by the flexible member are of comparable size. This situation often occurs in the positioning of a large radio telescope. In this case the transfer function includes a complex zero with almost no damping in addition to the lightly damped complex poles. A suitable transfer function is an extension of (1.4) given by

$$G_5(s) = \frac{(s^2/9) + 1}{s^2[(s^2/25) + 0.1(s/5) + 1]}.$$ (1.5)

Our fifth example also comes from the process industries and illustrates multivariable control of the liquid level and total pressure (head) in the head box of a paper machine. This example has two controls and two outputs so the transfer function is a matrix given by

$$G_6(s) = \frac{1}{\Delta(s)} \begin{bmatrix} s & (s + 1)(s + 0.07) \\ -0.05 & 0.7(s + 1)(s + 0.13) \end{bmatrix},$$ (1.6a)

where

$$\Delta(s) = (s + 1)(s + 0.1707)(s + 0.02929).$$ (1.6b)

1.3 OVERVIEW OF DESIGN APPROACH

An overview of the path we plan to take toward the design of digital controls will be useful before we begin the specific details. As mentioned above, we place systems of interest in three categories according to the nature of the signals present. These are discrete systems, sampled-data systems, and digital systems.

In discrete systems all signals vary at discrete times only. We will begin with an analysis of these in Chapter 2 and develop the z-transform of discrete signals and "pulse"-transfer functions for linear constant discrete systems. Of special interest will be the characterization of the dynamic response of discrete systems. As a special case of discrete systems we consider discrete equivalents to continuous systems which is one aspect of the currently popular field of *digital filters*. If quantization effects are ignored, digital filters are discrete systems which are designed to process discrete signals in such a fashion that the digital device (a digital computer program, for example) can be used to replace a continuous filter. Our treatment in Chapter 3 will concentrate on the use of discrete filtering techniques to find discrete equivalents of continuous-control compensator transfer functions.

A sampled-data system has both discrete and continuous signals and such situations are studied in Chapter 4. Here we are concerned with the question of data extrapolation to convert discrete signals as they might emerge from a digital computer into the continuous signals necessary for providing the input to one of the plants described above. This action typically occurs in conjunction with the D/A conversion. In addition to data extrapolation, we consider the analysis of discrete signals from the viewpoint of continuous analysis. For this purpose we introduce impulse modulation as a model of sampling and use Fourier analysis to give a clear picture of the problem of an ambiguity which can arise between continuous and discrete signals, also known as *aliasing*. A corollary of aliasing is the *sampling theorem* which specifies the conditions necessary if this ambiguity is to be removed and only one continuous signal allowed to correspond to a given set of samples.

Once we have developed the tools of analysis for discrete and sampled systems we can begin the design of feedback controls. Here we divide our techniques into two categories: transform[4] and state space[5] methods. In Chapter 5 we study the transform methods of the root locus and the frequency response as they are used to design digital control systems. Our state-space technique for design is mainly "pole assignment," a scheme for forcing the closed-loop poles to be in desirable locations. This technique is presented in Chapter 6 along with methods for estimating the "states" which do not have sensors directly on them. A study of quantization effects in Chapter 7 both presents a "worse-case" analysis and introduces the idea of random signals in order to describe a method for treating the "average" effects of this important nonlinearity.

The last four chapters cover more advanced topics which are essential for most complete designs. First of these topics is *identification* introduced in Chapter 8. Here the matter of model making is extended to the use of experimental data to verify and correct a theoretical model or to supply a dynamic description based only on input-output data. Only the most elementary of the concepts in this enormous field can be covered, of course. We present the method of least squares and some of the concepts of maximum likelihood. In Chapter 9 the topic of optimal control is introduced, with emphasis on the steady-state solution for linear constant discrete systems with quadratic loss func-

---

[4] Named because they use the Laplace or Fourier transform to represent systems.
[5] The state space is an extension of the space of displacement and velocity used in physics. Much that is called *modern control theory* uses differential equations in state-space form. We discuss this representation in Chapter 6 and use it extensively afterwards.

tions. The results are a valuable part of the designer's repertoire and are the only techniques presented here suitable for handling multivariable designs. Chapter 10 presents a topic with specific application to sampled-data and digital controls: the question of sampling-rate selection. In our earlier analysis we develop methods for examining the effects of different sampling rates, but in this chapter we consider for the first time the question of sample rate as a design parameter.

1.4 SUMMARY

In this chapter we introduced the subject of digital control design and defined the variables which characterize discrete, sampled-data, and digital control systems. Pointing out that we will be mainly concerned with simple single-loop linear constant systems, we gave the (highly simplified) transfer functions of five physical examples which will be used to illustrate the analysis and design techniques as they are presented in the book. These examples represent aspects of a space vehicle, a servomechanism, temperature control, a flexible structure, and a papermaking machine.

Finally, we gave an overview of our study from the analysis of discrete and sampled-data systems to the design of digital controls via transform and state space techniques. We pointed out that, following development of design methods for the most elementary case, we will consider the special topics of quantization, identification, optimal control, and selection of sampling rate.

SUGGESTIONS FOR FURTHER READING

Several histories of feedback control are readily available, including a *Scientific American Book* (1955), and the study of Mayr (1970). A good discussion of the historical developments of control is given by Dorf (1980) and by Fortmann and Hitz (1977), and many other references are cited by these authors for the interested reader. One of the earliest published studies of control systems operating on discrete time data (sampled-data systems in our terminology) is given by Hurewicz in Chapter 5 of the book by James, Nichols, and Phillips (1947).

The ideas of tracking and robustness embody many elements of control system design objectives. Within tracking are contained the requirements of system stability, good transient response, and good steady-state accuracy, all concepts fundamental to every control system. Robustness is a property essential to good performance in practical designs because real parameters are subject to change, and external unwanted signals invade every system. Discussion of control system performance specifications are given in most books on introductory control, including those of D'Azzo and Houpis (1975), Dorf (1980), Kuo (1975), and Ogata (1970). We will study these matters in later chapters with particular reference to digital control design.

A comprehensive treatment of conversion techniques between analog and digital signals is given by Hnatek (1976). A discussion relevant to low-cost microcomputer-based systems is given by Morrison (1978).

A comprehensive text concerned with writing equations of motion for systems in a form suitable for control studies is Cannon (1967). The topic is also the subject of a chapter in D'Azzo and Houpis (1975), Dorf (1980), Kuo (1975), and Ogata (1970).

PROBLEMS AND EXERCISES

1.1 Suppose a radar search antenna at San Francisco airport rotates at 6 rev/min and data points corresponding to the position of flight 1081 are plotted on the controller's screen once per antenna revolution. Flight 1081 is traveling directly toward the airport at 540 mi/hr. A feedback control system is established when the controller gives course corrections to the pilot. He wishes to do so each 9 mi of travel of the aircraft, and his instructions consist of course headings in integral degree values.

- a) What is the sampling rate, in seconds, of the range signal plotted on the radar screen?
- b) What is the sampling rate of the controller's instructions, in seconds?
- c) Identify the following signals as continuous, discrete, or digital:
  - i) the aircraft's range from the airport,
  - ii) the range data as plotted on the radar screen,
  - iii) the controller's instructions to the pilot,
  - iv) the pilot's actions on the aircraft control surfaces.
- d) Is this a continuous, sampled-data, or digital control system?
- e) Show that it is possible for the pilot of flight 1081 to fly a zigzag course which would show up as a straight line on the controller's screen. What is the (lowest) frequency of a sinusoidal zigzag course which will be hidden from the controller's radar?

1.2 From Truxal (1955), page 122.[6] An electronic designer has three component amplifiers with gain given by $K = + K_0 + \delta K$, where the magnitude of δK is less than 10% of K_0 and K_0 is 1000. He wishes to design an overall amplifier with a very precise gain of 100. Three topologies are suggested, as sketched in Fig. 1.3. (Topology I is certainly not practical as it stands but is useful to allow comparison of open-loop to closed-loop sensitivity.) We define the sensitivity of an overall gain G as the ratio of the relative change in G to the relative change in the parameter. The sensitivity of G to K is, for infinitesimal changes, given by

$$\mathscr{S}_G^K = \frac{\delta G/G}{\delta K/K} = \frac{K}{G}\frac{\partial G}{\partial K}.$$

- a) Compute the sensitivities of the three systems above to K if the β_i are selected to make $G_0 = 100$ in each case.
- b) Which has the lowest sensitivity?
- c) If each K changes by 10%, how close is this gain to 100?

---

[6] From *Automatic Feedback Control System Synthesis*, by J. G. Truxal (p. 122). Copyright © 1955 by McGraw-Hill. Used with permission of McGraw-Hill Book Company.

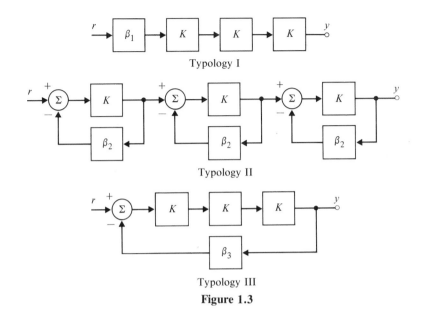

Typology I

Typology II

Typology III

Figure 1.3

d) Suppose each amplifier has an input noise w. Which system has the best disturbance rejection, i.e., the smallest output signal due to the disturbance alone, when $r = 0$?

e) Considering part (d) above, if one of the amplifiers has significantly *lower* input noise than the other two, which position should it occupy, i.e., should it be at the input or the output end of the three-amplifier chain?

f) Compute the sensitivity of the given systems with respect to changes in β, that is,

$$\mathscr{S}_G^\beta = \frac{\beta}{G} \frac{\partial G}{\partial \beta},$$

and compare the three topologies with respect to this figure of merit.

1.3 The assumptions of Problem 1.2 with respect to disturbance rejection are typical for electronic amplifiers but not for control systems. The typical control case is shown in Fig. 1.4. If the designer has the freedom to select the value of K_1, what value of K_1 gives the smallest error due to disturbance w?

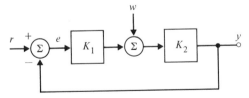

Figure 1.4

2 / Linear Discrete Dynamic Systems Analysis: The z-Transform

2.1 INTRODUCTION

The element in the structure of Fig. 1.1 that makes it different from other control models is the digital computer. The purpose of this chapter is to develop tools of analysis necessary to understand and to guide the design of programs for a computer acting as a linear dynamic control component. Needless to say, digital computers can do many things other than control linear dynamic systems; it is our purpose to examine their characteristics when doing this elementary control task and to provide the foundation needed to write programs for a real-time control computer.

2.2 LINEAR DIFFERENCE EQUATIONS

We assume that the analog-to-digital converter (A/D) in Fig. 1.1 takes samples of the signal e^1 at discrete times and passes them to the computer so that $m(kT) = e(kT)$. The job of the computer is to take these sample values and compute in some fashion the signals to be put out through the digital-to-analog converter (D/A). The characteristics of the A/D and D/A converters will be discussed later. First we consider the treatment of the data inside the computer. Suppose we call the input signals up to the kth sample $e_0, e_1, e_2, \ldots, e_k$, and the output signals prior to that time $u_0, u_1, u_2, \ldots, u_{k-1}$. Then, to get the next output, we have the machine compute some function which we can express in symbolic form as

$$u_k = f(e_0, \ldots, e_k; u_0, \ldots, u_{k-1}). \qquad (2.1)$$

Because we plan to emphasize the elementary and the dynamic possibilities, we

---

[1] We will drop the hat from $\hat{e}$ with no loss of generality.

assume that the function f in (2.1) is *linear* and depends on only a finite number of e's and u's. Thus we write

$$u_k = a_1 u_{k-1} + a_2 u_{k-2} + \cdots + a_n u_{k-n} + b_0 e_k + b_1 e_{k-1} + \cdots + b_m e_{k-m}. \quad (2.2)$$

Equation (2.2) is called a linear recurrence equation or difference equation and, as we shall see, has many similarities with a linear differential equation. The name "difference equation" derives from the fact that we could write (2.2) using u_k plus the differences in u_k which are defined as

$$\begin{aligned}
\nabla u_k &= u_k - u_{k-1} & &\text{(first difference),} \\
\nabla^2 u_k &= \nabla u_k - \nabla u_{k-1} & &\text{(second difference),} \quad (2.3) \\
\nabla^n u_k &= \nabla^{n-1} u_k - \nabla^{n-1} u_{k-1} & &\text{(nth difference).}
\end{aligned}$$

If we solve (2.3) for the values of u_k, u_{k-1}, u_{k-2} in terms of differences, we find

$$\begin{aligned}
u_k &= u_k, \\
u_{k-1} &= u_k - \nabla u_k, \\
u_{k-2} &= u_k - 2\nabla u_k + \nabla^2 u_k.
\end{aligned}$$

Thus, for a second-order equation with coefficients a_1, a_2, and b_0 (we let $b_1 = b_2 = 0$ for simplicity), we find the equivalent difference equation

$$-a_2 \nabla^2 u_k + (a_1 - 2a_2)\nabla u_k + (a_2 + a_1 - 1)u_k = b_0 e_k.$$

If the a's and b's in (2.2) are constant, then the computer is solving a *constant-coefficient difference equation* (CCDE). We plan to demonstrate later that with such equations the computer can control linear constant dynamic systems and perform most of the other tasks of linear constant dynamic systems, including performing the functions of electric filters. To do so, it is necessary first to examine methods of obtaining solutions to (2.2) and to study the general properties of these solutions.

To solve a specific CCDE is an elementary matter. We need a starting time (k-value) and some initial conditions to characterize the contents of the computer memory at this time. For example, suppose we take the case

$$u_k = u_{k-1} + u_{k-2} \quad (2.4)$$

and start at $k = 2$. Here there are no input values and to compute u_2 we need to know the (initial) values for u_0 and u_1. Let us take them to be $u_0 = u_1 = 1$. A plot of the values of u_k versus k is shown in Fig. 2.1. The results, the Fibonacci numbers, are named after the 13th-century mathematician[2] who studied them. For example, (2.4) has been used to model the growth of rabbits in a protected environment.[3] However that may be, the result as a computer program would

---

[2] Leonardo Fibonacci of Pisa, who introduced Arabic notation to the Latin world about 1200 A.D.
[3] Wilde (1964). Assume that u_k represents pairs of rabbits and that babies are born in pairs. Assume that no rabbits die and that a new pair begin reproduction after one period. Thus at time k, we have all the old rabbits, u_{k-1}, plus the newborn pairs born to the mature rabbits, which are u_{k-2}.

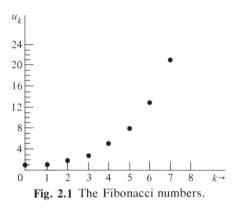

Fig. 2.1 The Fibonacci numbers.

seem to be unstable, to say the least. We would like to be able to examine equations like (2.2) and, without having to solve them, see if they are stable.

As a second elementary example of the origins of a difference equation, consider the discrete approximation to integration. Suppose we have a continuous signal, $e(t)$, of which a segment is sketched in Fig. 2.2, and we wish to compute an approximation to the integral

$$\mathcal{I} = \int_0^t e(t)\ dt, \tag{2.5}$$

using only the discrete values $e(0)$, ... , $e(t_{k-1})$, $e(t_k)$. We assume that we have an approximation for the integral from zero to the time t_{k-1} and we call it u_{k-1}. The problem is to obtain u_k from this information. Taking the view of the integral as the area under the curve $e(t)$, we see that this problem reduces to finding an approximation to the area under the curve between t_{k-1} and t_k. Three alternatives are sketched in Fig. 2.2. We can use the rectangle of height e_{k-1}, or the rectangle of height e_k, or the trapezoid formed by connecting e_{k-1} to e_k by a straight line. If we take the third choice, we find the area of the trapezoid to be

$$A = \frac{t_k - t_{k-1}}{2}\ (e_k + e_{k-1}). \tag{2.6}$$

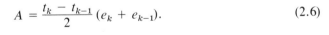
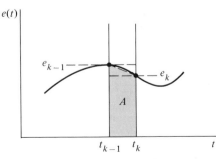

Fig. 2.2 Plot of a function and alternative approximations to the area under the curve over a single time interval.

Finally, if we assume that the sampling period, $t_k - t_{k-1}$, is a constant, T, we are led to a simple formula for discrete (trapezoid rule) integration:

$$u_k = u_{k-1} + \frac{T}{2}(e_k + e_{k-1}). \tag{2.7}$$

If $e(t) = t$, then $e_k = kT$ and substitution of $u_k = (T^2/2)k^2$ satisfies (2.7) and is exactly the integral of e. [It should be, because if $e(t)$ is a straight line, the trapezoid is the *exact* area.]

So we see that difference equations can be solved by a digital computer and that they may represent useful objects including models of physical processes and approximations to integration. It turns out that if the equations are linear with coefficients that are constant, we can describe the relation between u and e by a transfer function, and thereby gain a great aid to analysis and also to the design of linear, constant, discrete controls.

2.3 THE DISCRETE TRANSFER FUNCTION

We will obtain the transfer function of linear constant discrete systems by the method of z-transform analysis. An alternative viewpoint that requires a bit more mathematics but has some appeal, is given in the appendix to this chapter. The results are the same.

If a signal has discrete values $e_0, e_1, \ldots, e_k, \ldots$ we define the z-transform of the signal as the function[4,5]

$$E(z) \overset{\Delta}{=} \mathscr{z}\{e_k\}$$

$$\overset{\Delta}{=} \sum_{k=-\infty}^{\infty} e_k z^{-k}, \qquad r_0 \leq |z| \leq R_0, \tag{2.8}$$

and we assume we can find a range of values of the magnitude of the complex variable z for which the series (2.8) converges. A discussion of convergence is deferred until Section 2.5. The z-transform has the same role in discrete systems that the Laplace transform has in continuous systems analysis. For example, the z-transforms for e_k and u_k of the difference equations (2.2) and the trapezoid integration (2.7) are related in a simple way that permits the rapid solution of equations of this kind. To find the relation, we proceed by direct substitution. We

---

[4] We use the notation $\overset{\Delta}{=}$ to mean "is defined as."

[5] In (2.8) the lower limit is $-\infty$ so that values of e_k on both sides of $k = 0$ are included. The transform so defined is sometimes called the two-sided z-transform to distinguish it from the one-sided definition which would be $\Sigma_0^\infty e_k z^{-k}$. For signals which are zero for $k < 0$, the transforms obviously give identical results. To take the one-sided transform of u_{k-1}, however, we must handle the value of u_{-1}, and thus are initial conditions introduced by the one-sided transform. Examination of this property and other features of the one-sided transform are invited by the problems. We select the two-sided transform because we need to consider signals which extend into negative time when we come to Chapter 8.

take the definition given by (2.8) and, in the same way, we define the z-transform of the sequence $\{u_k\}$ as

$$U(z) \triangleq \sum_{k=-\infty}^{\infty} u_k z^{-k}. \tag{2.9}$$

Now we multiply (2.7) by z^{-k} and sum over k. We get

$$\sum_{k=-\infty}^{\infty} u_k z^{-k} = \sum_{k=-\infty}^{\infty} u_{k-1} z^{-k} + \frac{T}{2} \left(\sum_{k=-\infty}^{\infty} e_k z^{-k} + \sum_{k=-\infty}^{\infty} e_{k-1} z^{-k} \right). \tag{2.10}$$

From (2.9), we recognize the left-hand side as $U(z)$. In the first term on the right, we let $k - 1 = j$ to obtain

$$\sum_{k=-\infty}^{\infty} u_{k-1} z^{-k} = \sum_{j=-\infty}^{\infty} u_j z^{-(j+1)} = z^{-1} U(z). \tag{2.11}$$

By similar operations on the third and fourth terms we may reduce (2.10) to

$$U(z) = z^{-1} U(z) + \frac{T}{2} [E(z) + z^{-1} E(z)]. \tag{2.12}$$

Equation (2.12) is now simply an algebraic equation in z and the functions U and E. Solving it we obtain

$$U(z) = \frac{T}{2} \frac{1 + z^{-1}}{1 - z^{-1}} E(z). \tag{2.13}$$

We *define* the ratio of the transform of the input to the transform of the output as the *transfer function*, $H(z)$. Thus, in this case, the transfer function for trapezoid rule integration is

$$\frac{U(z)}{E(z)} \triangleq H(z) = \frac{T}{2} \frac{z + 1}{z - 1}. \tag{2.14}$$

For the more general relation given by (2.2), it is readily verified by the same techniques that

$$H(z) = \frac{b_0 + b_1 z^{-1} + \cdots + b_m z^{-m}}{1 - a_1 z^{-1} - a_2 z^{-2} - \cdots - a_n z^{-n}},$$

and if $n \geq m$, we can write this as a ratio of polynomials in z as

$$H(z) = \frac{b_0 z^n + b_1 z^{n-1} + \cdots + b_m z^{n-m}}{z^n - a_1 z^{n-1} - a_2 z^{n-2} - \cdots - a_n}$$

$$= \frac{b(z)}{a(z)}, \tag{2.15}$$

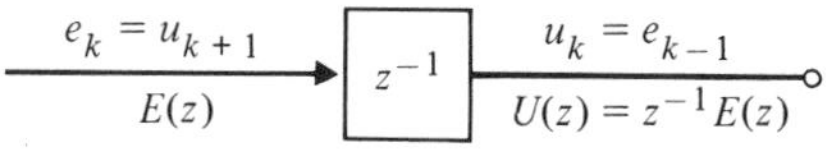

$$e_k = u_{k+1} \qquad \boxed{z^{-1}} \qquad u_k = e_{k-1}$$
$$E(z) \qquad\qquad U(z) = z^{-1} E(z)$$

Fig. 2.3 The unit delay.

and the general input-output relation between transforms is

$$U(z) = H(z)E(z). \tag{2.16}$$

Since $H(z)$ is a rational function of a complex variable, we use the terminology of that subject. Suppose we call the numerator polynomial $b(z)$ and the denominator $a(z)$. The places in z where $b(z) = 0$ are *zeros* of the transfer function and the places in z where $a(z) = 0$ are the *poles* of $H(z)$. If z_0 is a pole and $(z - z_0)^n H(z)$ has neither pole nor zero at z_0, we say that $H(z)$ has a pole of order n at z_0. If $n = 1$, the pole is simple.

We can now give a physical meaning to the variable z. Suppose we let all coefficients in (2.15) be zero except b_1 and we take b_1 to be 1. Then $H(z) = z^{-1}$. But $H(z)$ represents the transform of (2.2), and with these coefficient values the difference equation reduces to

$$u_k = e_{k-1}. \tag{2.17}$$

The present value of the output, u_k, equals the input *delayed by one period.* Thus we see that a transfer function of z^{-1} is a *delay* of one time unit. We can picture the situation as in Fig. 2.3 where both time and transform relations are shown.

Since the relations of (2.7), (2.14), and (2.15) are all composed of delays, they can be expressed in terms of z^{-1}! Consider (2.7). In Fig. 2.4 we illustrate the difference equation (2.7) using the transfer function z^{-1} as the symbol for a unit delay.

We can follow the operations of the discrete integrator by tracing the signals through Fig. 2.4. For example, the present value of e_k is passed to the first summer, where it is added to the previous value e_{k-1}, and the sum is multiplied by $T/2$ to compute the area of the trapezoid between e_{k-1} and e_k. This is the signal marked "a_k." After this, there is another sum where the previous output, u_{k-1}, is added to the new area to form the next value of the integral, u_k. The discrete integration occurs in the loop with one delay, z^{-1}, and unity gain.

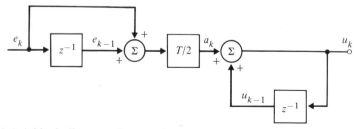

Fig. 2.4 A block diagram of trapezoid integration as represented by Eq. (2.7).

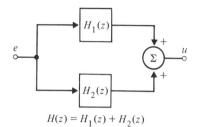

$$H(z) = H_1(z) + H_2(z)$$

Fig. 2.5 Block diagram of parallel blocks.

2.3.1 Block Diagram Reduction and Mason's Rule

Since Eq. (2.16) is linear and constant, we can use block-diagram analysis to manipulate discrete-transfer-function relationships. There are only four important cases:

1. The transfer function of paths in parallel is the sum of the single-path transfer functions (Fig. 2.5).

2. The transfer function of paths in series is the *product* of the path transfer functions (Fig. 2.6).

3. The transfer function of a single loop of paths is the transfer function of the forward path divided by one minus the loop transfer function (Fig. 2.7).

4. The transfer function of a multipath diagram or graph is given by Mason's rule.[6] If each forward path touches every loop path, the graph is tightly con-

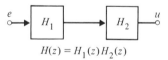

$$H(z) = H_1(z) H_2(z)$$

Fig. 2.6 Block diagram of cascade blocks.

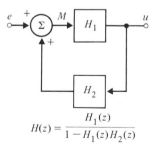

$$H(z) = \frac{H_1(z)}{1 - H_1(z) H_2(z)}$$

Fig. 2.7 Feedback transfer function.

[6] Mason (1956).

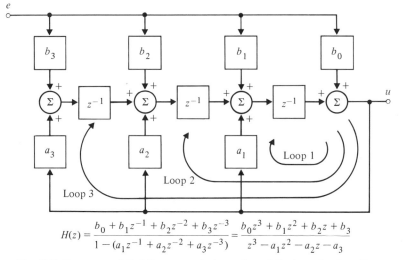

$$H(z) = \frac{b_0 + b_1 z^{-1} + b_2 z^{-2} + b_3 z^{-3}}{1 - (a_1 z^{-1} + a_2 z^{-2} + a_3 z^{-3})} = \frac{b_0 z^3 + b_1 z^2 + b_2 z + b_3}{z^3 - a_1 z^2 - a_2 z - a_3}$$

Fig. 2.8 One general tightly connected graph and its transfer function.

nected and the rule reduces to: The transfer function of a tightly connected graph is the sum of the forward-path transfer functions divided by one minus the sum of all loop transfer functions (Fig. 2.8).

We will demonstrate the reduction of the block diagram of Fig. 2.8. First we note that the signal u is unchanged if we move the b_0-block back to be parallel with the b_1-block, provided we multiply b_0 by z. Thus the diagram is reduced to Fig. 2.9. We have merely used the fact that $(b_0 z)z^{-1} = b_0$. This argument can, in an obvious way, be continued so that $b_1 z + b_0 z^2$ are in parallel with b_2, and these can then again be moved to be in parallel with b_3, with the final effect shown in Fig. 2.10.

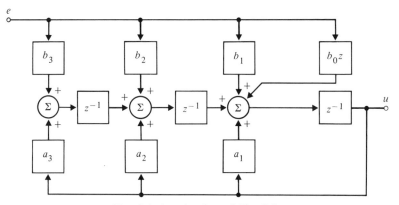

Fig. 2.9 A reduction of Fig. 2.8.

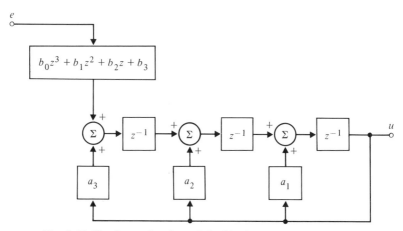

Fig. 2.10 Further reduction of the block diagram of Fig. 2.8.

With Fig. 2.10 we have reduced the graph to a single forward path and we can begin reducing the feedback loops. First we note that loop 1 is a simple loop with forward path gain z^{-1} and loop gain $a_1 z^{-1}$. Thus the next reduction by the formula of Fig. 2.7 is as shown in Fig. 2.11.

And now we are almost done because loop 2 is a simple loop which has equivalent gain:

$$\frac{z^{-2}/(1 - a_1 z^{-1})}{1 - a_2 z^{-2}/(1 - a_1 z^{-1})} = \frac{z^{-2}}{1 - a_1 z^{-1} - a_2 z^{-2}}.$$

The last loop has thus been reduced to simple-loop status and reveals a gain of:

$$\frac{z^{-3}}{1 - a_1 z^{-1} - a_2 z^{-2} - a_3 z^{-3}}.$$

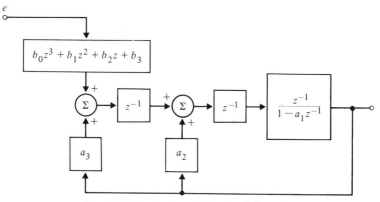

Fig. 2.11 Reduction of feedback loop 1.

The product of this factor, now in series with the terms in b_i, gives the final result:

$$H(z) = \frac{b_0 + b_1 z^{-1} + b_2 z^{-2} + b_3 z^{-3}}{1 - a_1 z^{-1} - a_2 z^{-2} - a_3 z^{-3}}.$$

Because the feedback loops of Fig. 2.8 all come from the output or "observed" signal, we call this structure the *observer canonical* form. It is a direct canonical form because the gains of the blocks in the structure are obtained directly from the coefficients of the numerator and denominator polynomials in the transfer function, $H(z)$. Another useful direct canonical form is shown in Fig. 2.12.

In this case all the feedback loops return to the point of application of the input or "control" variable, and hence the form is referred to as the *control canonical form*. Reduction of the structure by Mason's rule or by elementary operations, beginning first with moving the b_0-block to the right and multiplying it by z, then moving $b_0 z + b_1$, and so on, followed by reduction of the feedback loops one at a time, shows that this structure also has the transfer function given by $H(z)$.

The block diagrams of Figs. 2.8 and 2.12 are called *direct canonical* realizations of the transfer function $H(z)$ because the gains of the realizations are coefficients in the transfer-function polynomials. If we realize a transfer function by placing several first- or second-order direct forms in series with each other, we have a *cascade canonical* form. In this case, the poles and zeros of the transfer function are isolated and the $H(z)$ is represented as a product of factors.

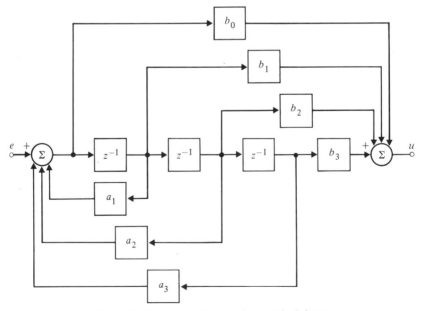

Fig. 2.12 Structure of control canonical form.

2.3.2 Relation of Transfer Function to Pulse Response

We have shown that a transfer function of z^{-1} is a unit delay.   We can also give a time-domain meaning to an arbitrary transfer function.   Recall that the z-transform is defined by (2.8) to be $E(z) = \Sigma e_k z^{-k}$, and the transfer function is defined from (2.16) as $H(z)$ when the input and output are related by $U(z) = H(z)E(z)$.   Now suppose we deliberately select $e(k)$ to be the unit discrete pulse defined by

$$e_k = \begin{cases} 1 & (k = 0) \\ 0 & (k \neq 0) \end{cases}$$
$$\overset{\Delta}{=} \delta_k.$$

(2.18)

Then it follows that $E(z) = 1$ and therefore that

$$U(z) = H(z).$$

(2.19)

Thus the transfer function $H(z)$ is seen to be the *transform* of the signal which is the response of the system to a unit pulse input.   For example, let us look at the system of Fig. 2.4 and put a unit pulse in at the e_k-node (with no signals in the system beforehand).[7]   We can readily follow the pulse through the block and build Table 2.1.   Thus the unit pulse response is zero for negative k, is $T/2$ at $k = 0$, and equals T thereafter.   The z-transform of this sequence is

$$H(z) = \sum_{-\infty}^{\infty} h_k z^{-k}.$$

If we add $T/2$ to the z^0-term and subtract $T/2$ from the whole series, we have a simpler sum, as follows:

$$H(z) = \sum_{k=0}^{\infty} Tz^{-k} - \frac{T}{2}$$

$$= \frac{T}{1 - z^{-1}} - \frac{T}{2} \quad (1 < |z|)$$

$$= \frac{2T - T(1 - z^{-1})}{2(1 - z^{-1})}$$

$$= \frac{T + Tz^{-1}}{2(1 - z^{-1})}$$

$$= \frac{T}{2} \frac{z + 1}{z - 1} \quad (1 < |z|).$$

(2.20)

Of course, this is the transfer function we obtained in (2.13) from direct analysis of the difference equation.

---

[7] In this development we assume that (2.7) is intended to be used as a formula for computing values of e_k as k *increases*.   There is no reason why we could not also solve for e_k as k takes on negative values. The direction of time comes from the application and not from the recurrence equation.

Table 2.1

k	e_{k-1}	e_k	a_k	u_{k-1}	u_k
0	0	1	$T/2$	0	$T/2$
1	1	0	$T/2$	$T/2$	T
2	0	0	0	T	T
3	0	0	0	T	T

A final point of view useful in the interpretation of the discrete transfer function is obtained by multiplying the infinite polynomials of $E(z)$ and $H(z)$ as required in (2.16). For simplicity, we will assume that the unit pulse response, h_k, is zero for $k < 0$.[8] Likewise, we will take $k = 0$ to be the starting time for e_k. Then the product that produces $U(z)$ is the polynomial product given in Fig. 2.13. Since this product has been shown to be $U(z) = \Sigma u_k z^{-k}$, it must therefore follow that the coefficient of z^{-k} in the product is u_k. Listing these coefficients we have the relations

$$u_0 = e_0 h_0,$$
$$u_1 = e_0 h_1 + e_1 h_0,$$
$$u_2 = e_0 h_2 + e_1 h_1 + e_2 h_0,$$
$$u_3 = e_0 h_3 + e_1 h_2 + e_2 h_1 + e_3 h_0.$$

By extrapolation of this simple pattern, we derive the result

$$u_k = \sum_{j=0}^{k} e_j h_{k-j}.$$

By extension, we let the lower limit of the sum be $-\infty$ and the upper limit be $+\infty$:

$$u_k = \sum_{j=-\infty}^{\infty} e_j h_{k-j}. \tag{2.21}$$

This is the discrete convolution sum and is the analog of the convolution integral which relates input and impulse response to output in linear constant continuous

e_0	$+ e_1 z^{-1}$	$+ e_2 z^{-2}$	$+ e_3 z^{-3}$	$+ \cdots$
h_0	$+ h_1 z^{-1}$	$+ h_2 z^{-2}$	$+ h_3 z^{-3}$	$+ \cdots$
$e_0 h_0$	$+ e_1 h_0 z^{-1}$	$+ e_2 h_0 z^{-2}$	$+ e_3 h_0 z^{-3}$	
	$+ e_0 h_1 z^{-1}$	$+ e_1 h_1 z^{-2}$	$+ e_2 h_1 z^{-3}$	
		$+ e_0 h_2 z^{-2}$	$+ e_1 h_2 z^{-3}$	
			$+ e_0 h_3 z^{-3}$	

$e_0 h_0 + (e_0 h_1 + e_1 h_0)z^{-1} + (e_0 h_2 + e_1 h_1 + e_2 h_0)z^{-2} + (e_0 h_3 + e_1 h_2 + e_2 h_1 + e_3 h_0)z^{-3} + \cdots$

Fig. 2.13 Representation of the product $E(z)H(z)$ as a product of polynomials.

---

[8] If this is not the case, it means that the system response begins *before* the input that causes it appears, and the system is said to be noncausal.

systems.   To verify (2.21) we take the z-transform of both sides:

$$\sum_{k=-\infty}^{\infty} u_k z^{-k} = \sum_{k=-\infty}^{\infty} z^{-k} \sum_{j=-\infty}^{\infty} e_j h_{k-j}.$$

Interchanging the sum on j with the sum on k leads to

$$U(z) = \sum_{j=-\infty}^{\infty} e_j \sum_{k=-\infty}^{\infty} z^{-k} h_{k-j}.$$

Now let $k - j = \ell$ in the second sum:

$$U(z) = \sum_{j=-\infty}^{\infty} e_j \sum_{\ell=-\infty}^{\infty} h_\ell z^{-(\ell+j)},$$

but $z^{-(\ell+j)} = z^{-\ell} z^{-j}$, which leads to

$$U(z) = \sum_{j=-\infty}^{\infty} e_j z^{-j} \sum_{\ell=-\infty}^{\infty} h_\ell z^{-\ell},$$

and we recognize these two separate sums as

$$U(z) = E(z)H(z). \qquad \text{QED}$$

We can also derive the convolution sum from the properties of linearity and stationarity.   First we need more formal definitions of "linear" and "stationary."

A system with input e and output u is *linear* if superposition applies.   Thus if $u_1(k)$ is the response to $e_1(k)$, and $u_2(k)$ is the response to $e_2(k)$, then the system is linear if and only if, for every scalar α and β, the response to $\alpha e_1 + \beta e_2$ is $\alpha u_1 + \beta u_2$.

A system is *stationary* or time invariant if the properties of the system do not change with time.   For example, if we put the system at rest (no internal energy in the system) and apply a certain signal $e(k)$, we observe a response $u(k)$.   If we repeat this experiment at a later time (N periods later) and the system is again at rest and we apply $e(k - N)$, we should see $u(k - N)$.

These properties may be used to derive the convolution in (2.21) as follows. If response to a unit pulse at $k = 0$ is $h(k)$, then response to a pulse of intensity e_0 is $e_0 h(k)$ if the system is linear.   Furthermore, if the system is *constant*, then a delay of the input will delay the response.   Thus, if

$$e = \begin{cases} e_\ell, & k = \ell, \\ 0, & k \neq \ell, \end{cases}$$

then the response will be $e_\ell h_{k-\ell}$.

Finally, the total response at time k to a sequence of these pulses is the *sum* of the responses, namely

$$u_k = e_0 h_k + e_1 h_{k-1} + \cdots + e_\ell h_{k-\ell} + \cdots + e_k h_0$$

or

$$u_k = \sum_{\ell=0}^{k} e_\ell h_{k-\ell}.$$

Now note that if the input sequence began in the distant past, we must include terms for $\ell < 0$, perhaps back to $\ell = -\infty$. Similarly, if the system should be non-causal, future values of e where $\ell > k$ may also come in. The general case is thus

$$u_k = \sum_{\ell=-\infty}^{\infty} e_\ell h_{k-\ell}. \tag{2.21}$$

2.3.3 External Stability

A very important qualitative property of some dynamic systems is stability. A system (or living organism, for that matter) may be said to be stable if its response is appropriate for the given stimulus. For discrete dynamic systems we sometimes consider the responses at all the internal variables that might appear at the delay elements in a canonical block diagram as in Fig. 2.8 or Fig. 2.12. Such studies identify *internal stability*. Often we are satisfied to consider only the *external stability* as given by the study of the input-output relation described for the linear stationary case by (2.21).

For external stability, the most common definition of "appropriate response" is that for every *Bounded Input*, we should have a *Bounded Output*. If this is true we say the system is BIBO stable. A test for BIBO stability may be given directly in terms of the unit pulse response, h_k. First we consider a sufficient condition. Suppose the input e_k is bounded, i.e., there is an M such that

$$|e_\ell| \leq M < \infty \qquad \text{for all} \quad \ell. \tag{2.22}$$

If we consider the magnitude of the response given by (2.21), it is easy to see that

$$|u_k| \leq \left| \sum e_\ell h_{k-\ell} \right|,$$

which is surely less than the sum of the magnitudes as given by

$$\leq \sum_{-\infty}^{\infty} |e_\ell| \, |h_{k-\ell}|.$$

But, since we assume (2.22), this result is in turn bounded by

$$\leq M \sum_{-\infty}^{\infty} |h_{k-\ell}|. \tag{2.23}$$

Thus the output will be bounded for every bounded input if

$$\sum_{\ell-\infty}^{\infty} |h_{k-\ell}| < \infty. \tag{2.24}$$

This condition is also necessary, for if we consider the bounded (by 1!) input

$$e_\ell = \frac{h_{-\ell}}{|h_{-\ell}|} \qquad (h_{-\ell} \neq 0)$$

$$= 0 \qquad (h_{-\ell} = 0)$$

and apply it to (2.21), the output at $k = 0$ is

$$u_0 = \sum_{\ell=-\infty}^{\infty} e_\ell h_{-\ell}$$

$$= \sum_{\ell=-\infty}^{\infty} \frac{(h_{-\ell})^2}{|h_{-\ell}|}$$

$$= \sum_{\ell=-\infty}^{\infty} |h_{-\ell}|. \qquad (2.25)$$

Thus, unless the condition given by (2.24) is true, the system is not BIBO stable.

The test given by (2.24) can be applied to the unit pulse response used to compute (2.20) and given as the u_k-column in Table 2.1:

$$h_0 = T/2,$$
$$h_k = T, \qquad k > 0,$$

$$\sum |h_k| = T/2 + \sum_1^{\infty} T = \text{unbounded}. \qquad (2.26)$$

Thus the approximation to integration is not (BIBO) stable! As a second example, we consider the difference equation (2.2) with all coefficients except a_1 and b_0 equal to zero:

$$u_k = a_1 u_{k-1} + b_0 e_k. \qquad (2.27)$$

The unit pulse response is easily developed from the first few terms to be

$$u_0 = b_0, \qquad u_1 = a_1 b_0, \qquad u_2 = a_1^2 b_0, \ldots$$
$$u_k = h_k = b_0 a^k, \qquad k \geq 0. \qquad (2.28)$$

Applying the test, we have

$$\sum_{-\infty}^{\infty} |h_\ell| = \sum_{\infty=0}^{\infty} b_0 |a^\ell| = b_0 \frac{1}{1 - |a|} \qquad (|a| < 1)$$

$$= \text{unbounded} \qquad (|a| \geq 1).$$

Thus we conclude that the system described by this equation is BIBO stable if $|a| < 1$, and unstable otherwise. This is exactly the kind of result we want—a test for stability in terms of the parameters of the system. This result will be developed further as we learn more about the properties of transfer functions.

2.4 SIGNAL ANALYSIS AND DYNAMIC RESPONSE

In Section 2.3 we demonstrated that if two variables are related by a linear constant difference equation, then the ratio of the z-transform of the output signal to that of the input is a function of the system equation alone, and the ratio is called the transfer function. A method for study of linear constant discrete systems is thereby indicated, consisting of the following steps:

1. Compute the transfer function of the system $H(z)$.
2. Compute the transform of the input signal, $E(z)$.
3. Form the product, $E(z)H(z)$, which is the transform of the output signal, u.
4. Invert the process of transformation to obtain $u(kT)$.

If the system description is available in difference-equation form, and the input signal is elementary, then the first three steps of this process require very little effort or computation. The final step, however, is often tedious and, because we will later be preoccupied with design of transfer functions to give desirable responses, we attach great benefit to avoiding the inversion step, if possible. Our approach to this problem is to present a repertoire of elementary signals with known features and to learn their representation in the transform or z-domain. Thus, when given an unknown transform, we will be able, by reference to these known solutions, to infer the major features of the time-domain signal and thus to determine whether the unknown is of sufficient interest to warrant the effort of detailed time-response computation.

To begin this process of attaching a connection between the time domain and the z-transform domain, we compute the transforms of a few elementary signals. The first case is the unit pulse.

We have already seen that the unit pulse is defined by

$$
\begin{aligned}
e_1(k) &= 1 \quad (k = 0) \\
&= 0 \quad (k \neq 0) \\
&= \delta_k;
\end{aligned}
$$

then

$$E_1(z) = \sum_{-\infty}^{\infty} \delta_k z^{-k} = z^0 = 1. \tag{2.29}$$

This result is much like the continuous case, wherein the Laplace transform of the unit impulse is the constant 1.0.

The quantity $E_1(z)$ gives us an instantaneous method to relate signals to systems: to characterize the system $H(z)$, consider the signal $u(k)$ which is the unit pulse response; then $U(z) = H(z)$.

As a second example, we consider the unit step function defined by[9]

---

[9] We will usually refer to this frequently occurring signal as $1(k)$. We have shifted notation here to use $e(k)$ rather than e_k for the kth sample, to allow use of subscripts to identify different signals.

$$e_2(k) = 1 \quad (k \geq 0)$$
$$= 0 \quad (k < 0). \tag{2.30}$$

In this case, the z-transform is

$$E_2(z) = \sum_{k=-\infty}^{\infty} e_2(k)z^{-k} = \sum_{k=0}^{\infty} z^{-k}$$

$$= \frac{1}{1 - z^{-1}} \quad (|z^{-1}| < 1)$$

$$= \frac{z}{z - 1} \quad (|z| > 1). \tag{2.31}$$

Here the transform is characterized by a zero at $z = 0$ and a pole at $z = 1$. The significance of the convergence being restricted to $|z| > 1$ will be explored later when we consider the inverse transform operation. The Laplace transform of the unit step is $1/s$; we may thus keep in mind that a pole at $s = 0$ for a continuous signal corresponds in some way to a pole at $z = 1$ for discrete signals. More about this later. In any event, we record that a pole at $z = 1$ with convergence outside the unit circle, $|z| = 1$ will correspond to a constant for positive time and zero for negative time.

To emphasize the connection between the time domain and the z-plane, we sketch in Fig. 2.14 the z-plane with the unit circle shown and the pole of $E_2(z)$ marked $\times$ and the zero marked $\bigcirc$. Beside the z-plane we sketch the time plot of $e_2(k)$.

For our third example, we consider the one-sided exponential

$$e_3(k) = r^k \quad (k \geq 0)$$
$$= 0 \quad (k < 0), \tag{2.32}$$

which is the same as $r^k 1(k)$, using the symbol $1(k)$ for the unit step function. Now

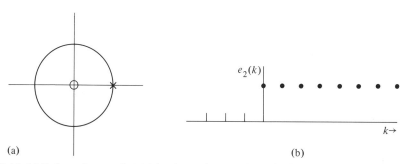

(a)   (b)

Fig. 2.14 (a) Pole and zero of $E_2(z)$ in the z-plane. The unit circle is shown for reference. (b) Plot of $e_2(k)$.

we get

$$E_3(z) = \sum_{k=0}^{\infty} r^k z^{-k}$$

$$= \sum_{k=0}^{\infty} (rz^{-1})^k$$

$$= \frac{1}{1 - rz^{-1}} \qquad (|rz^{-1}| < 1)$$

$$= \frac{z}{z - r} \qquad (|z| > |r|). \qquad (2.33)$$

The pole of $E_3(z)$ is at $z = r$. From (2.32) we know that $e_3(k)$ grows without bound if $|r| > 1$. From (2.33) we conclude that a z-transform which converges for large z and has a real pole *outside* the circle $|z| = 1$ corresponds to a growing signal. If such a signal were the unit pulse response of our system, such as our digital control program, we would say the program was *unstable* as we saw in (2.28). Again we plot in Fig. 2.15 the z-plane and the corresponding time history of $E_3(z)$ and $e_3(k)$ for $r \approx 0.6$.

Our next example considers the modulated sinusoid $e_4(k) = [r^k \cos k\theta]1(k)$, where we assume $r > 0$. Actually, we can decompose $e_4(k)$ into the sum of two complex exponentials as

$$e_4(k) = r^k \left(\frac{e^{jk\theta} + e^{-jk\theta}}{2} \right) 1(k)$$

and because the z-transform is linear,[10] we need only compute the transform of each single complex exponential and add the results later. We thus take first

$$e_5(k) = r^k e^{jk\theta} 1(k) \qquad (2.34)$$

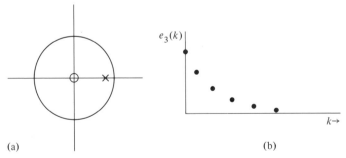

(a)   (b)

Fig. 2.15 (a) Pole and zero of $E_3(z)$ in the z-plane.   (b) Plot of $e_3(k)$.

---

[10] We have not shown this formally.   The demonstration, using the definition of linearity given above, is simple and given in Section 2.5.

and compute

$$E_5(z) = \sum_{k=0}^{\infty} r^k e^{j\theta k} z^{-k}$$

$$= \sum_{j=0}^{\infty} (re^{j\theta} z^{-1})^k$$

$$= \frac{1}{1 - re^{j\theta} z^{-1}}$$

$$= \frac{z}{z - re^{j\theta}} \qquad (|z| > r). \tag{2.35}$$

The signal $e_5(k)$ grows without bound as k gets large if and only if $r > 1$, and a system with this pulse response is BIBO stable if and only if $|r| < 1$. The boundary of stability is the unit circle. To complete the argument given above for $e_4(k) = r^k \cos k\theta 1(k)$, we see immediately that the other half is found by replacing θ by $-\theta$ in (2.35),

$$\mathscr{z}\{r^k e^{-j\theta k} 1(k)\} = \frac{z}{z - re^{-j\theta}} \qquad (|z| > r), \tag{2.36}$$

and thus that

$$E_4(z) = \frac{1}{2} \left\{ \frac{z}{z - re^{j\theta}} + \frac{z}{z - re^{-j\theta}} \right\}$$

$$= \frac{z(z - r\cos\theta)}{z^2 - 2r(\cos\theta)z + r^2} \qquad (|z| > r). \tag{2.37}$$

The z-plane pole-zero pattern of $E_4(z)$ and the time plot of $e_4(k)$ are shown in Fig. 2.16 for $r = 0.7$ and $\theta = 45°$.

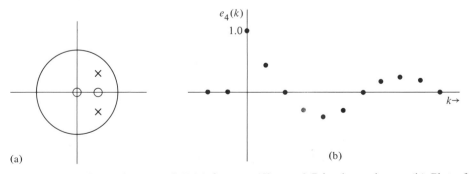

(a)

(b)

Fig. 2.16 (a) Poles and zeros of $E_4(z)$ for $\theta = 45°$, $r = 0.7$ in the z-plane. (b) Plot of $e_4(k)$.

We note in passing that if $\theta = 0$, then e_4 reduces to e_3 and, with $r = 1$, to e_2 so that three of our signals are special cases of e_4. By exploiting the features of $E_4(z)$ we can draw a number of conclusions about the relation between pole locations in the z-plane and the time-domain signals to which the poles correspond. We collect these for later reference.

1. The duration of a transient is set mainly by the value of the radius, r, of the poles.

   a) $r > 1$ corresponds to a growing signal and is unstable as a pulse response.
   b) $r = 1$ corresponds to a signal with nongrowing amplitude but which is *not* BIBO stable as a pulse response.
   c) For $r < 1$, the closer r is to 0 the shorter the duration of the signal. The corresponding system is BIBO stable.
   d) A pole at $r = 0$ corresponds to a transient of finite duration.

2. The number of samples per oscillation of a sinusoidal signal is determined by θ. If we require $\cos \theta k = \cos(\theta(k + N))$, we find that a period of 2π rad contains N samples, where

$$N = \left.\frac{2\pi}{\theta}\right|_{\text{rad}} = \left.\frac{360}{\theta}\right|_{\text{deg}} \text{ samples/cycle.}$$

For $\theta = 45°$, we have $N = 8$, and the plot of $e_4(k)$ given above shows the eight samples in the first cycle very clearly.

From the calculation of these few z-transforms we have established that the duration of a time signal is related to the radius of the pole locations and the number of samples per cycle is related to the angle, θ.

Another set of very useful relationships can be established by considering the signals to be samples from a continuous signal, $e(t)$, with Laplace transform $E(s)$. With this device we can exploit our knowledge of s-plane features by transferring them to equivalent z-plane properties. For the specific numbers represented in the illustration of e_4, we take the continuous signal

$$y(t) = e^{-at} \cos bt \tag{2.38}$$

with
$$aT = 0.3567,$$
$$bT = \pi/4.$$

And, taking samples one second apart ($T = 1$), we have

$$y(kT) = (e^{-0.3567})^k \cos \frac{\pi k}{4}$$

$$= (0.7)^k \cos \frac{\pi k}{4}$$

$$= e_4(k).$$

The poles of the Laplace transform of $y(t)$ (in the s-plane) are at

$$s_{1,2} = -a + jb, \ -a - jb.$$

From (2.29), the z-transform of $E_4(z)$ has poles at

$$z_{1,2} = re^{j\theta}, \ re^{-j\theta},$$

but since $y(kT)$ equals $e_4(k)$, it follows that

$$r = e^{-aT}, \qquad \theta = bT,$$

and

$$z_{1,2} = e^{s_1 T}, \ e^{s_2 T}.$$

If $E(z)$ is a ratio of polynomials in z, which will be the case if $e(k)$ is generated by a linear difference equation with constant coefficients, then by partial fraction expansion, $E(z)$ can be expressed as a sum of elementary terms like E_4 and E_3.[11] In all such cases, the discrete signal can be generated by samples from continuous signals where the relation between the s-plane poles and the corresponding z-plane poles is given by

$$z = e^{sT}. \tag{2.39}$$

If we know what it means to have a pole in a certain place in the s-plane, then (2.39) shows us where to look in the z-plane to find a representation of discrete samples having the *same time features*. It is useful to sketch several major features from the s-plane to the z-plane according to (2.39) to help fix these ideas. Such a sketch is shown in Fig. 2.17. Each feature should be traced in the mind to assure a good grasp is made of the relation. These features are given in Table 2.2. We note in passing that the map $z = e^{sT}$ of (2.39) is many-to-one. There are many values of s for each value of z. In fact, if

$$s_2 = s_1 + j\frac{2\pi}{T} N,$$

then $e^{s_1 T} = e^{s_2 T}$. The (great) significance of this fact will be explored in Chapter 4.

Our eventual purpose, of course, is to design digital controls and our interest in the relation between z-plane poles and zeros and time-domain response comes from our need to know how a proposed design will respond in a given dynamic situation. The generic dynamic test for controls is the step response and we will conclude this discussion of discrete system dynamic response with an examination of the relationships between the pole-zero patterns of elementary systems and the corresponding step responses for a discrete transfer function from u to y of a hypothetical plant. Our attention will be restricted to the step responses of the discrete system shown in Fig. 2.18 for a selected set of values of the parameters.

---

[11] Unless a pole of $E(z)$ is repeated. We have yet to compute the discrete version of a signal corresponding to a higher-order pole.

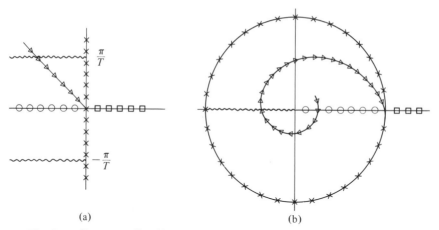

(a)   (b)

Fig. 2.17 Corresponding lines in s-plane and z-plane according to $z = e^{sT}$.

Table 2.2

s-plane	Symbol	z-plane
$\begin{cases} s = j\omega \\ \text{Real frequency axis} \end{cases}$	$\times\times\times$	$\begin{cases} \lvert z \rvert = 1 \\ \text{Unit circle} \end{cases}$
$s = \sigma \geq 0$	▢▢▢	$z = r \geq 1$
$s = \sigma \leq 0$	◯◯◯	$z = r, \quad 0 \leq r \leq 1$
$\begin{cases} s = -\zeta\omega_n + j\omega_n \sqrt{1 - \zeta^2} \\ \quad = -a + jb \\ \text{Constant damping ratio} \\ \text{if } \zeta \text{ is fixed and } \omega_n \\ \text{varies} \end{cases}$	△△△	$\begin{cases} z = re^{j\theta} \text{ where } r = \exp\{-\theta\zeta/\sqrt{1-\zeta^2}\} = e^{-aT}, \\ \qquad \theta = \omega_n T \sqrt{1-\zeta^2} = bT \\ \text{Logarithmic spiral} \end{cases}$
$s = \pm j(\pi/T)$	〰〰	$z = -r$

Note that if $z_1 = p_1$, the members of the one pole-zero pair cancel out, and that if $z_2 = r \cos \theta$, $a_1 = 2r \cos \theta$, $a_2 = r^2$, the system has a unit pulse response, $U(z) = 1$, whose transform is

$$Y(z) = \frac{z - r \cos \theta}{z^2 - 2r \cos \theta z + r^2}. \qquad (2.40)$$

This transform, when compared with the transform $E_4(z)$ given in (2.37), is seen to be

$$Y(z) = z^{-1}E_4(z),$$

$$U(z) \longrightarrow \boxed{K \dfrac{(z - z_1)(z - z_2)}{(z - p_1)(z^2 - a_1 z + a_2)}} \longrightarrow Y(z)$$

Fig. 2.18 Definition of the parameters of the system whose step responses are to be catalogued.

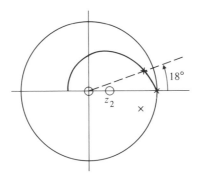

Fig. 2.19 Pole-zero pattern of $Y(z)$ for the system of Fig. 2.18, with $z_1 = P_1$, $U(z) = z/(z - 1)$, a_1 and a_2 selected for $\theta = 18°$, $\zeta = 0.5$.

and we conclude that under these circumstances the system pulse response is a delayed version of $e_4(k)$, a typical second-order system pulse response.

For our first study we consider the effect of zero location. We let $z_1 = p_1$ and explore the effect of the (remaining) zero location, z_2, on the step response overshoot for three sets of values of a_1 and a_2. We select a_1 and a_2 so that the poles of the system correspond to a response with damping ratio $\zeta = 0.5$ and consider values of θ of 18, 45, and 72 degrees. In every case, we will take the gain K to be such that the steady-state output value equals the step size. The situation in the z-plane is sketched in Fig. 2.19 for $\theta = 18°$. The curve for $\zeta = 0.5$ is also shown for reference. In addition to the two poles and one zero of $H(z)$, we show the pole at $z = 1$ and the zero at $z = 0$ which come from the transform of the input step, $U(z)$, given by $z/(z - 1)$.

The major effect of the zero z_2 on the step response $y(k)$ is to change the percent overshoot, as may be seen from the four-step responses for this case plotted in Fig. 2.20. To summarize all these data, we plot the percent overshoot versus

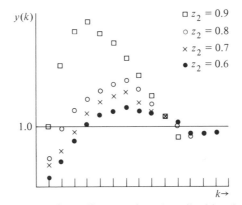

Fig. 2.20 Plot of step responses for a discrete plant described by the pole-zero pattern of Fig. 2.19 for various values of z_2.

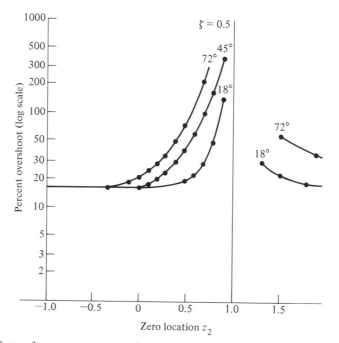

Fig. 2.21 Effects of an extra zero on a discrete second-order system, $\zeta = 0.5$; $\theta = 18°, 45°,$ and $72°$.

zero location in Fig. 2.21 for $\zeta = 0.5$ and in Fig. 2.22 for $\zeta = 0.707$. The major feature of these plots is that the zero has very little influence when on the negative axis, but its influence is dramatic as it comes near $+1$. Also included on the plots of Fig. 2.21 are overshoot figures for a zero in the unstable region on the positive real axis. These responses go in the *negative* direction at first and for the zero very near $+1$, the negative peak is larger than one![12]

Our second class of step responses corresponds to a study of the influence of a third pole on a basically second order response. For this case we again consider the system of Fig. 2.18, but this time we fix $z_1 = z_2 = -1$ and let p_1 vary from near -1 to near $+1$. In this case, the major influence of the moving singularity is on the *rise time* of the step response. We plot this effect for $\theta = 18, 45,$ and 72 degrees and $\zeta = 0.5$ on Fig. 2.23. In the figure we define the rise time as the time required for the response to rise to 0.95, which is to 5% of its final value. We see here that the extra pole causes the rise time to get very much longer as the location of p_1 moves toward $z = +1$ and comes to dominate the response.

Our conclusions from these plots are that the addition of a pole or zero to a

---

[12] Such systems are called nonminimum phase by Bode because the phase shift they impart to a sinusoidal input is greater than the phase of a system whose *magnitude* response is the same but which has a zero in the stable rather than the unstable region.

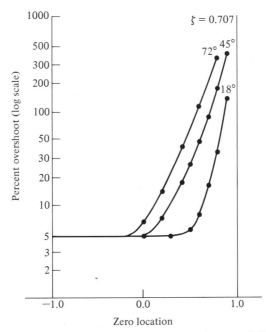

Fig. 2.22 Effects of extra zero on second-order system when $\zeta = 0.707$; $\theta = 18°, 45°, 72°$. Percent overshoot versus zero location.

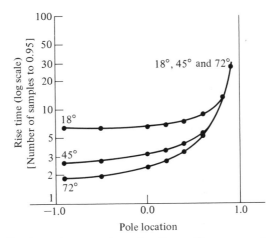

Fig. 2.23 Effects of extra pole on system rise time.   Two zeros at -1, one zero at ∞. $\zeta = 0.5$; $\theta = 18°, 45°, 72°$.

given system has only a small effect if the added singularities are in the range from 0 to -1. However, a zero moving toward $z = +1$ greatly increases the system overshoot. A pole placed toward $z = +1$ causes the response to slow down and thus primarily affects the rise time which is being progressively increased.

2.5 PROPERTIES OF THE z-TRANSFORM

We have used the z-transform to show that linear constant discrete systems can be described by a transfer function which is the z-transform of the system's unit pulse response, and we have studied the relationship between the pole-zero patterns of transfer functions in the z-plane and the corresponding time responses. We turn now to consideration of some of the properties of the z-transform which are essential to the effective and correct use of this important tool.

As with the Laplace transform, the z-transform is actually one of a pair of transforms which connect functions of time to functions of the complex variable s or z. The z-transform computes a function of z from a sequence in k. (We identify the sequence number k with time in our analysis of dynamic systems but there is nothing in the transform *per se* which requires this.) The inverse z-transform is a means to compute a sequence in k from a given function of z. There is an integral for the inverse transform which we will study presently, but our understanding of it will be enhanced if we first examine two more elementary schemes for inversion of a given $F(z)$ which can be used if we know beforehand that $F(z)$ is rational in z and converges as z approaches infinity. For a sequence $f(k)$, the z-transform has been defined as

$$F(z) = \sum_{k=-\infty}^{\infty} f(k)z^{-k}, \qquad r_0 < |z| < R_0. \tag{2.41}$$

If any value of $f(k)$ for negative k is nonzero, then there will be a term in (2.41) with a positive power of z. This term will be unbounded if the magnitude of z is unbounded and thus if $F(z)$ converges as $|z|$ approaches infinity, we know that $f(k)$ is zero for $k < 0$. In this case, (2.41) is one-sided and we can write

$$F(z) = \sum_{k=0}^{\infty} f(k)z^{-k}, \qquad r_0 < |z|. \tag{2.42}$$

The right-hand side of (2.42) is a series expansion of $F(z)$ about infinity or about $z^{-1} = 0$. Such an expansion is especially easy if $F(z)$ is the ratio of two polynomials in z^{-1}. We need only divide the numerator by the denominator in the correct way, and the division once done, the coefficient of z^{-k} is automatically the sequence value $f(k)$. An example we have worked out before will illustrate the process. Suppose we take our system to be the trapezoid rule integration with transfer function given by (2.14)

$$H(z) = \frac{T}{2} \frac{z + 1}{z - 1}, \qquad |z| > 1. \tag{2.14}$$

We will take the input to be the geometric series represented by $e_3(k)$ with $r = 0.5$. Then we have

$$E_3(z) = \frac{z}{z - 0.5}, \qquad |z| > 0.5, \tag{2.43}$$

and

$$U(z) = E_3(z)H(z)$$

$$= \frac{z}{z - 0.5} \frac{T}{2} \frac{z + 1}{z - 1}, \quad |z| > 1. \tag{2.44}$$

Equation (2.44) represents the transform of the system output, $u(k)$. Keeping out the factor of $T/2$, we write $U(z)$ as a ratio of polynomials in z^{-1},

$$U(z) = \frac{T}{2} \frac{1 + z^{-1}}{1 - 1.5z^{-1} + 0.5z^{-2}}, \tag{2.45}$$

and divide as follows:

$$\frac{T}{2}[1 + 2.5z^{-1} + 3.25z^{-2} + 3.625z^{-3} + \cdots$$

$$1 - 1.5z^{-1} + 0.5z^{-2}\overline{)1 + z^{-1}}$$
$$\underline{1 - 1.5z^{-1} + 0.5z^{-2}}$$
$$2.5z^{-1} - 0.5z^{-2}$$
$$\underline{2.5z^{-1} - 3.75z^{-2} + 1.25z^{-3}}$$
$$3.25z^{-2} - 1.25z^{-3}$$
$$\underline{3.25z^{-2} - 4.875z^{-3} + 1.625z^{-4}}$$
$$3.625z^{-3} - 1.625z^{-4}$$
$$\underline{3.625z^{-3} - \cdots}$$

By direct comparison with $U(z) = \sum_0^\infty u(k)z^{-k}$, we conclude that

$$u_0 = T/2,$$
$$u_1 = (T/2)2.5,$$
$$u_2 = (T/2)3.25, \tag{2.46}$$
$$\cdot$$
$$\cdot$$
$$\cdot$$

Clearly, the use of a computer will greatly aid the speed of this process in all but the simplest of cases.[13]

The second special method for the inversion of z-transforms is to decompose $F(z)$ by partial fraction expansion and look up the components of the sequence $f(k)$

---

[13] Or, as the authors find, a programmable hand-held calculator for transforms to 6th order. Some may prefer to use synthetic division and omit copying over all the extraneous z's in the division. The process is identical to converting $F(z)$ to the equivalent difference equation and solving for the unit pulse response.

in a previously prepared table. We consider again (2.44) and expand $U(z)$ as a function of z^{-1} as follows,

$$U(z) = \frac{T}{2} \frac{1 + z^{-1}}{1 - z^{-1}} \frac{1}{1 - 0.5z^{-1}} = \frac{A}{1 - z^{-1}} + \frac{B}{1 - 0.5z^{-1}}.$$

We multiply both sides by $1 - z^{-1}$, let $z^{-1} = 1$, and compute

$$A = \frac{T}{2} \frac{2}{0.5} = 2T.$$

Similarly, at $z^{-1} = 2$, we evaluate

$$B = \frac{T}{2} \frac{1 + 2}{1 - 2} = -\frac{3T}{2}.$$

Looking back now at e_2 and e_3, which constitute our "table" for the moment, we can copy down that

$$u_k = Ae_2(k) + Be_3(k)$$

$$= 2Te_2(k) - \frac{3T}{2} e_3(k)$$

$$= 2T - \frac{3T}{2} \left(\frac{1}{2}\right)^k$$

$$= \frac{T}{2} \left[4 - \frac{3}{2^k}\right], \qquad k \geq 0. \tag{2.47}$$

Evaluation of (2.47) for $k = 0, 1, 2, \ldots$ will, naturally, give the same values for $u(k)$ as we found in (2.46). A table of z-transforms is given in Appendix B, at the back of the book, for general reference.

In order to make maximum use of the table, one must be able to use a few simple properties of the z-transform which follow directly from the definition. Some of these, such as linearity, we have already used without making a formal statement of it, and others such as the transform of the convolution we have previously derived. For reference, we will demonstrate a few properties here and collect them into Appendix B for future reference. In all the properties listed below, we assume that $F_i(z) = \mathscr{z}\{f_i(kT)\}$.

1. *Linearity:* A function $f(x)$ is linear if $f(\alpha x_1 + \beta x_2) = \alpha f(x_1) + \beta f(x_2)$. Applying this result to the definition of the z-transform, we find immediately that

$$\mathscr{z}\{\alpha f_1(kT) + \beta f_2(kT)\} = \sum_{k=-\infty}^{\infty} \{\alpha f_1(k) + \beta f_2(k)\}z^{-k}$$

$$= \alpha\mathscr{z}\{f_1(k)\} + \beta\mathscr{z}\{f_2(k)\}$$

$$= \alpha F_1(z) + \beta F_2(z). \tag{2.48}$$

Thus the z-transform is a linear function.   It is the linearity of the transform which makes the partial fraction technique work.

2. *Convolution of Time Sequences:*

$$\mathscr{z}\left\{\sum_{\ell=-\infty}^{\infty} f_1(\ell)f_2(k-\ell)\right\} = F_1(z)F_2(z). \tag{2.49}$$

We have already developed this result in connection with (2.21).   It is this result which makes the transform so useful in linear constant-system analysis because the combination of such dynamic systems is done by linear algebra on the transfer functions.

3. *Time Shift:*

$$\mathscr{z}\{f(k+n)\} = z^{+n}F(z). \tag{2.50}$$

We demonstrate this result also by direct calculation:

$$\mathscr{z}\{f(k+n)\} = \sum_{k=-\infty}^{\infty} f(k+n)z^{-k}.$$

If we let $k + n = j$, then

$$\mathscr{z}\{f(k+n)\} = \sum_{j=-\infty}^{\infty} f(j)z^{-(j-n)}$$

$$= z^n F(z). \qquad \text{QED}$$

This property is the essential tool in solving linear constant-coefficient difference equations by transforms.   We should note here that the transform of the time shift is not the same for the one-sided transform since a shift may introduce terms with negative argument which are not included in the one-sided transform and must be treated separately.   This effect causes initial conditions for the difference equation to be introduced when solution is done with the one-sided transform.   See Problem 2.5.

4. *Scaling in the z-plane:*

$$\mathscr{z}\{r^{-k}f(k)\} = F(rz). \tag{2.51}$$

By direct substitution,

$$\mathscr{z}\{r^{-k}f(k)\} = \sum_{k=-\infty}^{\infty} r^{-k}f(k)z^{-k}$$

$$= \sum_{k=-\infty}^{\infty} f(k)(rz)^{-k}$$

$$= F(rz). \qquad \text{QED}$$

As an illustration of this property, we consider the z-transform of the unit step, $1(k)$, which we have computed before:

$$\mathscr{z}\{1(k)\} = \sum_{k=0}^{\infty} z^{-k} = \frac{z}{z-1}.$$

By property 4 we have immediately that

$$\mathscr{z}\{r^{-k}1(k)\} = \frac{rz}{rz-1} = \frac{z}{z-(1/r)}.$$

5. *Final-Value Theorem:* If $F(z)$ converges for $|z| > 1$ and all poles of $(1 - z)F(z)$ are inside the unit circle, then

$$\lim_{k\to\infty} f(k) = \lim_{z\to 1} (z-1)F(z). \tag{2.52}$$

The conditions on $F(z)$ assure that the only possible pole of $F(z)$ not strictly inside the unit circle is a simple pole at $z = 1$ which is removed in $(z - 1)F(z)$. Furthermore, the fact that $F(z)$ converges as the magnitude of z gets arbitrarily large ensures that $f(k)$ is zero for negative k. Therefore, all components of $f(k)$ tend to zero as k gets large, with the possible exception of the constant term due to the pole at $z = 1$. The size of this constant is given by the coefficient of $1/(z - 1)$ in the partial fraction expansion of $F(z)$, namely

$$C = \lim_{z\to 1} (z-1)F(z).$$

However, since all other terms in $f(k)$ tend to zero, the constant C is the final value of $f(k)$ and (2.52) results. QED

As an illustration of this property, we consider the signal whose transform is given by $U(z)$ in (2.44):

$$U(z) = \frac{z}{z-0.5}\frac{T}{2}\frac{z+1}{z-1}, \qquad |z| > 1. \tag{2.44}$$

Since $U(z)$ satisfies the conditions of (2.52), we have

$$\lim_{k\to\infty} u(k) = \lim_{z\to 1} (z-1)\frac{z}{z-0.5}\frac{T}{2}\frac{z+1}{z-1}$$

$$= \lim_{z\to 1} \frac{z}{z-0.5}\frac{T}{2}(z+1)$$

$$= \frac{1}{1-0.5}\frac{T}{2}(1+1)$$

$$= 2T.$$

This result may be checked against the closed form for $u(k)$ given by (2.47).

2.6 REGIONS OF CONVERGENCE AND THE INVERSE INTEGRAL[14]

We must now examine more closely the role of the region of convergence of the z-transform and present the inverse transform integral. We begin with another example. The sequence

$$f(k) = \begin{cases} -1, & k < 0, \\ 0, & k \geq 0 \end{cases}$$

has the transform

$$F(z) = \sum_{k=-\infty}^{-1} - z^{-k}$$

$$= - \left[\sum_{0}^{\infty} z^k - 1 \right]$$

$$= \frac{z}{z - 1}, \qquad |z| < 1. \tag{2.53}$$

This transform is exactly the same as the transform of the unit step $1(k)$, (2.31), except that this transform converges *inside* the unit circle and the transform of the $1(k)$ converges outside the unit circle. Knowledge of the region of convergence is obviously essential to the proper inversion of the transform to obtain the time sequence. The inverse z-transform is the closed complex integral[15]

$$f(k) = \frac{1}{2\pi j} \oint F(z) z^k \frac{dz}{z}, \tag{2.54}$$

where the contour is a circle in the region of convergence of $F(z)$. To demonstrate the correctness of the integral and to use it to compute inverses it is useful to apply Cauchy's residue calculus [see Rosenbrock and Storey (1970)]. The result is that a closed integral of a function of z which is analytic on and inside a closed contour except at a finite number of isolated singularities z_i is given by

$$\frac{1}{2\pi j} \oint_C F(z) \, dz = \sum_i \mathrm{Res}(z_i). \tag{2.55}$$

In (2.55), $\mathrm{Res}(z_i)$ means the residue of $F(z)$ at the singularity at z_i. We will be considering only rational functions, and these have only poles as singularities. If $F(z)$ has a pole of order n at z_1, then $(z - z_1)^n F(z)$ is regular at z_1 and may be expanded

---

[14] Contains material which may be omitted without loss of continuity.
[15] If it is known that $f(k)$ is causal, that is $f(k) = 0$ for $k < 0$, then the region of convergence is outside the smallest circle which contains all the poles of $F(z)$ for rational transforms. It is this property which permits inversion by partial-fraction expansion and long division.

in a Taylor series near z_1 as

$$(z - z_1)^n F(z) = A_{-n} + A_{-n+1}(z - z_1) + \cdots + A_{-1}(z - z_1)^{n-1}$$
$$+ A_0(z - z_1)^n + \cdots \quad (2.56)$$

The residue of $F(z)$ at z_1 is A_{-1}.

First we will use Cauchy's formula to verify (2.54). If $F(z)$ is the z-transform of $f(k)$, then we write

$$\mathcal{I} = \frac{1}{2\pi j} \oint \sum_{\ell=-\infty}^{\infty} f(\ell) z^{-\ell} z^k \frac{dz}{z}.$$

We assume that the series for $F(z)$ converges uniformly on the contour of integration, so the series may be integrated term by term. Thus

$$\mathcal{I} = \frac{1}{2\pi j} \sum_{\ell=-\infty}^{\infty} f(\ell) \oint z^{k-\ell} \frac{dz}{z}.$$

The argument of the integral has no pole inside the contour if $k - \ell \geq 1$, and has zero residue at the pole at $z = 0$ if $k - \ell < 0$. Only if $k = \ell$, does the integral have a residue and that is 1. By (2.55), the integral is zero if $k \neq \ell$ and is $2\pi j$ if $k = \ell$. Thus $\mathcal{I} = f(k)$, which demonstrates (2.54).

To illustrate the use of (2.54) to compute the inverse of a z-transform, we will use the function $z/(z - 1)$ and consider first the case of convergence for $|z| > 1$ and second the case of convergence for $|z| < 1$. For the first case,

$$f_1(k) = \frac{1}{2\pi j} \oint_{|z|=R>1} \frac{z}{z - 1} z^k \frac{dz}{z}, \quad (2.57)$$

where the contour is a circle of radius greater than 1. Suppose $k < 0$. In this case, the argument of the integral has two poles inside the contour: one at $z = 1$ with residue

$$\lim_{z \to 1} (z - 1) \frac{z^k}{z - 1} = 1,$$

and one pole at $z = 0$ with residue found as in (2.56) (if $k < 0$, then z^{-k} removes the pole):

$$z^{-k} \frac{z^k}{z - 1} = -\frac{1}{1 - z}$$
$$= -(1 + z + z^2 + \cdots + z^{k-1} + \cdots).$$

The residue is thus -1 for all k, and the sum of the residues is zero, and

$$f_1(k) = 0, \quad k < 0. \quad (2.58)$$

For $k \geq 0$, the argument of the integral in (2.57) has only the pole at $z = 1$ with residue 1. Thus

$$f_1(k) = 1, \qquad k \geq 0. \qquad (2.59)$$

Equations (2.58) and (2.59) correspond to the unit step function, as they should. We would write the inverse transform symbolically as $\mathscr{z}^{-1}\{\cdot\}$ as, in this case,

$$\mathscr{z}^{-1}\left\{\frac{z}{z-1}\right\} = 1(k) \qquad (2.60)$$

when $z/(z-1)$ converges for $|z| > 1$.

If, on the other hand, convergence is inside the unit circle, then for $k \geq 0$, there are no poles of the integrand contained in the contour, and

$$f_2(k) = 0, \qquad k \geq 0.$$

At $k < 0$, there is a pole at the origin of z, and as before, the residue is equal to -1 there, so

$$f_2(k) = -1, \qquad k < 0.$$

In symbols, corresponding to (2.60), we have

$$\mathscr{z}^{-1}\left\{\frac{z}{z-1}\right\} = 1(k) - 1 \qquad (2.61)$$

when $z/(z-1)$ converges for $|z| < 1$.

Although, as we have just seen, the inverse integral can be used to compute an expression for a sequence to which a transform corresponds, a more effective use of the integral is in more general manipulations. We consider one such case of some interest. First, we consider an expression for the transform of a product of two sequences. Suppose we have

$$f_3(k) = f_1(k)f_2(k), \qquad (2.62)$$

and f_1 and f_2 are such that the transform of the product exists. An expression for $F_3(z)$ in terms of $F_1(z)$ and $F_2(z)$ may be developed as follows. By definition

$$F_3(z) = \sum_{k=-\infty}^{\infty} f_1(k)f_2(k)z^{-k}. \qquad (2.63)$$

From the inversion integral, (2.54), we can replace $f_2(k)$ by an integral:

$$F_3(z) = \sum_{k=-\infty}^{\infty} f_1(k)z^{-k} \frac{1}{2\pi j} \oint_{C_2} F_2(\zeta)\zeta^k \frac{d\zeta}{\zeta}.$$

We assume that we can find a region where we may exchange the summation with the integration. The contour will be called C_3 in this case:

$$F_3(z) = \frac{1}{2\pi j} \oint_{C_3} F_2(\zeta) \sum_{k=-\infty}^{\infty} f_1(k) \left(\frac{z}{\zeta}\right)^{-k} \frac{d\zeta}{\zeta}.$$

The sum may now be recognized as $F_1(z/\zeta)$ and, when we substitute this,

$$F_3(z) = \frac{1}{2\pi j} \oint_{C_3} F_2(\zeta)F_1\left(\frac{z}{\zeta}\right) \frac{d\zeta}{\zeta}, \tag{2.64}$$

the contour C_3 may be in the overlap of the convergence regions of $F_2(\zeta)$ and $F_1(z/\zeta)$. Then $F_3(z)$ will converge for the range of values of z for which C_3 can be found.

If we let $f_1 = f_2$ and $z = 1$ in (2.64), we have the discrete version of Parseval's theorem, where convergence is on the unit circle:

$$F_3(1) = \sum_{k=-\infty}^{\infty} f_1^2 = \frac{1}{2\pi j} \oint F_1(\zeta)F_1\left(\frac{1}{\zeta}\right) \frac{d\zeta}{\zeta}. \tag{2.65}$$

2.7 IMPLEMENTATION OF DIFFERENCE EQUATIONS IN REAL TIME

We have presented an analysis of linear constant difference equations and implied that we expect the equations to be solved in real time on a digital computer. While almost every real-time facility is unique, there are certain important features which are independent of implementation which we would like to describe. To do so, we will expand the digital part of Fig. 1.1, especially the A/D converter. Actually this device usually has three major components, as shown in Fig. 2.24: a multiplexer, a sample-and-hold, and the converter proper.

As suggested by the figure, the multiplexer is a switch (electronic, usually) which selects from among several (n) lines the one that will supply the signal for the current conversion. Most multiplexers have a random-address mode whereby any given line can be selected. Some have a "burst" mode in which a number k between 1 and n is given, and with each clock pulse, the multiplexer will automatically step from 1, 2, . . . , k.

The sample-and-hold is an analog circuit which may look like drawing (b) and

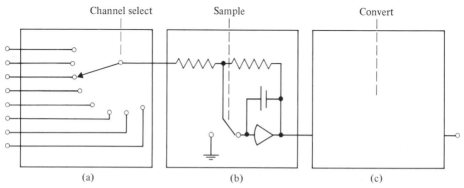

Fig. 2.24 Components of analog-to-digital converter system.   (a) Multiplexer. (b) Sample-and-hold.   (c) A/D converter.

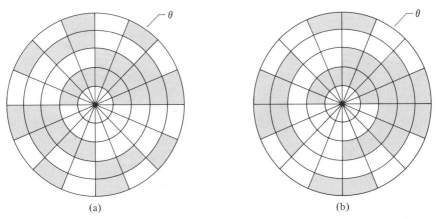

Fig. 2.25 Sketch of shaft-encoded disks. (a) Binary code. (b) Gray code.

causes the charge on a capacitor to track the input voltage until the "sample" signal is given, at which time the input is disconnected and the capacitor voltage holds constant for a specified time long enough to allow the A/D converter to complete its conversion. The purpose of the sample-and-hold is to guarantee that the A/D converter produces a number which accurately represents the input at the sampling "instant."

Analog-to-digital converters take many forms. One common class is used in servomechanisms which control the rotation of a shaft. The angular position of the shaft can be coded into digital values by a disk attached to the shaft and composed of bands of transparent and dark segments as sketched in Fig. 2.25. If a light is behind each band and a light detector (photo diode, perhaps) is in front of each band, then the shaft angle code is given by the pattern of signals from the light detectors. The common binary radix code[16] corresponds to the signals generated by the binary disk. These signals have the serious problem that the transition of angle by one sector may cause the code to shift many bits. For example, the transition from $\theta \sim 0000$ to $\theta \sim 1111$ has every bit in the code change. A minor misalignment of the photo detectors could cause a major error in angle reading. The Gray code avoids this ambiguity for it arranges the clear and opaque areas so that a movement of angle by one sector causes only one bit in the code to change.

Even more common than shaft encoders are A/D converters whose inputs are electric voltages. Of the many types of voltage converters which exist, we will describe two which are very common. The first technique uses a digital counter as the device which generates the digital number. If the input voltage is converted into a train of pulses with variable frequency, then for a fixed period, these pulses are counted and the result is a number proportional to the voltage. A vari-

---

[16] The number represented by the code $b_m b_{m-1}, \ldots, b_1 b_0 = b_0 + b_1 2 + b_2 2^2 + \cdots + b_m 2^m$.

ation on this theme is to count a train of pulses of fixed frequency for a time which is modulated by the size of the voltage to be converted. The variable time can be generated by starting a linear (in time) voltage at the start of the count and stopping the count when the linear voltage has reached the magnitude of the input signal. A technique which is much faster than the counter-based converters is the method of successive approximation. For this technique the unknown voltage is first compared to a reference which is half the maximum. If the unknown is larger, the most significant bit is set and the unknown is compared to three-quarters of the maximum to determine the next bit, and so on. Such converters set one bit each clock cycle so they need n cycles to generate n bits. At the same clock rate, the counter-based converters may require as many as 2^n cycles.

The digital-to-analog converter may also take many forms. The basic idea is that the given digital code can be used to cause switches (electronic gates) to be opened or closed, and these, in turn, generate an output voltage corresponding to the code. A standard circuit is shown in Fig. 2.26. The output of this circuit is

$$V_0 = \pm\, V_{\text{Ref}}\, \frac{R}{2}\left\{\frac{b_4}{R} + \frac{b_3}{2R} + \frac{b_2}{4R} + \frac{b_1}{8R} + \frac{b_0}{16R}\right\},$$

which is a multiple of the number corresponding to the binary code $b_4 b_3 b_2 b_1 b_0$.

Once the word is inside the machine, it is necessary to have a program which solves the difference equation and generates the correct number to be converted at the output. The logic of a typical program for a single-input–single-output system is shown in Fig. 2.27. To be concrete, we show the computation for the third-order observer canonical form shown in Fig. 2.8. The output variables of the delay elements are named x_1, x_2, and x_3 from right to left. The program has two flags, STOP and FLAG. The program assumes that the machine has a pro-

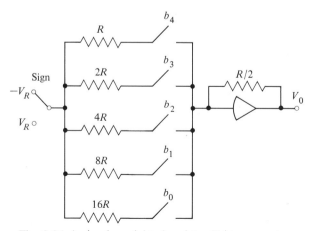

Fig. 2.26 A simple weighted resistor D/A converter.

PROGRAM

1. Store the initial values in variables e, u, x_1, x_2, x_3 and read values for parameters a_1, a_2, a_3, b_0, b_1, b_2, b_3.

2. Reset the FLAG to zero, select the sample period, and start the clock.

3. If the STOP flag is set, halt the machine (or return to the monitor program).

4. If the FLAG is set to one, then print the message "sampling period too short" and halt the machine (or return to the monitor program).

5. If the FLAG is reset to zero, then wait at step 5. ELSE start the A/D converter, store the result from channel C in the location of variable e and reset the FLAG to zero.

6. $u \leftarrow b_0 e + x_1$

7. Send u to the D/A.

8. $x_1 \leftarrow b_1 e + a_1 u + x_2$

9. $x_2 \leftarrow b_2 e + a_2 u + x_3$

10. $x_3 \leftarrow b_3 e + a_3 u$

11. Go to step 3.

Fig. 2.27 Logic of a real-time control program.   (The arrow pointing left is an assignment operator to be read "is replaced by.")

grammable real-time clock such that a clock interrupt will occur once every T seconds where T may be set under program control.   *When the clock interrupt occurs, the variable identified as FLAG is set to one.*

In step 1 the program parameters are set up, and the initial conditions of the difference equation are established.   In step 2 the computer is set up to respond to the clock interrupt, and the clock is started.   The STOP flag tested in step 3 allows the operator (or another program) to turn off this real-time loop.   In a small system, the STOP flag is often set by a front panel switch.   In step 4 the FLAG is tested to see if the clock has interrupted the program before the equations have been solved.   Obviously if we try to solve complicated equations in a very short sampling period, our computer may not complete the calculations before another output is due.   In this case we must either ask for simpler (faster) computations or else increase the sampling period.   In normal operation, the program sits at step 5 waiting for the clock to trigger the next A/D conversion which it does by setting the FLAG.   When the FLAG is set by the clock, the program goes to the A/D conversion.   When the A/D conversion is done, the FLAG is reset, and we move on to the difference equations.   If the parameter b_0 is zero, step 6 is eliminated and step 7 changed to "send x_1 to the D/A."   The idea is to send out the new value of u as soon as possible because any time delay between the clock interrupt and the generation of the control output interferes with achieving good dynamic response.   In steps 8, 9, and 10, the main computations are done for the next cycle.   Note that if the clock interrupt occurs anywhere in these latter steps, the FLAG will be set by the clock program and step 4 will identify the error.

The program logic of Fig. 2.27 does not show the number system used for calculation or the effects of data overflow that may occur in steps 6, 8, 9, or 10.   A common practice is to turn on an overflow light (or flag), substitute the positive or negative limiting value in the available number system for the overflow value, and continue with the computation.   The result is like the saturation of an analog operational amplifier.

2.8 SUMMARY

In this chapter we have shown how systems described by linear difference equations with constant coefficients may be described by transfer functions if the signals are represented by z-transforms.   The transfer function was shown to be the z-transform of the unit-pulse response of the system, and furthermore the system output was shown to be the convolution of the input with the unit-pulse response.   We introduced the observer and the control canonical forms for transfer functions and gave rules for block-diagram reduction of transfer functions.

We studied the dynamic response of discrete systems, including especially the step response of a second-order system.   The effects of the location of the zero and of a third pole were plotted, largely for future reference in design.

Several of the properties of the z-transform were demonstrated, and the calculation of the inverse of a z-transform was presented by long division, by partial fraction expansion, and by evaluation of the inverse transform integral.

In the final section, we discussed some practical issues which arise when one wishes to implement a difference equation in real time on a digital computer and analyzed a simplified program logic for this purpose.

Appendix to Chapter 2

Another Derivation of the Discrete Transfer Function.[17]

Let $\mathscr{D}$ be a discrete system which maps an input sequence, $\{e(k)\}$, into an output sequence, $\{u(k)\}$.   Then, expressing this as an operator on $e(k)$, we have

$$u(k) = \mathscr{D}\{e(k)\}.$$

If $\mathscr{D}$ is linear, then

$$\mathscr{D}\{\alpha e_1(k) + \beta e_2(k)\} = \alpha\mathscr{D}\{e_1(k)\} + \beta\mathscr{D}\{e_2(k)\}. \tag{2.66}$$

If the system is constant, a shift in $e(k)$ to $e(k + j)$ must result in no other effects

---

[17] Suggested by L. A. Zadeh (1952) at Columbia University.

but a shift in the response, u. We write

$$\mathscr{D}\{e(k + j)\} = u(k + j) \qquad \text{for all} \quad j \tag{2.67}$$

if

$$\mathscr{D}\{e(k)\} = u(k).$$

Theorem   If $\mathscr{D}$ is linear and constant and is given an input z^k for a value of z for which the output is finite at time k, then the output will be of the form $H(z)z^k$.

Proof.   In general, if $e(k) = z^k$, then an arbitrary finite response may be written

$$u(k) = H(z, k)z^k.$$

Consider $e_2(k) = z^{k+j} = z^j z^k$ for some fixed j. From (2.66), if we let $\alpha = z^j$, it must follow that

$$\begin{aligned}
u_2 &= z^j u(k) \\
&= z^j H(z, k)z^k \\
&= H(z, k)z^{k+j}.
\end{aligned} \tag{2.68}$$

From (2.67), we must have

$$\begin{aligned}
u_2(k) &= u(k + j) \\
&= H(z, j + k)z^{k+j} \qquad \text{for all} \quad j.
\end{aligned} \tag{2.69}$$

From a comparison of (2.68) and (2.69), it follows that

$$H(z, k) = H(z, k + j) \qquad \text{for all} \quad j;$$

that is, H does not depend on the second argument and may be written $H(z)$. Thus for the elemental signal $e(k) = z^k$, we have a solution $u(k)$ of the same (exponential) shape but modulated by a ratio $H(z)$.

Can we represent a general signal as a *linear sum* (integral) of such elements? We can, by the integral

$$e(k) = \frac{1}{2\pi j} \oint E(z)z^k \frac{dz}{z}, \tag{2.70}$$

where

$$E(z) = \sum_{-\infty}^{\infty} e(k)z^{-k}, \qquad r < |z| < R, \tag{2.71}$$

for signals with $r < R$ for which (2.71) converges. We call $E(z)$ the z-transform of $e(k)$, and the (closed) path of integration is in the annular region of convergence of (2.71). If $e(k) = 0$, $k < 0$, then $R \to \infty$, and this region is the whole z-plane *out-*

side a circle of finite radius.   Signals which are zero for $k < 0$ and bounded by an exponential for $k > 0$ may be represented by (2.70).

The consequences of linearity are that the response to a sum of signals is the sum of the responses as given in (2.66).   Although (2.70) is the limit of a sum, the result still holds, and we can write

$$u(k) = \frac{1}{2\pi j} \oint E(z) \frac{dz}{z} \,[\text{response to } z^k],$$

but, by the theorem, the response to z^k is $H(z)z^k$.   Therefore

$$u(k) = \frac{1}{2\pi j} \oint E(z) \frac{dz}{z} \,[H(z)z^k].$$

$$= \frac{1}{2\Pi j} \oint H(z)E(z)z^k \frac{dz}{z}. \tag{2.72}$$

We may define $U(z) = H(z)E(z)$ by comparison with (2.70) and note that

$$U(z) = \sum_{k=-\infty}^{\infty} u(k)z^{-k} = H(z)E(z). \tag{2.73}$$

Thus $H(z)$ is the *transfer function,* which is the ratio of the transforms of $e(k)$ and $u(k)$ as well as the amplitude response to inputs of the form z^k.

This derivation begins with linearity and stationarity and derives the z-transform as the natural tool of analysis from the fact that input signals in the form z^k produce an output which has the same shape.[18]   It is somewhat more satisfying to derive the necessary transform than to start with the transform and see what systems it is good for.   Better to start with the problem and find a tool than start with a tool and look for a problem.   Unfortunately, the direct approach requires extensive use of the inversion integral and more sophisticated analysis to develop the main result, which is (2.73).   Chacun à son goût.

PROBLEMS AND EXERCISES

2.1 a) Derive the difference equation corresponding to the approximation of integration found by fitting a parabola to the points e_{k-2}, e_{k-1}, e_k and taking the area under this parabola between $t = kT - T$ and $t = kT$ as the approximation to the integral of $e(t)$ over this range.

  b) Find the transfer function of the resulting discrete system and plot the poles and zeros in the z-plane.

2.2 Verify that the transfer function of the system of Fig. 2.12 is given by the same $H(z)$ as the system of Fig. 2.8.

---

[18] Since z^k is unchanged in shape by passage through the linear constant system, we say that z^k is an eigenfunction of such systems.

2.3 a) Compute and plot the unit pulse response of the system derived in Exercise 2.1.
b) Is this system BIBO stable?

2.4 The first-order system $(z - \alpha)/(1 - \alpha)z$ has a zero at $z = \alpha$.

a) Plot the step response for this system for $\alpha = 0.8, 0.9, 1.1, 1.2, 2$.
b) Plot the overshoot of this system on the same coordinates as those appearing in Fig. 2.21 for $-1 < \alpha < 1$.
c) In what way is the step response of this system unusual for $\alpha > 1$?

2.5 The one-sided z-transform is defined as

$$F(z) = \sum_0^\infty f(k)z^{-k}.$$

a) Show that the one-sided transform of $f(k + 1)$ is

$$\mathscr{z}\{f(k + 1)\} = zF(z) - zf(0).$$

b) Use the one-sided transform to solve for the transforms of the Fibonacci numbers by writing (2.4) as $u_{k+2} = u_{k+1} + u_k$. Let $u_0 = u_1 = 1$. [You will need to compute the transform of $f(k + 2)$.]
c) Compute the location of the poles of the transform of the Fibonacci numbers.
d) Compute the inverse transform of the numbers.
e) Show that if u_k is the kth Fibonacci number, then the ratio u_{k+1}/u_k will go to $(1 + \sqrt{5})/2$, the golden ratio of the Greeks.
f) Show that if we add a forcing term, $e(k)$, to (2.4) we can generate the Fibonacci numbers by a system which can be analyzed by the two-sided transform; i.e., let $u_k = u_{k-1} + u_{k-2} + e_k$ and let $e_k = \delta_0(k)[\delta_0(k) = 1$ at $k = 0$ and zero elsewhere]. Take the two-sided transform and show that the same $U(z)$ results as in part (b).

2.6 Consider the transfer function

$$H(z) = \frac{(z + 1)(z^2 - 1.3z + 0.81)}{(z^2 - 1.2z + 0.5)(z^2 - 1.4z + 0.81)}.$$

Draw a *cascade* realization, using observer canonical forms for second-order blocks and in such a way that the coefficients as shown in $H(z)$ above are the parameters of the block diagram.

2.7 a) Write the $H(z)$ of Exercise 2.6 in partial fractions in two terms of second order each and draw a *parallel* realization, using the observer canonical form for each block and showing the coefficients of the partial fraction expansion as the parameters of the realization.
b) Suppose the two factors in the denominator of $H(z)$ were identical (say we change the 1.4 to 1.2 and the 0.81 to 0.5). What would the parallel realization be in this case?

2.8 For a second-order system with damping ratio 0.5 and poles at an angle in the z-plane of $\theta = 30°$, what percent overshoot to a step would you expect if the system had a zero at $z_2 = 0.6$?

P39
Time Shift

2.9 Consider a signal with the transform (which converges for $|z| > 2$)

$$U(z) = \frac{z}{(z - 1)(z - 2)}.$$

a) What value is given by the formula (final-value theorem) of (2.52) applied to this $U(z)$?

b) Find the final value of $u(k)$ by taking the inverse transform of $U(z)$, using partial-fraction expansion and the tables.

c) Explain why the two results of (a) and (b) differ.

2.10 a) Find the z-transform and be sure to give the region of convergence for the signal

$$u(k) = r^{+|k|}, \qquad r < 1.$$

[*Hint:* Write u as the sum of two functions, one for $k \geq 0$ and one for $k < 0$, find the individual transforms, and determine values of z for which *both* terms converge.]

b) If a rational function $U(z)$ is known to converge on the unit circle $|z| = 1$, show how partial-fraction expansion may be used to compute the inverse transform. Apply your result to the transform you found in part (a).

2.11 Compute the inverse transform, $f(k)$, for each of the following transforms:

a) $F(z) = \dfrac{1}{1 + z^{-2}}, \qquad |z| > 1;$

b) $F(z) = \dfrac{z(z - 1)}{z^2 - 1.25z + 0.25}, \qquad |z| > 1;$

c) $F(z) = \dfrac{z}{z^2 - 2z + 1}, \qquad |z| > 1;$

d) $F(z) = \dfrac{z}{(z - \frac{1}{2})(z - 2)}, \qquad \dfrac{1}{2} < |z| < 2.$

2.12 Use the z-transform to solve the difference equation

$$y(k) - 3y(k - 1) + 2y(k - 2) = 2u(k - 1) - 2u(k - 2),$$
$$\begin{aligned} u(k) &= k & k \geq 0 \\ &= 0, & k < 0; \\ y(k) &= 0, & k < 0. \end{aligned}$$

2.13 a) Write a program in BASIC to implement the program flow of Fig. 2.27. Assume we have routines CALL(1, T, F) to do step 2, CALL(2, E, C1, F) to do step 5, and CALL(3, U, C2) to do step 7. In each case, T is the sampling period in milliseconds, F is the FLAG variable, E is the variable where the A/D converter stores its value, C1 is the channel of the A/D multiplexer, U is the variable to be converted by the D/A, and C2 is the channel where the output is to appear.

b) Do part (a) in FORTRAN.

3 / Discrete Equivalents to Continuous Transfer Functions: The Digital Filter

3.1 INTRODUCTION

One of the exciting fields of application of digital systems[1] is in signal processing and digital filtering. A filter is a device designed to pass desirable elements and hold back or reject undesirable ones; in signal processing it is common to represent signals as a sum of sinusoids and to define the "desirable elements" as those signals whose frequency components are in a specified band. Thus a radio receiver filter passes the band of frequencies transmitted by the station we want to hear and rejects all others. We would call such a filter a *bandpass filter*. In electrocardiography it often happens that power-line frequency signals are strong and unwanted, so we design a filter to pass signals between 1 and 500 Hz, but to eliminate those at 60 Hz. The magnitude of the transfer function for this purpose may look like Fig. 3.1 on a log-frequency scale, where the amplitude response between 59.5 and 60.5 Hz might reach 10^{-3}. Here we have a band reject filter with a 60 dB rejection ratio in a 1-Hz band centered at 60 Hz. In long-distance telephony some

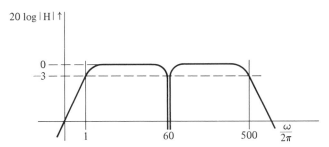

Fig. 3.1 Magnitude of a low-frequency bandpass filter with a narrow rejection band.

---

[1] Including microprocessors and special-purpose large-scale integration (LSI) digital chips.

filters play a conceptually different role.   There the issue is that transmission media—wires or microwaves—introduce distortion in the amplitude and phase of a sinusoid which must be removed.   Filters to accomplish this correction are called *equalizers*.   And in control we must control systems whose dynamics require modification in order that the complete system have satisfactory dynamic response.   We call the devices which make these changes *compensators.*

Whatever the name—filter, equalizer, or compensator—many fields have found use for devices having specified amplitude and phase transmission characteristics, and the trend is to perform these functions by digital means.   The design of electronic filters is a well-established subject which includes not only very sophisticated techniques, but also well-tested computer programs [see Daniels (1974)].   Much of the effort in digital filter design has been directed toward the design of digital filters which have the same characteristics (as nearly as possible) as those of a satisfactory continuous design.   For digital control systems we have much the same motivation: since continuous control designs are well established, we should like to know how to take advantage of a good continuous design and cause a digital computer to produce a discrete equivalent to the continuous compensator.

Thus we are led to the specific problem of this chapter: given a transfer function, $H(s)$, what discrete transfer function will have approximately the same characteristics?   We will present here three approaches to this task:

Method 1: *numerical integration.*
Method 2: *pole-zero mapping.*
Method 3: *hold equivalence.*

3.2 DESIGN OF DIGITAL FILTERS BY NUMERICAL INTEGRATION

The topic of numerical integration of differential equations is quite complex, and only the most elementary techniques are presented here.   The fundamental concept is to represent the given filter transfer function $H(s)$ as a differential equation and to derive a difference equation whose solution is an approximation to that of the differential equation.   For example, the system

$$\frac{U(s)}{E(s)} = H(s) = \frac{a}{s + a} \tag{3.1}$$

is equivalent to the differential equation

$$\dot{u} + au = ae. \tag{3.2}$$

Now, if we write (3.2) in integral form, we have a development much like that of

(2.5) in Chapter 2, except that the integral is more complex here:

$$u(t) = \int^t [-au(\tau) + ae(\tau)] \, d\tau,$$

$$u(kT) = \int^{kT-T} [-au + ae] \, d\tau + \int_{kT-T}^{kT} [-au + ae] \, d\tau$$

$$= u(kT - T) + \left\{ \begin{array}{l} \text{area of } -au + ae \\ \text{over } kT - T \le \tau < kT \end{array} \right\}. \tag{3.3}$$

We can now develop many rules based on our selection of the approximation of the incremental area term. The first approximation leads to the forward rectangular rule[2]: we approximate the area by the rectangle looking forward from $kT - T$ and take the amplitude of the rectangle to be the value of the integrand at $kT - T$. The width of the rectangle is T. The result is an equation in the first approximation, u_1:

$$u_1(kT) = u_1(kT - T) + T[-au_1(kT - T) + ae(kT - T)]$$
$$= (1 - aT)u_1(kT - T) + aTe(kT - T). \tag{3.4a}$$

The transfer function corresponding to the forward rectangular rule is

$$H_F(z) = \frac{aTz^{-1}}{1 - (1 - aT)z^{-1}}$$

$$= \frac{a}{(z - 1)/T + a} \qquad \text{(forward rectangular rule)}. \tag{3.4b}$$

A second rule follows from taking the amplitude of the approximating rectangle to be the value looking backward from kT toward $kT - T$, namely $-au(kT) + ae(kT)$. The equation for u_2, the second approximation, is

$$u_2(kT) = u_2(kT - T) + T[-au_2(kT) + ae(kT)]$$

$$= \frac{u_2(kT - T)}{1 + aT} + \frac{aT}{1 + aT} e(kT). \tag{3.5a}[3]$$

Again we can take the z-transform and compute the transfer function of the backward rule:

$$H_B(z) = \frac{aT}{1 + aT} \frac{1}{1 - z^{-1}/(1 + aT)} = \frac{aTz}{z(1 + aT) - 1}$$

$$= \frac{a}{(z - 1)/Tz + a} \qquad \text{(backward rectangular rule)}. \tag{3.5b}$$

---

[2] Also known as *Euler's rule*.
[3] It is worthy of attention that in order to solve for (3.5a) we had to eliminate $u(kT)$ from the right-hand side where it entered from the integrand. Had Eq. (3.2) been nonlinear, the result would have been an implicit equation requiring an iterative solution. This topic is the subject of predictor-corrector rules, which are beyond our scope of interest. A discussion is found in most books on numerical analysis, for example in Hamming (1962).

Our final version of integration rules is the *trapezoid rule* found by taking the area approximated in (3.3) to be that of the trapezoid formed by the average of the previously selected rectangles. The approximating difference equation is

$$u_3(kT) = u_3(kT - T) + \frac{T}{2}[-au_3(kT - T) + ae(kT - T) - au_3(kT) + ae(kT)]$$

$$= \frac{1 - (aT/2)}{1 + (aT/2)} u_3(kT - T) + \frac{aT/2}{1 + (aT/2)} [e_3(kT - T) + e_3(kT)]. \qquad (3.6a)$$

The corresponding transfer function from the trapezoid rule is

$$H_T(z) = \frac{aT(z + 1)}{(2 + aT)z + aT - 2}$$

$$= \frac{a}{\dfrac{2}{T}\dfrac{z - 1}{z + 1} + a} \qquad \text{(trapezoid rule)} \qquad (3.6b)$$

Suppose we tabulate our results obtained thus far:

$H(s)$	Method	Transfer function	
$\dfrac{a}{s + a}$	Forward rectangular rule	$H_F = \dfrac{a}{(z - 1)/T + a}$	
$\dfrac{a}{s + a}$	Backward rectangular rule	$H_B = \dfrac{a}{(z - 1)/Tz + a}$	(3.7)
$\dfrac{a}{s + a}$	Trapezoid rule	$H_T = \dfrac{a}{(2/T)[(z - 1)/(z + 1)] + a}$	

What is obvious about this tabulation is that the effect of each of our methods is to present a discrete transfer function which can be obtained from the given Laplace transfer function $H(s)$ by substitution of an approximation for the frequency variable as shown below:

Method	Approximation		
Forward rectangular rule	$s \sim \dfrac{z - 1}{T}$	(i)	
Backward rectangular rule	$s \sim \dfrac{z - 1}{Tz}$	(ii)	(3.8)
Trapezoid rule or Tustin's bilinear rule	$s \sim \dfrac{2}{T}\dfrac{z - 1}{z + 1}$	(iii)	

The trapezoid rule substitution is also known, especially in digital and sampled-data control circles, as *Tustin's method* [Tustin (1947)] after the British engineer whose work stimulated a great deal of interest in this approach. The transformation is also called the *bilinear transformation* from consideration of its mathematical form. The design method may be summarized by stating the rule: Given a continuous transfer function (filter), $H(s)$, a discrete equivalent may be found by the substitution

$$H_T(z) = H(s)\big|_{s=(2/T)[(z-1)/(z+1)]} \tag{3.9}$$

Each of the approximations given in (3.8) may be viewed as a map from the s-plane to the z-plane. A further understanding of the maps may be obtained by considering them graphically. For example, since the $(s = j\omega)$-axis is the boundary between poles of stable systems and poles of unstable systems, it would be interesting to know how the $j\omega$-axis is mapped by the three rules, and where the left (stable) half of the s-plane appears in the z-plane. For this purpose we must solve the relations in (3.8) for z in terms of s. We find

$$\begin{array}{lll}
\text{i)} & z = 1 + Ts, & \text{forward rectangular rule,} \\[2mm]
\text{ii)} & z = \dfrac{1}{1 - Ts}, & \text{backward rectangular rule,} \\[2mm]
\text{iii)} & z = \dfrac{1 + Ts/2}{1 - Ts/2}, & \text{Tustin's rule.}
\end{array} \tag{3.10}$$

If we let $s = j\omega$ in these equations, we obtain the boundaries of the shaded regions sketched in the z-plane in Fig. 3.2 for each case. To show that rule (ii) results in a circle, $\frac{1}{2}$ is added to and subtracted from the right-hand side to yield

$$\begin{aligned}
z &= \frac{1}{2} + \left[\frac{1}{1 - Ts} - \frac{1}{2}\right] \\
&= \frac{1}{2} - \frac{1}{2}\frac{1 + Ts}{1 - Ts}.
\end{aligned} \tag{3.11}$$

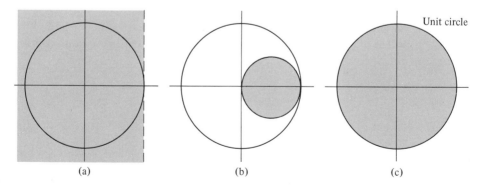

(a)   (b)   (c)

Fig. 3.2 Maps of the left half s-plane to the z-plane by the integration rules of Eq. (3.8). Stable s-plane poles map into the shaded regions in the z-plane. The unit circle is shown for reference. (a) Forward rectangular rule. (b) Backward rectangular rule. (c) Trapezoid rule.

Now it is easy to see that with $s = j\omega$, the magnitude of $z - \frac{1}{2}$ is constant,

$$|z - \tfrac{1}{2}| = \tfrac{1}{2},$$

and the curve is thus a circle as drawn in Fig. 3.2(b). Since the unit circle is the stability boundary in the z-plane, it is obvious from Fig. 3.2 that the forward rectangular rule could cause a stable continuous filter to be mapped into an unstable digital filter.

It is especially interesting to notice that Tustin's rule maps the stable region of the s-plane exactly into the stable region of the z-plane although the entire $j\omega$-axis of the s-plane is stuffed into the 2π-length of the unit circle! Obviously a great deal of distortion takes place in the mapping in spite of the congruence of the stability regions. As our final rule deriving from numerical integration ideas, we discuss a formula which extends Tustin's rule one step in an attempt to correct for the inevitable distortion of real frequencies mapped by the rule. We begin with our elementary transfer function (3.1) and consider the Tustin rule approximation

$$H_T(z) = \frac{a}{(2/T)[(z - 1)/(z + 1)] + a}. \tag{3.6b}$$

The original $H(s)$ had a pole at $s = -a$, and for real frequencies, $s = j\omega$, the magnitude of $H(j\omega)$ is given by

$$|H(j\omega)|^2 = \frac{a^2}{\omega^2 + a^2}$$

$$= \frac{1}{\omega^2/a^2 + 1}. \tag{3.12}$$

Thus our reference filter has a half-power point, $|H|^2 = \frac{1}{2}$, at $\omega = a$. It may be interesting to know where $H_T(z)$ has a half-power point.

As we saw in Chapter 2, signals with poles on the imaginary axis in the s-plane (sinusoids) map into signals on the unit circle of the z-plane. A sinusoid of frequency ω_1 corresponds to $z_1 = e^{j\omega_1 T}$ and the response of $H_T(z)$ to a sinusoid of frequency ω_1 is $H_T(z_1)$. We consider now (3.6b) for $H_T(z_1)$ and manipulate it into a more convenient form for our present purposes:

$$H_T(z_1) = \frac{a}{\dfrac{2}{T}\dfrac{e^{j\omega_1 T} - 1}{e^{j\omega_1 T} + 1} + a}$$

$$= \frac{a}{\dfrac{2}{T}\dfrac{e^{j\omega_1 T/2} - e^{-j\omega_1 T/2}}{e^{j\omega_1 T/2} + e^{-j\omega_1 T/2}} + a}$$

$$= \frac{a}{\dfrac{2}{T}j \tan \dfrac{\omega_1 T}{2} + a}. \tag{3.13}$$

The magnitude squared of H_T will be $\frac{1}{2}$ when

frequency distortion or warping

$$\frac{2}{T} \tan \frac{\omega_1 T}{2} = a$$

or

$$\tan \frac{\omega_1 T}{2} = \frac{aT}{2}. \tag{3.14}$$

Equation (3.14) is a measure of the frequency distortion or warping caused by Tustin's rule. Whereas we wanted to have a half-power point at $\omega = a$, we realized a half-power point at $\omega_1 = (2/T) \tan^{-1} (aT/2)$. We can turn the intentions around and suppose that we really want the half-power point to be at ω_1. Equation (3.14) can be made into an equation of prewarping: if we select "a" according to (3.14), then, using Tustin's bilinear rule for the design, the half-power point will be at ω_1. A statement of a complete set of rules for filter design via bilinear transformation with prewarping is:

a) Write the desired filter characteristic with transform variable s and critical frequency ω_1[4] in the form $H(s/\omega_1)$.

b) Replace ω_1 by a such that

$$a = \frac{2}{T} \tan \frac{\omega_1 T}{2}$$

and in place of $H(s/\omega_1)$, consider the prewarped function $H(s/a)$.

c) Substitute

$$s = \frac{2}{T} \frac{z-1}{z+1}$$

in $H(s/a)$ to obtain $H_p(z)$.

As a frequency substitution the result may be expressed as

$$H_p(z) = H(s/\omega_1) \Bigg|_{s = \dfrac{\omega_1}{\tan \dfrac{\omega_1 T}{2}} \dfrac{z-1}{z+1}} \tag{3.15}$$

It is clear from (3.15) that when $\omega = \omega_1$, $H_p(z_1) = H(j1)$ and the discrete filter has exactly the same transmission at ω_1 as the continuous filter has at this frequency. This is the consequence of prewarping. We also note that as the sampling period gets small, $H_p(z)$ approaches $H(j\omega/\omega_1)$.

A comparison of the four rules studied in this section is illustrated in Fig. 3.3

---

[4] The critical frequency need not be the band edge. We can use the band center of a bandpass filter or the crossover frequency of a Bode plot compensator. However, we must have $\omega_1 < \pi/T$.

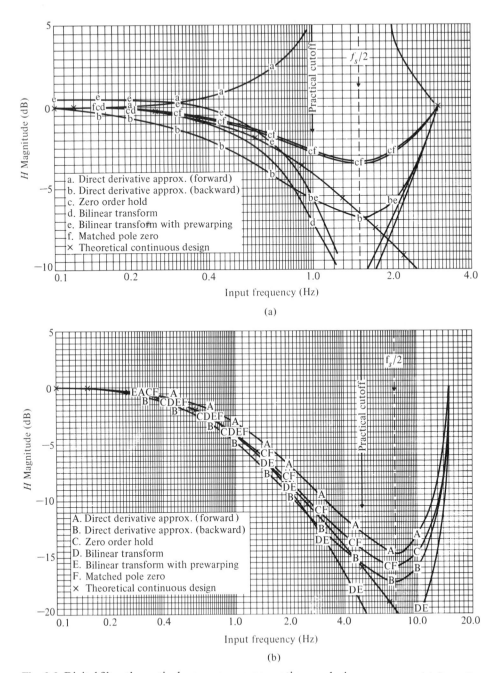

Fig. 3.3 Digital filter theoretical response versus continuous design response.   (a) Sampling frequency $f_s = 3$ Hz.   (b) Sampling frequency $f_s = 15$ Hz.

along with a zero-order hold equivalent and a matched pole-zero approximation, two techniques to be discussed in following sections.[5]

Figure 3.3 shows plots of $H(j\omega) = 5/(j\omega + 5)$ and the frequency responses of the discrete equivalents for $T = \frac{1}{3}$ in Fig. 3.3(a) and $T = \frac{1}{15}$ in Fig. 3.3(b). The major features are:

a) At $T = \frac{1}{3}$, the forward rectangular rule gives a totally unacceptable approximation.

b) The bilinear transformation gives a zero at $f = 1.5$ Hz, which is half the sampling frequency (or at $f = 7.5$ in Fig. 3.3b).

c) With prewarping, curve e confirms that the approximation has the same gain at $f_1 = \frac{5}{2}\pi$, the half-power point which was selected as the critical frequency in the design.

d) All the digital equivalents have frequency response characteristics which are periodic in ω with period $2\pi/T$. Thus the plots have symmetry about the "Nyquist" frequency π/T, which is 1.5 Hz in Fig. 3.3(a) and 7.5 Hz in Fig. 3.3(b).

3.3 POLE-ZERO MAPPING

A very simple but effective method of obtaining a discrete equivalent to a continuous transfer function is to be found by extrapolation of the relation derived in Chapter 2 between the s- and z-planes. If we take the z-transform of samples of a continuous signal $e(t)$, then the poles of the discrete transform $E_1(z)$ are related to the poles of $E(s)$ according to $z = e^{sT}$. We must go through the z-transform process to locate the zeros of $E_1(z)$, however. The idea of the pole-zero mapping technique is that the map $z = e^{sT}$ could be applied to the zeros also. The technique consists of a set of heuristic rules for locating the zeros and gain of a z-transform which will describe a discrete equivalent transfer function which approximates the given $H(s)$. The rules are as follows:

1. All poles of $H(s)$ are mapped according to $z = e^{sT}$. If $H(s)$ has a pole at $s = -a$, then $H_{pz}(z)$ has a pole at $z = e^{-aT}$.

2. All *finite* zeros are also mapped by $z = e^{sT}$. If $H(s)$ has a zero at $s = -b$, then $H_{pz}(z)$ has a zero at $z = e^{-bT}$.

3. All zeros of $H(s)$ at $s = \infty$ are mapped in $H_{pz}(z)$ to the point $z = -1$.

3a. If a unit delay in the digital filter unit pulse response is desirable for any reason, i.e., computation time is necessary to process each sample, one zero of $H(s)$ at $s = \infty$ is mapped into $z = \infty$. $H_{pz}(z)$ is left with the number of zeros one less than the number of poles in the finite plane. The series expansion of

---

[5] These curves were obtained by E. Freeman in Course E207 at Stanford in February 1978.

$H(z)$ in powers of z^{-1} will have no constant term, and thus the $h(k)$ has a one-unit delay in response to the unit pulse.

4. The gain of the digital filter is selected to match the gain of $H(s)$ at the band center or a similar critical point. In most control applications, the critical frequency is $s = 0$, and hence we select the gain so that

$$H(s)\,|_{s=0} = H_{pz}(z)\,|_{z=1}.$$

With respect to the first two of these rules, one must map complex conjugate poles or zeros together with the result that a pair at $s = -a \pm jb$ will be mapped to $z = r \exp \pm j\theta$, where $r = e^{-aT}$ and $\theta = bT$. The rationale behind rule 3 is that the map of real frequencies from $j\omega = 0$ to increasing ω is onto the unit circle at $z = e^{j0} = 1$ until $z = e^{j\pi} = -1$. Thus the point $z = -1$ represents, in a real way, the highest frequency possible in the discrete transfer function, so it is appropriate that if $H(s)$ is zero at the highest (continuous) frequency, $|H_{pz}(z)|$ should also have this property.

Application of these rules to $H(s) = a/(s + a)$ gives

$$H_{pz}(z) = \frac{(z + 1)(1 - e^{-aT})}{2(z - e^{-aT})}, \qquad (3.16)$$

or, using rule 3(a),

$$H_{pz}(z) = \frac{1 - e^{-aT}}{z - e^{-aT}}. \qquad (3.17)$$

This function is plotted on Fig. 3.3 for purposes of comparison.

3.4 HOLD EQUIVALENCE[6]

For this technique, we assume the situation to be as sketched in Fig. 3.4. The purpose of the samplers in Fig. 3.4(b) is to require that H_{ho} have only samples to work on and produce only samples of its output, $\hat{u}$, and thus H_{ho} can be realized as a discrete transfer function. The philosophy of the design is the following. We are asked to design a system which, with an input consisting of samples of $e(t)$, has an output which approximates the output of $H(s)$ whose input is the continuous $e(t)$. We generate the discrete equivalent by first approximating $e(t)$ from the samples $e(k)$ and then putting this $\hat{e}(t)$ through $H(s)$. First consider the possibilities for approximation. These are techniques for taking a sequence of samples and extrapolating or holding them to produce a continuous signal. Suppose we have the $e(t)$ as sketched in Fig. 3.5. This figure also shows a sketch of a piecewise constant approximation to $e(t)$ obtained by the operation of holding $\hat{e}(t)$ constant at $e(k)$ over the interval from kT to $(k + 1)T$. This operation uses a

---

[6] May be omitted on first reading or may be considered after Chapter 4.

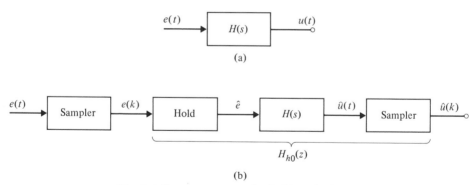

Fig. 3.4 Systems as seen for hold equivalence.

zero-order polynomial (constant) to approximate $e(t)$ and is called the *zero-order hold*.[7] If we use a first-order polynomial for extrapolation, we have a *first-order hold*, and so on for second-, and *n*th-order holds. Let us first show how to obtain $H_{h0}(z)$ from this picture: $H_{h0}(z)$ must relate the $e(k)$ samples to the $\hat{u}(k)$ samples, so we need to compute $\hat{u}(k)$.

From Fig. 3.4, we see that $\hat{u}(t)$ is the response of $H(s)$ to the sequence of pulses from the hold circuit. If we had only $e(0)$, then $\hat{e}(t)$ would be a step of size $e(0)$ followed T sec later by a negative step of the same size. The samples at $\hat{u}$ in response to a step of size $e(0)$ at $\hat{e}$ are the samples of the signal having Laplace transform $H(s)/s$. We can symbolize the z-transform of these samples as

$$\mathcal{Z}\{H(s)/s\}.$$

Likewise, the samples due to the step delayed one period have the transform

$$z^{-1}\mathcal{Z}\left\{\frac{H(s)}{s}\right\},$$

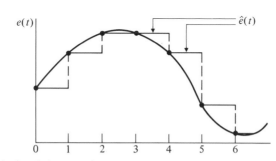

Fig. 3.5 A signal, its samples, and its approximations by zero-order hold.

---

[7] Note that the signal $\hat{e}$ is, on the average, delayed from e by $T/2$ sec. This delay is one measure of the quality of the approximation and a guide to the selection of T.

and the total contribution of $e(0)$ to the transform of the $\hat{u}(k)$ samples is

$$e(0)(1 - z^{-1})\mathcal{Z}\left\{\frac{H(s)}{s}\right\}. \tag{3.18}$$

If $e(1)$ were applied alone, the dynamic effects would be the same as those of $e(0)$ but would be delayed a further period to give the contribution

$$e(1)z^{-1}(1 - z^{-1})\mathcal{Z}\left\{\frac{H(s)}{s}\right\}. \tag{3.19}$$

Continuation of this line of reasoning and addition of all the components lead us to

$$\hat{U}(z) = \sum_{k=0}^{\infty} e(k)z^{-k}(1 - z^{-1})\mathcal{Z}\left\{\frac{H(s)}{s}\right\} \tag{3.20}$$

$$= E(z)(1 - z^{-1})\mathcal{Z}\left\{\frac{H(s)}{s}\right\}. \tag{3.21}$$

Therefore, the zero-order-hold equivalent to $H(s)$ is given by $\hat{U}(z)/E(z)$ or

$$H_{h0}(z) = (1 - z^{-1})\mathcal{Z}\left\{\frac{H(s)}{s}\right\}. \tag{3.22}$$

An example will fix ideas. Suppose we again take the first-order filter

$$H(s) = \frac{a}{s + a}.$$

Then

$$\frac{H(s)}{s} = \frac{a}{s(s + a)} = \frac{1}{s} - \frac{1}{s + a}$$

and

$$\mathcal{Z}\left\{\frac{H(s)}{s}\right\} = \mathcal{Z}\left\{\frac{1}{s}\right\} - \mathcal{Z}\left\{\frac{1}{s + a}\right\}, \tag{3.23}$$

and, by definition of the operation given in (3.23),

$$\mathcal{Z}\left\{\frac{H(s)}{s}\right\} = \sum_{0}^{\infty} z^{-k} - \sum_{0}^{\infty} z^{-k}e^{-akT}$$

$$= \frac{1}{1 - z^{-1}} - \frac{1}{1 - e^{-aT}z^{-1}}$$

$$= \frac{(1 - e^{-aT}z^{-1}) - (1 - z^{-1})}{(1 - z^{-1})(1 - e^{-aT}z^{-1})}. \tag{3.24}$$

Finally, substituting (3.24) in (3.22), we get the zero-order-hold equivalent of $H(s)$, namely

$$H_{h0}(z) = \frac{(1 - e^{-aT})}{z - e^{-aT}}. \tag{3.25}$$

This function is plotted, for $z = e^{j\omega T}$, in Fig. 3.3 for comparison with the other techniques of deriving discrete equivalents to continuous transfer functions. We note that for the trivial example given, the zero-order-hold equivalent of (3.25) is identical to the matched pole-zero equivalent given by (3.17). We will see in Chapter 6 that the zero-order-hold equivalent is readily computed by an algorithm well suited to implementation on a digital computer, and thus either of these techniques is attractive from a computational point of view.

Visual examination of Fig. 3.3 shows that curves "c" and "f" which correspond to the zero-order-hold and the pole-zero mapping equivalents, respectively, give good approximation to the frequency response of the simple first-order filter. Comparison curves for a more complex example are plotted in Fig. 3.6. For this case,

$$H(s) = \frac{s}{s^2 + s + 25},$$

corresponding to a bandpass structure with a center frequency at $\omega_0 = 5$ ($f_0 =$

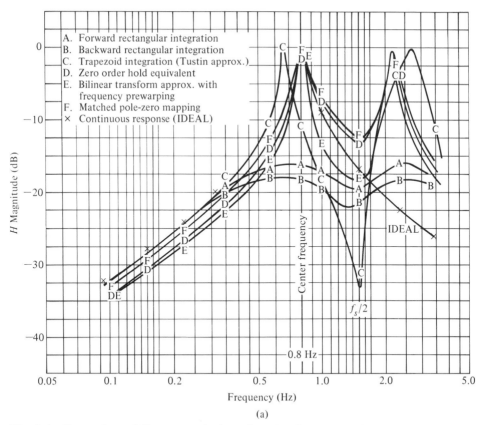

(a)

Fig. 3.6a Comparison of discrete equivalents for a bandpass example sampling frequency of 3 Hz.

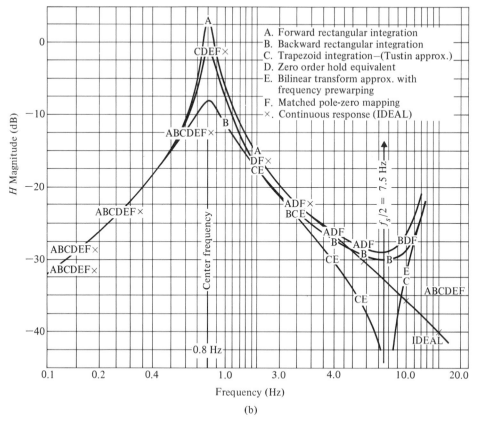

Fig. 3.6b Comparison for frequency of 15 Hz.

0.795 Hz) and a bandwidth[8] of 1 rad (0.159 Hz). The frequency ω_0 was used as the "critical" frequency for prewarping and pole-zero mapping. These curves show, first of all, the extremely poor quality of the approximations generated by the rectangular rules. Furthermore the curves obtained at a 3-Hz sampling also show very clearly the error in the center frequency of the filter designed by the bilinear transformation (curve C) and the correction of this error by prewarping (curve E).

3.5 BUTTERWORTH AND ITAE EQUIVALENTS

We have developed several techniques for constructing discrete transfer functions which are equivalent to a given continuous transfer function and we have compared results of applying the techniques to two elementary examples. The easiest

---

[8] The bandwidth of a bandpass filter is defined as the width of the amplitude response at the amplitude of 0.707.

method to apply, if the poles and zeros of the continuous transfer function are known, is the pole-zero mapping technique. We would like to explore briefly the application of this technique to more complex transfer functions. Our purpose is partly to demonstrate the simplicity of the method and partly to present two classes of transfer functions which can be used as models for later designs. The classes to be studied are the Butterworth (1930) and the minimum Integral of Time-multiplied Absolute Error, or ITAE, transfer functions. [Graham and Lathrop (1953)].

The Butterworth filter is designed to meet frequency-response amplitude specifications such that the transfer function, $H(j\omega)$, is taken to have n poles, to have all its zeros at $\omega = \infty$, to have a bandwidth ranging from zero to ω_0,[9] and to have an amplitude characteristic which has the value 1.0 at $\omega = 0$ and is as flat as possible at $\omega = 0$, which is the band center. The magnitude squared function which meets these requirements is

$$|H|^2 = \frac{1}{1 + (\omega/\omega_0)^{2n}}. \tag{3.26}$$

In (3.26) ω_0 is the bandwidth, and the larger n is, the more complicated the filter is, but the cutoff of signals having frequency components above ω_0 is much sharper. The unit step responses of the Butterworth and (for later reference) the ITAE transfer functions are shown in Fig. 3.7. We can use (3.26) to find the poles of $H(s)$. The magnitude squared for any $H(j\omega)$ is $H(s)H(-s)|_{s=j\omega}$, and thus the poles of $H(s)H(-s)$ in this case are the zeros of

$$1 + (s/j\omega_0)^{2n} = 0$$
$$(s/j\omega_0)^{2n} = -1 = e^{j2\pi k}e^{j\pi}, \qquad k = 0, 1, 2, \ldots ;$$

taking the $2n$th root and multiplying by ω_0, we have

$$s = j\omega_0 e^{j(2\pi k/2n)}e^{j(\pi/2n)}$$
$$= \omega_0 e^{j(\pi k/n)}e^{j(\pi/2n)}e^{j(\pi/2)}. \tag{3.27}$$

All $2n$ poles of $H(s)H(-s)$ lie on a circle of radius ω_0 since the magnitude of s in (3.27) is ω_0. Furthermore, the poles are equally spaced at π/n rad and are in conjugate pairs. For $n = 3$, the poles are as shown in Fig. 3.8.

Since we want a filter transfer function $H(s)$ which is causal and stable, all the poles of $H(s)$ must be in the left half-plane. Therefore, the poles of $H(s)$ are at $s_1 = -\omega_0$ and $s_{2,3} = -\omega_0/2 \pm j(\sqrt{3}/2)\omega_0$. The transfer function is found by multiplying the corresponding factors, that is

$$H_3(s) = \frac{1}{(s/\omega_0 + 1)\left(s^2/\omega_0^2 + \dfrac{s}{\omega_0} + 1\right)} = \frac{1}{B_3\left(\dfrac{s}{\omega_0}\right)}. \tag{3.28}$$

---

[9] The limit of the band of frequencies passed by the filter is defined to be that frequency where the magnitude squared of the transfer function is one-half or the magnitude is 0.707.

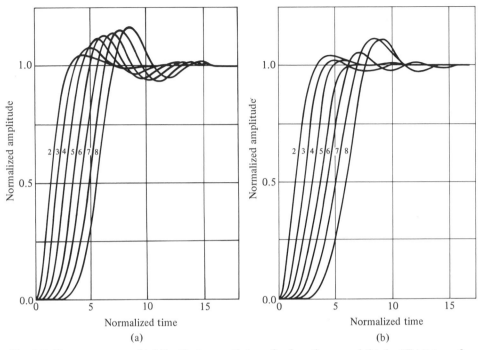

Fig. 3.7 Step responses of (a) the Butterworth transfer functions, and (b) the ITAE transfer functions. (From *Linear Control System Analysis and Design,* by D'Azzo and Houpis (pp. 489, 490). Copyright © 1975 by McGraw-Hill Book Company. Used with permission of McGraw-Hill Book Company.)

The denominator of (3.28) is, by definition, a *Butterworth polynomial* in s of third degree. In general, using the geometry of Fig. 3.8, we can write $H(s) = 1/B_n(s)$, where for n odd,

$$B_n\left(\frac{s}{\omega_0}\right) = \left(\frac{s}{\omega_0} + 1\right) \prod_{k=1}^{(n-1)/2} \left(\frac{s^2}{\omega_0^2} + 2(\cos\theta_k)\frac{s}{\omega_0} + 1\right), \qquad \theta_k = \frac{\pi k}{n}, \quad (3.29)$$

and for n even,

$$B_n(s) = \prod_{k=1}^{n/2} \left(\frac{s^2}{\omega_0^2} + 2(\cos\theta_k)\frac{s}{\omega_0} + 1\right), \qquad \theta_k = \frac{\pi}{2n}(2k - 1).$$

In rectangular coordinates, the complex poles of the Butterworth filter are at $-a_k \pm jb_k$, where

$$a_k = \omega_0 \cos\theta_k, \qquad b_k = \omega_0 \sin\theta_k. \qquad (3.30)$$

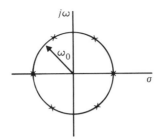

Fig. 3.8 Poles of the third-order Butterworth filter magnitude squared.

Using the pole-zero mapping technique, a discrete equivalent to the Butterworth filter has complex poles in the z-plane at

$$z_k = r_k e^{j\phi_k},$$

where

$$r_k = e^{-a_k T}, \qquad \phi_k = b_k T, \tag{3.31}$$

and a real pole at $z = e^{-\omega_0 T}$ if n is odd. According to the rules for pole-zero mapping, the zeros of the discrete equivalent will all be at $z = -1$ with the possible exception of one zero at $z = \infty$ if a unit delay is desired.

To compute a discrete equivalent to a Butterworth transfer function by pole-zero mapping, we need to follow the steps of Section 3.3, beginning with location of the poles and zeros of $H(s)$ and selection of a sampling period. For illustrative purposes, we will develop the equivalent of a fourth-order system and select the sampling period to satisfy $T\omega_0 = 1$. The poles of the continuous filter are found from (3.29) for $n = 4$, and the poles of the discrete equivalent are from (3.31). These are given in Table 3.1 with the parameters of the second-order factors of $H_{pz}(z)$. The block diagram of the discrete equivalent is shown in Fig. 3.9, in a cascade structure with plots of the s-plane poles of $H(s)$, the z-plane poles and zeros of $H_{pz}(z)$, and the step response of $H_{pz}(z)$. Since we are consider-

Table 3.1 Design of Fourth-Order Butterworth Equivalent via Pole-Zero Mapping

Continuous poles	Discrete poles	Parameters of canonical form
$a_1 = \omega_0 \cos(\pi/8)$ $= \omega_0(0.92388)$	$r_1 = 0.39698$	$-r_1^2 = -0.15759$
$b_1 = \omega_0 \sin(\pi/8)$ $= \omega_0(0.38268)$	$\phi_1 = 0.38268$ $= 21.9°$	$+2r_1 \cos \phi_1 = +0.73653$
$a_2 = \omega_0 \cos(\pi/8)3$ $= 0.38268$	$r_2 = 0.68203$	$-r_2^2 = -0.465166$
$b_2 = \omega_0 \sin(\pi/8)3$ $= \omega_0 0.92388$	$\phi_2 = 0.60273$ $= 52.9°$	$+2r_2 \cos \phi_2 = +0.82216$

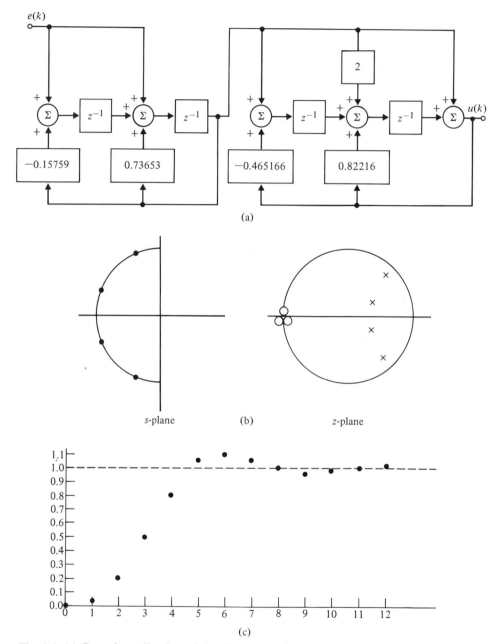

Fig. 3.9 (a) Cascade realization of the discrete equivalent of a fourth-order Butterworth transfer function for $\omega_0 T = 1$. (b) Poles and zeros of fourth-order Butterworth transfer function and a discrete equivalent designed by pole-zero mapping with $\omega_0 T = 1$. (c) Step response of the discrete equivalent of the fourth-order Butterworth transfer function.

Table 3.2 Pole Locations for the ITAE Transfer Functions

k	
1	$s + 1$
2	$s + 0.707 \pm j0.707$
3	$(s + 0.7081)(s + 0.521 \pm j1.068)$
4	$(s + 0.424 \pm j1.263)(s + 0.626 \pm j0.4141)$
5	$(s + 0.8955)(s + 0.376 \pm j1.292)(s + 0.5758 \pm j0.5339)$
6	$(s + 0.3099 \pm j1.263)(s + 0.5805 \pm j0.7828)(s + 0.7346 \pm j0.2873)$

ing only the discrete transfer functions, the step response is shown at sampling times only. A control system designed to match these sample-time values would fill in the curve in a manner depending on the specific plant dynamics.

The coefficients of the ITAE transfer functions were computed to minimize the integral of the product of time and the absolute values of the error to a step. The result is a series of prototype transfer functions which have significantly lower overshoot than the Butterworth transfer functions. The poles of the first six ITAE transfer functions are given in Table 3.2 in normalized form.[10] The design values for a discrete equivalent to the fourth-order ITAE form via pole-zero mapping are given in Table 3.3. The block diagram, poles and zeros, and step response are given in Fig. 3.10. Again, we select the sampling period so that $\omega_0 T = 1$.

Comparison of Figs. 3.10 and 3.9 confirms the qualitative rules developed in Chapter 2 to the effect that moving poles along the real axis toward the stability boundary tends to lower the overshoot and increase the rise time. In the transition from the Butterworth to the ITAE transfer functions the roots at angles of about 20° have been moved toward the unit circle. However, the effect is complicated by the fact that the higher-frequency poles are also moved. Certainly we can conclude that the ITAE parameters provide a basis for suitable pole locations for control system design and that simple pole-zero mapping is adequate to obtain a quite acceptable discrete equivalent.

Table 3.3 Design Parameters for Fourth-Order ITAE Digital Equivalent via Pole-Zero Mapping

$a_1 = 0.424$	$r_1 = 0.654$	$-r_1^2 = -0.42827$
$b_1 = 1.263$	$\phi_1 = 1.263 = 72.36°$	$+2r_1 \cos \phi_1 = +0.39653$
$a_2 = 0.626$	$r_2 = 0.53473$	$-r_2^2 = -0.28593$
$b_2 = 0.4141$	$\phi_2 = 0.4141 = 23.73°$	$+2r_2 \cos \phi_2 = +0.979$

---

[10] The factors $(s + a + jb)(s + a - jb)$ are written as $s + a \pm jb$ to conserve space. To convert the results to a bandwidth ω_0, replace s by s/ω_0 throughout.

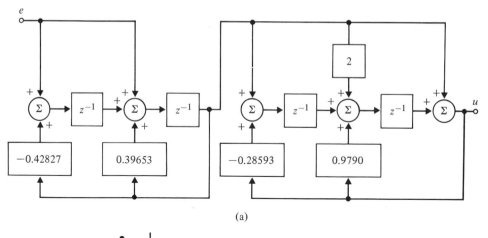

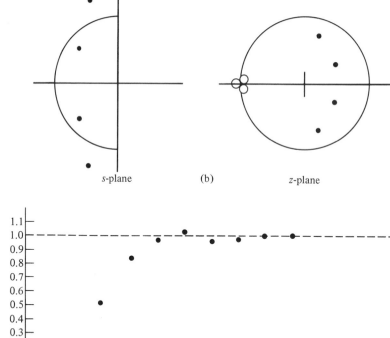

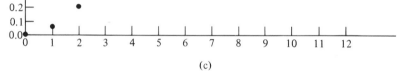

Fig. 3.10 Characteristics of discrete equivalent to fourth-order ITAE transfer function.

3.6 SUMMARY

In this chapter we have presented several techniques for the construction of discrete equivalents to continuous transfer functions so that known design methods for continuous systems—controls and filters—may be used as a basis for discrete system design. The methods presented were:

1. *Numerical integration*
   a) Forward rectangular rule
   b) Backward rectangular rule
   c) Trapezoid or Tustin's rule
   d) Bilinear transformation with prewarping
2. *Pole-zero mapping*
3. *Hold equivalence*

All methods, except the forward rectangular rule, guarantee a stable discrete system from a stable continuous prototype. The bilinear transformation with prewarping affords exact control over the transmission at a selected critical frequency which must be less than $\frac{1}{2}T$. Pole-zero mapping is the simplest to apply computationally and gives good accuracy to transient responses, as was illustrated by examples from Butterworth and ITAE designs. The hold-equivalence method lends itself to computer implementation, as we will see in Chapter 6.

PROBLEMS AND EXERCISES

3.1 Sketch the zone in the z-plane where poles corresponding to the left half of the s-plane will be mapped by the pole-zero mapping technique and the zero-order-hold technique.

3.2 a) Prove that Eq. (3.15) is true.
 b) Suppose that an nth-order Butterworth transfer function with magnitude squared given by (3.26) is used to construct a discrete equivalent by bilinear transformation with prewarping. Give a compact expression for the magnitude squared of the discrete transfer function.

3.3 a) The following transfer function is a lead network designed to add about 60° lead at $\omega_1 = 3$ rad:

$$H(s) = \frac{s + 1}{0.1s + 1}.$$

For each of the following design methods compute and plot in the z-plane the pole and zero locations and compute the amount of phase lead given by the network at $z_1 = e^{j w_1 T}$. Let $T = 0.25$ sec.

   i) Forward rectangular rule
   ii) Backward rectangular rule
   iii) Trapezoid rule
   iv) Bilinear with prewarping (use ω_1 as the warping frequency)

v) Pole-zero mapping

vi) Zero-order-hold equivalent

b) Plot on log-log paper over the frequency range $\omega = 0.1 \rightarrow \omega = 100$ the amplitude Bode plots on each of the above equivalents.

3.4 a) The following transfer function is a lag network designed to introduce gain attenuation of a factor of 10 (20 dB) at $\omega = 3$:

$$H(s) = \frac{10s + 1}{100s + 1}.$$

For each of the following design methods, compute and plot on the z-plane the pole-zero pattern of the resulting discrete equivalent and give the gain attenuation at $z_1 = e^{j\omega_1 T}$. Let $T = 0.25$ sec.

i) Forward rectangular rule

ii) Backward rectangular rule

iii) Trapezoid rule

iv) Bilinear with prewarping (use $\omega_1 = 3$ rad as the warping frequency)

v) Pole-zero mapping

vi) Zero-order-hold equivalent

b) For each case computed, plot the Bode amplitude curves over the range $\omega_\ell = 0.01 \rightarrow \omega_h = 10$ rad.

3.5 The Butterworth transfer function described by (3.26) is a low-pass characteristic since the gain is near one for $\omega < \omega_0$ and approaches zero for $\omega > \omega_0$. A high-pass characteristic can be obtained by replacing s/ω_0 by ω_0/s in $B_n(s)$. More subtly, a bandpass characteristic can be found by the replacement

$$\frac{\omega}{\omega_0} \leftarrow Q \left(\frac{\omega}{\omega_0} - \frac{\omega_0}{\omega} \right) \triangleq Q\gamma,$$

where the parameter Q comes from a consideration of the quality of an inductance used at radio frequencies to construct bandpass filters. (The term was apparently introduced by F.E. Terman (1943).) The magnitude squared of the Butterworth bandpass filter is

$$|H|^2 = \frac{1}{1 + (\gamma Q)^{2n}}.$$

a) Show that if the band edges ω_1 and ω_2 are defined so that $|\gamma_1 Q| = |\gamma_2 Q| = 1$, then the bandwidth $(\omega_2 - \omega_1)$ is

$$(\omega_2 - \omega_1) = \omega_0/Q$$

and $\omega_1 \omega_2 = \omega_0^2$.

b) If we use the bilinear transformation with prewarping, show that the equivalent change is

$$Q\gamma \leftarrow Q^* \gamma^* \triangleq \left[\frac{\tan \omega T/2}{\tan \omega_0 T/2} - \frac{\tan \omega_0 T/2}{\tan \omega T/2} \right] Q^*.$$

c) Give the expression for the bandwidth of the discrete equivalent Butterworth bandpass filter in terms of ω_0 and Q^*.

d) Show that $1/B_n(1/\gamma Q)$ is a band reject characteristic.

e) Select ω_0 and Q to give the band reject characteristics of the 60-Hz elimination filter shown as part of Fig. 3.1. $Bw = 1 \text{ Hz}; f_0 = 60 \text{ Hz} = \omega_0/2\pi$. Let $n = 1$ and $T = .0005$.

f) Compute the poles and zeros of the discrete equivalent to the band reject filter of part (e), using pole-zero mapping, and plot its frequency response over the range $50 \le f \le 70$.

3.6 Write a computer program to implement the cascade realization shown in Fig. 3.9(a). Assume that the samples of the input are stored in location e by the statement CALL A/D(e) and that the number in location u is put out by the statement CALL D/A(u). Use (a) BASIC, or (b) FORTRAN, or (c) the language used in your control lab. Minimize the delay between sampling e and updating u.

3.7 Write a program to compute the frequency response of a digital filter. Allow either poles and zeros or polynomial coefficients to specify the filter and have the program ask for the minimum and maximum frequencies to be computed as powers of 10, that is, $f_{min} = 0.01 \text{ Hz}; f_{max} = 1 \text{ Hz}$. Compute 10 points per decade at points 1, 1.5, 2, 2.5, 3, 4, 5, 6, 7.5, 9.

4 / Sampled Data Systems

4.1 INTRODUCTION

We have thus far introduced linear difference equations and used the z-transform for their solutions. We have also shown how, by associating the discrete sample values with samples from a continuous signal, one may interpret the difference equations as digital filters and design specific discrete transfer functions which are equivalent to a given continuous transfer function. In all this we have taken essentially a digital or discrete centered point of view. We must now study dynamic systems which include discrete parts but in the process we must take a continuous system point of view and develop techniques necessary to allow us to analyze discrete components when they are embedded within a continuous system. We are going to show how to analyze discrete signals by the continuous tool, the Laplace transform.

4.2 SAMPLING AS IMPULSE MODULATION

We consider first the sampling operation, shown symbolically in Fig. 4.1. The idea of the sampler is to give a mathematical representation of the operation of taking periodic samples from $r(t)$ to produce $r(kT)$ in such a way that we can model the discrete signals generated by the A/D converter and can analyze them simultaneously with continuous signals using the same tool (Laplace transform). The technique is to use *impulse modulation* as the mathematical representation of sampling. Thus, from Fig. 4.1, we picture the output of the sampler as a string of impulses,

$$r^*(t) = \cdots + \delta(t + T)r(-T) + \delta(t)r(0) + \delta(t - T)r(T) + \delta(t - 2T)r(2T) + \cdots$$

In general,

$$r^*(t) = \sum_{-\infty}^{\infty} r(t)\delta(t - kT). \tag{4.1}$$

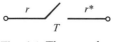

Fig. 4.1 The sampler.

The impulse may be taken as the limit of a pulse of unit area which has growing amplitude and shrinking duration. It has the sifting property that

$$\int_{-\infty}^{\infty} f(t)\delta(t - a)\, dt = f(a) \tag{4.2}$$

for all f that are continuous at a. The impulse is related to the unit step by the integral

$$\int_{-\infty}^{t} \delta(\tau)\, d\tau = 1(t) \tag{4.3}$$

and has the Laplace transform

$$\mathcal{L}\{\delta(t)\} = \int_{-\infty}^{\infty} \delta(t)e^{-st}\, dt = 1. \tag{4.4}$$

Using these properties we can see that $r^*(t)$ depends only on the discrete sample values $r(kT)$ and furthermore has the Laplace transform

$$\mathcal{L}\{r^*(t)\} \triangleq \int_{-\infty}^{\infty} r^*(t)e^{-st}\, dt$$

$$= \int_{-\infty}^{\infty} \sum_{-\infty}^{\infty} r(t)\delta(t - kT)e^{-st}\, dt$$

$$R^*(s) = \sum_{-\infty}^{\infty} r(kT)e^{-skT}. \tag{4.5}$$

The notation $R^*(s)$ is used to symbolize the (Laplace) transform of $r^*(t)$, the sampled or impulse-modulated $r(t)$.[1] Now note that (4.5) is the z-transform of the sample sequence $r(k)$ if we let $e^{sT} = z$. We might have anticipated this result since

---

[1] It will be necessary, from time to time, to consider sampling a signal which is not continuous. The only case we will consider will be equivalent to applying a step function, $1(t)$, to a sampler. For the purposes of this book we will define the unit step to be continuous from the right and assume that the impulse, $\delta(t)$, picks up the full value of unity. By this convention and (4.1) we compute

$$1^*(t) = \sum_{k=0}^{\infty} \delta(t - kT), \tag{a}$$

and, using (4.3), we obtain

$$\mathcal{L}\{1^*(t)\} = 1/(1 - e^{-Ts}). \tag{b}$$

The reader should be warned that the Fourier integral converges to the *average* value of a function at a discontinuity and not the value approached from the right as we assume. Because our use of the transform theory is elementary and the convenience of equation (b) above is substantial, we have selected the continuous-from-the-right convention. In case of doubt, the discontinuous term should be separated and treated by special analysis, perhaps in the time domain.

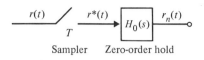

Fig. 4.2 Linear model of the A/D converter.

we obtained the z-transform of samples taken from a continuous signal and found this substitution related the s-plane characteristics of the continuous signal to the z-plane characteristics of the discrete signal.

The use of the impulse modulation model for the sampler is well illustrated by the model of the linear effects (ignoring amplitude quantization) of the A/D converter. As we saw earlier, the effect of the A/D converter is to take a sample in negligible time (the sample-and-hold circuit is often used to avoid the effects of finite conversion time), and the sample is stored as a constant until the next sampling instant. The problem of modeling this process requires that we represent two processes: (1) extracting the samples, and (2) holding the result fixed for one period. The impulse modulator effectively extracts the sample in the form of $r(kT)\delta(t - kT)$. The remaining problem is to construct a linear constant system which will convert this impulse into a pulse of height $r(kT)$ and width T. Thus the impulse response of the second half of the A/D model is a unit pulse of width T sec. The Laplace transform of this impulse response is the transfer function of the hold operation, which we call the zero-order hold, namely,

$$
\begin{aligned}
H_0(s) &= \mathscr{L}\{p(t)\} \\
&= \int_0^\infty [1(t) - 1(t - T)]e^{-st}\, dt \\
&= (1 - e^{-sT})/s.
\end{aligned}
\tag{4.6}
$$

Thus the linear behavior of the A/D converter may be modeled by Fig. 4.2. We must emphasize that the signal $r*(t)$ in Fig. 4.2 is not expected to represent a physical signal in the A/D converter circuit, but rather is introduced to allow us to obtain a transfer-function model of the hold operation and to make a reasonable input-output model of the converter action.

4.3 SAMPLED SPECTRA AND ALIASING

We can get further insight into the process of sampling by an alternative representation of the transform of $r*(t)$, using Fourier analysis. From (4.1) we see that $r*(t)$ is a product of $r(t)$ and the train of impulses, $\Sigma\delta(t - kT)$. This latter series, being periodic, can be represented by a Fourier series[2]; namely,

$$
\sum_{k=-\infty}^{\infty} \delta(t - kT) = \sum_{n=-\infty}^{\infty} C_n e^{j(2\pi n/T)t},
\tag{4.7}
$$

---

[2] A good reference for Fourier analysis is Bracewell (1978).

where

$$C_n = \frac{1}{T} \int_{-T/2}^{T/2} \sum_{k=-\infty}^{\infty} \delta(t - kT)e^{-jn(2\pi t/T)} \, dt = \frac{1}{T}.$$

We define $\omega_s = 2\pi/T$ as the radian sampling frequency. We now substitute (4.7) into (4.1) using ω_s, and take the Laplace transform

$$\mathcal{L}\{r^*(t)\} = \int_{-\infty}^{\infty} r(t) \left\{ \frac{1}{T} \sum_{n=-\infty}^{\infty} e^{jn\omega_s t} \right\} e^{-st} \, dt,$$

$$R^*(s) = \frac{1}{T} \sum_{n=-\infty}^{\infty} \int_{-\infty}^{\infty} r(t) \, e^{jn\omega_s t} \, e^{-st} \, dt$$

$$= \frac{1}{T} \sum_{n=-\infty}^{\infty} \int_{-\infty}^{\infty} r(t) \, e^{-(s-jn\omega_s)t} \, dt$$

$$= \frac{1}{T} \sum_{n=-\infty}^{\infty} R(s - jn\omega_s), \tag{4.8}$$

where $R(s)$ is the transform of $r(t)$. In communication or radio engineering terms, (4.7) expresses the fact that the impulse train corresponds to an infinite sequence of carrier frequencies at integral values of $2\pi/T$, and (4.8) shows that when $r(t)$ modulates all these carriers, it produces a never-ending train of sidebands. A sketch of the elements in the sum given in (4.8) is shown in Fig. 4.3.

An important feature of sampling is shown in Fig. 4.3 and illustrated at the frequency marked ω_1. Two curves are drawn comprising the elements of the sum at ω_1. One of these, the larger amplitude in the figure, is the value of $R(j\omega_1)$. The other comes from the spectrum centered at $2\pi/T$ and is $R(j\omega_0)$, where ω_0 is such that $\omega_0 = \omega_1 - 2\pi/T$. This frequency, ω_0, which shows up at ω_1 after sampling,

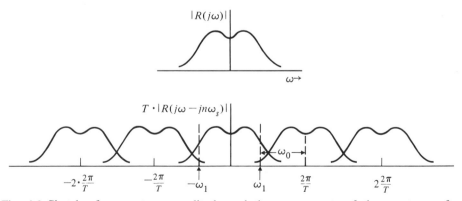

Fig. 4.3 Sketch of a spectrum amplitude and the components of the spectrum after sampling, showing aliasing.

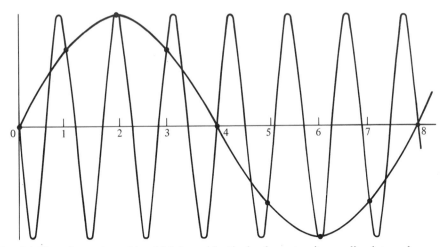

Fig. 4.4 Plot of two sinusoids which have identical values at unit sampling intervals: an example of aliasing.

is called in the trade an "alias" of ω_1 and the process is called *aliasing*. When experimental data are to be sampled for later analysis it is essential that one include an alias guard filter or "prefilter" to reduce the spectrum overlap if a clear picture of the signal spectrum is to be seen.[3]

The phenomenon of aliasing has a clear meaning in time. Two specific sinusoids of different frequencies have identical samples. We cannot, therefore, distinguish between them from their samples. A plot of a cycle of a sinusoid at $\frac{1}{8}$ Hz and of several cycles of a sinusoid at $\frac{7}{8}$ Hz are shown in Fig. 4.4. If we sample these waves at 1 Hz, as indicated by the dots, then we get the same sample values from both signals and would continue to get the same sample values for all time. Note that the sampling frequency is 1, and, if $f_1 = \frac{1}{8}$, then

$$f_0 = \tfrac{1}{8} - 1 = -\tfrac{7}{8}.$$

The significance of the negative frequency is that the $\frac{7}{8}$-Hz sinusoid in Fig. 4.4 is the negative of the sine.

A corollary to the aliasing problem is the sampling theorem. Suppose we ask the following question: What is the highest frequency we can allow in $r(t)$ so that aliasing will not occur? From inspection of Fig. 4.3, it is clear that if $R(j\omega)$ has components above $\omega_s/2$ or π/T, then overlap and aliasing will occur. Conversely, we can say that if $R(j\omega)$ is zero for $|\omega| \geq \pi/T$, then sampling at intervals of T sec will produce no aliasing and the original signal can be recovered exactly from R^*, the spectrum of the samples. This is the sampling theorem: To recover

---

[3] If you were to design a Butterworth guard filter for sampling at 100 Hz, what should the cutoff frequency be? For more extensive discussion, see Chapter 10.

a signal from its samples, you must sample ($\omega_s = 2\pi/T$) *at least twice* the highest frequency (π/T) in the signal.

A phenomenon closely related to aliasing is that of *hidden oscillations*. If signal frequencies only up to π/T may be sampled without confusion, there is the possibility that a signal could contain some frequencies that the samples do not show *at all*. Show how to construct such a signal, even one that grows in an unstable fashion but whose sample values are zero. Such signals, when they show up in a digital control, are called "hidden oscillations."

4.4 DATA EXTRAPOLATION AND IMPOSTORS

The sampling theorem states that under the right conditions, it is possible to recover the signal from the samples; this section discusses practical techniques for doing so. Looking at Fig. 4.3 we can see that the spectrum of $R(j\omega)$ is contained in the low-frequency part of $R^*(j\omega)$. Therefore, to recover $R(j\omega)$ we need only process $R^*(j\omega)$ through a low-pass filter to regain R. As a matter of fact, if $R(j\omega)$ has zero energy for frequencies in the bands above π/T (such an R is said to be band-limited), then an *ideal low-pass filter* with gain T for $-\pi/T \leq \omega \leq \pi/T$ and zero elsewhere would recover $R(j\omega)$ from $R^*(j\omega)$ exactly. Suppose we define this ideal low-pass filter characteristic as $L(j\omega)$. Then we have the result

$$R(j\omega) = L(j\omega)R^*(j\omega). \tag{4.9}$$

The signal $r(t)$ is the inverse transform of $R(j\omega)$ and since by (4.9), $R(j\omega)$ is the *product* of two transforms, its inverse transform $r(t)$ must be the convolution of the time functions $\ell(t)$ and $r^*(t)$. The form of the filter impulse response can be computed by using the definition of $L(j\omega)$ from which the inverse transform gives

$$\ell(t) = \frac{1}{2\pi} \int_{-\pi/T}^{\pi/T} Te^{j\omega t}\, d\omega$$

$$= \frac{T}{2\pi} \frac{e^{j\omega t}}{jt} \Bigg|_{-\pi/T}^{\pi/T}$$

$$= \frac{T}{2\pi jt} (e^{j(\pi t/T)} - e^{-j(\pi t/T)})$$

$$= \frac{1}{\pi t/T} \sin \frac{\pi t}{T}$$

$$\overset{\Delta}{=} \operatorname{sinc} \frac{\pi t}{T}. [4] \tag{4.10}$$

Using (4.1) for $r^*(t)$ and (4.10) for $\ell(t)$, we find that their convolution is

$$r(t) = \int_{-\infty}^{\infty} r(\tau) \sum_{-\infty}^{\infty} \delta(\tau - kT) \operatorname{sinc} \frac{\pi(t - \tau)}{T}\, d\tau.$$

---

[4] "Sinc" is the name given to the function defined by $\operatorname{sinc}(\theta) = \sin(\theta)/\theta$.

Using the sifting property of the impulse, we have

$$r(t) = \sum_{k=-\infty}^{\infty} r(kT) \operatorname{sinc} \frac{\pi(t - kT)}{T}. \tag{4.11}$$

Equation (4.11) is a constructive statement of the sampling theorem: it shows explicitly how to construct the (by assumption) band-limited function $r(t)$ from its samples. The sinc function is the interpolator that fills in the time gaps between samples with a wave that has no frequencies above π/T.

There is one serious drawback to the extrapolating signal given by (4.10). Since $\ell(t)$ is the impulse response of the low-pass filter $L(j\omega)$, it follows that this filter is noncausal because $\ell(t)$ is nonzero for $t < 0$. $\ell(t)$ starts at $t = -\infty$ when the impulse that triggers it does not occur until $t = 0$! In many communications problems it is possible to approximate $\ell(t)$ closely by adding a phase lag, $e^{-j\phi}$, to $L(j\omega)$ which adds a *delay* to $\ell(t)$. In feedback control systems, a large lag is usually disastrous for stability, so we avoid approximations to this function and use something else, like the polynomial holds.

In Section 4.2 we introduced the zero-order hold as a model for the storage register in an A/D converter that maintains a constant signal value between samples and has the transfer function

$$H_0(j\omega) = \frac{1 - e^{-j\omega T}}{j\omega}. \tag{4.12}$$

This, too, is a low-pass filter. In fact, the magnitude response is

$$|H_0(j\omega)| = T \left| \operatorname{sinc} \frac{\omega T}{2} \right|, \tag{4.13}$$

which slowly gets smaller as ω gets larger until it is zero for the first time at $\omega = \omega_s = 2\pi/T$. The phase is

$$\angle H_0(j\omega) = \frac{-\omega T}{2}. \tag{4.14}$$

A plot of the magnitude and phase of the zero-order hold is shown in Fig. 4.5. A more complex circuit which may be preferred to the zero-order hold is the first-order hold, which, as suggested by its name, extrapolates data between sampling periods by a first-order polynomial, a straight line. We can, as with the zero-order hold, compute the transfer function of the filter which, acting with impulse sampling, produces the action of a first-order hold. A sketch of the response to a unit pulse will be helpful and is shown in Fig. 4.6. The first line, rising from 1 to 2 over the period from $t = 0$ to $t = T$ is an extrapolation of the line between the points at $t = -T$ (where the sample was, by assumption, equal to zero) and at $t = 0$, where the sample was unity. Likewise the line going negative from 0 at $t = T$ to -1 at $t = 2T$ is the extrapolation of the line between the points

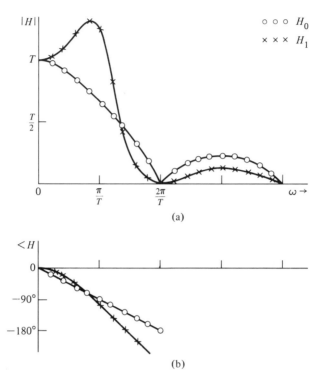

(a)

(b)

Fig. 4.5 Magnitude (a) and phase (b) of polynomial hold filters.

at $t = 0$ and $t = T$. The Laplace transform of this $h(t)$ is

$$H_1(s) = T \left(\frac{1 - e^{-Ts}}{Ts} \right)^2 (Ts + 1).$$ (4.15)

The magnitude and phase of $H_1(j\omega)$ are plotted in Fig. 4.5 to permit comparison with the characteristics of H_0. Note that for low frequencies (below $\pi/2T$) the

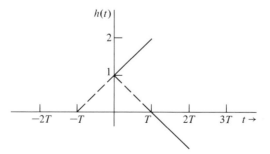

Fig. 4.6 Impulse response of filter, which produces first-order hold action.

first-order hold has significantly less phase lag than does the zero-order hold. However, no clear guidelines seem to exist indicating that one is preferred to the other in any particular circumstances.

Obviously many other, more sophisticated, data extrapolators may and could be designed. However, rarely is additional complexity justified in terms of improved performance in feedback control. As to pure data extrapolation for information purposes, the designs will be best accomplished when the design methods of control systems have been developed.

Associated with the construction of hold circuits and data extrapolators is a problem which is dual to aliasing. Suppose we wish to use the computer as a signal generator and cause it to produce a continuous sinusoidal output signal. Because the digital signals are constructed by samples only, we will be required to include a data extrapolator of some kind—most typically a zero-order hold. The spectrum of the sampled sinusoid will be repeated as shown in Fig. 4.3 and will thus have components not only at the desired frequency ω_0 but also at $\omega_0 + k2\pi/T$, where T is the sampling period. When this composite spectrum is filtered by the zero-order hold characteristic of Fig. 4.5, these higher harmonics appear in the output—impostors at $\omega_0 + k2\pi/T$ pretending to be components of a signal at ω_0 as it were.

4.5 BLOCK DIAGRAM
ANALYSIS OF SAMPLED DATA SYSTEMS

We have thus far talked mainly about discrete, continuous, and sampled signals. To analyze a feedback system which contains a digital computer, we need to be able to compute the transforms of output signals of systems which contain sampling operations in various places, including feedback loops, in the block diagram. The technique for doing this is a simple extension of the ideas of block-diagram analysis of systems which are all continuous or all discrete, but one or two rules need to be carefully observed to assure success. First, we should review the facts of sampled-signal analysis.

We represent the process of sampling a continuous signal and holding it by impulse modulation followed by low-pass filtering. For example, the system of Fig. 4.7 leads to

$$E(s) = R^*(s)H(s),$$
$$U(s) = E^*(s)G(s). \qquad (4.16)$$

The result of impulse modulation of continuous-time signals like $e(t)$ and $u(t)$ is to produce a series of sidebands as in (4.8) and Fig. 4.3, which add up to a periodic function of frequency. If the transform of the signal to be sampled is a product of a transform that is already periodic of period $2\pi/T$, and one that is not, as $U(s) = E^*(s)G(s)$, where $E^*(s)$ is periodic and $G(s)$ is not, we can show that $E^*(s)$ comes

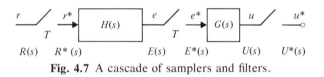

Fig. 4.7 A cascade of samplers and filters.

out as a factor of the result. This is the most important relation for the block diagram analysis of sampled-data systems, namely[5]

$$U^*(s) = (E^*(s)G(s))^* = E^*(s)G^*(s). \tag{4.17}$$

We can prove (4.17) either in the frequency domain, using (4.8) or in the time domain, using (4.1) and convolution. We will use (4.8) here. If $U(s) = E^*(s)G(s)$, then

$$U^*(s) = \frac{1}{T} \sum_{n=-\infty}^{\infty} E^*(s - jn\omega_s)G(s - jn\omega_s), \tag{4.18}$$

but $E^*(s)$ is

$$E^*(s) = \frac{1}{T} \sum_{k=-\infty}^{\infty} E(s - jk\omega_s),$$

so that

$$E^*(s - jn\omega_s) = \frac{1}{T} \sum_{k=-\infty}^{\infty} E(s - jk\omega_s - jn\omega_s). \tag{4.19}$$

Now in (4.19) we can let $k = \ell - n$ to get

$$E^*(s - jn\omega_s) = \frac{1}{T} \sum_{\ell=-\infty}^{\infty} E(s - j\ell\omega_s)$$

$$= E^*(s). \tag{4.20}$$

In other words, because E^* is already periodic, shifting it an integral number of periods leaves it unchanged. Substituting (4.20) into (4.18) yields

$$U^*(s) = E^*(s) \frac{1}{T} \sum_{-\infty}^{\infty} G(s - jn\omega_s)$$

$$= E^*(s)G^*(s). \qquad \text{QED} \tag{4.21}$$

Note especially what is *not* true. If $U(s) = E(s)G(s)$, then $U^*(s) \neq E^*(s)G^*(s)$ but rather $U^*(s) = (EG)^*(s)$. The periodic character of E^* in (4.17) is crucial.

The final result we require is that given a sampled-signal transform such as

---

[5] We of course assume the existence of $U^*(s)$, which is assured if $G(s)$ tends to zero as s tends to infinity at least as fast as $1/s$. We must be careful to avoid impulse modulation of impulses, for $\delta(t)\delta(t)$ is undefined.

$U^*(s)$, we may find the corresponding z-transform simply by letting $e^{sT} = z$ or

$$U(z) = U^*(s)\big|_{e^{sT}=z}. \tag{4.22}$$

There is an important time-domain reflection of (4.22). The inverse Laplace transform of $U^*(s)$ is the sequence of *impulses* with intensities given by $u(kT)$; the inverse z-transform of $U(z)$ is the sequence of values $u(kT)$. Conceptually, sequences of values and the corresponding z-transforms are easy to think about as being processed by a computer program, while the model of sampling as a sequence of impulses is what allows us to analyze a discrete system embedded in a continuous world. Of course, the impulse modulator must *always* be followed by a low-pass circuit (hold circuit) in the physical world. Note that (4.22) can also be used in the other direction to obtain $U^*(s)$, the Laplace transform of the train of impulses, from a given $U(z)$.

These rules of analysis can be illustrated by example. Consider the block diagram given in Fig. 4.7 taken from Fig. 1.1. In Fig. 4.8 we have modeled the A/D converter plus computer program plus D/A converter as an impulse modulator [which takes the samples from $e(t)$], a computer program which processes these samples, and a zero-order hold which constructs the piecewise constant output of the D/A converter from the impulses of m^*. In the actual computer we assume that the samples of $e(t)$ are manipulated by a difference equation whose input-output effect is described by the z-transform $D(z)$. These operations are represented in Fig. 4.8 *as if* they were performed on impulses, and hence the transfer function is $D^*(s)$ according to (4.22). Finally, the manipulated impulses, $m^*(t)$, are applied to the zero-order hold from which the piecewise constant-control signal $u(t)$ comes. In reality, of course, the computer operates on the sample values of $e(t)$ and the piecewise constant output is generated via a storage register and D/A converter. The impulses provide us with a convenient, consistent, and effective model of the processes to which Laplace-transform methods can be applied.

From the results given thus far, we can write relations among Laplace transforms as

$$E(s) = R - Y, \tag{4.23}$$

$$M^*(s) = E^*D^*, \tag{4.24}$$

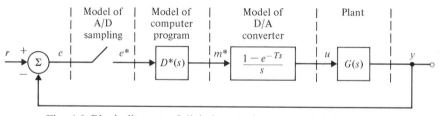

Fig. 4.8 Block diagram of digital control as a sampled-data system.

$$U = M^* \left[\frac{1 - e^{-Ts}}{s}\right],$$ (4.25)

$$Y = GU.$$ (4.26)

The usual idea is to relate the discrete output, Y^*, to the discrete input, R^*. Suppose we sample each of these equations by using the results of Eq. (4.5) to "star" each transform. The equations are[6]

$$E^* = R^* - Y^*$$ (4.27)

$$M^* = E^*D^*$$ (4.28)

$$U^* = M^*$$ (4.29)

$$Y^* = [GU]^*$$ (4.30)

Now (4.30) indicates that we need U, not U^*, to compute Y^* so we must back up to substitute (4.25) into (4.30):

$$Y^* = \left[GM^* \left(\frac{1 - e^{-Ts}}{s}\right)\right]^*.$$ (4.31)

Taking out the periodic parts, which are those in which s appears only as e^{sT} [which include $M^*(s)$], we have

$$Y^* = (1 - e^{-Ts})M^* \left(\frac{G}{s}\right)^*.$$ (4.32)

Substituting from (4.28) for M^* gives

$$Y^* = (1 - e^{-Ts})E^*D^*(G/s)^*.$$ (4.33)

And substituting (4.27) for E^* yields

$$Y^* = (1 - e^{-Ts})D^*(G/s)^*[R^* - Y^*].$$ (4.34)

If we call

$$(1 - e^{-Ts})D^*(G/s)^* = H^*,$$

then we can solve (4.34) for Y^* obtaining

$$Y^* = \frac{H^*}{1 + H^*} R^*.$$ (4.35)

---

[6] In sampling (4.25) we obtain (4.29) by use of the convention given on p. 78 for impulse modulation of discontinuous functions. From the time domain operation of the zero-order hold, it is clear that the samples of u and m are the same and from this (4.29) follows.

These equations can be illustrated with a simple example.   Suppose our plant has the first-order transfer function

$$G(s) = \frac{a}{s + a}, \tag{4.36}$$

the computer program corresponds to a discrete integrator like the backward rectangular-rule equivalent, namely

$$u(kT) = u(kT - T) + K_0 e(kT), \tag{4.37}$$

and the computer D/A holds the output constant so that the zero-order hold is the correct model.   Suppose we select the sampling period T so that $e^{-T} = \frac{1}{2}$.   We wish to compute the components of H^* given in (4.35).   For the computer program we have the transfer function of (4.37) which in terms of z is

$$D(z) = \frac{U(z)}{E(z)} = \frac{K_0}{1 - z^{-1}} = \frac{K_0 z}{z - 1}.$$

Using (4.22), we get

$$D^*(s) = \frac{K_0 e^{sT}}{e^{sT} - 1}. \tag{4.38}$$

For the plant and zero-order hold we require

$$(1 - e^{-Ts})(G(s)/s)^* = (1 - e^{-Ts})\left(\frac{a}{s(s + a)}\right)^*$$

$$= (1 - e^{-Ts})\left(\frac{1}{s} - \frac{1}{s + a}\right)^*.$$

Using (4.5), we have

$$(1 - e^{-Ts})(G(s)/s)^* = (1 - e^{-Ts})\left(\frac{1}{1 - e^{-Ts}} - \frac{1}{1 - e^{-aT}e^{-Ts}}\right).$$

Since we assumed (for simplicity) that $e^{-aT} = \frac{1}{2}$, this reduces to

$$(1 - e^{-Ts})(G(s)/s)^* = \frac{\frac{1}{2}e^{-Ts}}{1 - \frac{1}{2}e^{-Ts}}$$

$$= \frac{\frac{1}{2}}{e^{Ts} - \frac{1}{2}}. \tag{4.39}$$

Combining (4.39) and (4.38), then, in this case, we obtain

$$H^*(s) = \frac{K_0}{2} \frac{e^{sT}}{(e^{sT} - 1)(e^{sT} - \frac{1}{2})}. \tag{4.40}$$

Equation (4.40) can now be used in (4.35) to find the closed-loop transfer function from which the dynamic and static responses can be studied, as a function of K_0, the program gain. We note also that beginning with (4.26), we can readily calculate that

$$Y(s) = R^* \frac{D^*}{1 + H^*} \frac{(1 - e^{-Ts})}{s} G(s). \tag{4.41}$$

Equation (4.41) shows how to compute the response of this system in between sampling instants. For a given $r(t)$, the terms in (4.41) which are starred and the $(1 - e^{-Ts})$-term correspond to a train of impulses whose individual values can be computed by expanding in powers of e^{-Ts}. These impulses are applied to $G(s)/s$, which is the step response of the plant. Thus, between sampling instants, we will see segments of the plant step response.

With the exception of the odd-looking forward transfer function, (4.35) looks like the familiar feedback formula: forward-over-one-plus-feedback. Unfortunately, the sequence of equations by which (4.35) was computed was a bit haphazard and such an effort may not always succeed. Another example will further illustrate the problem. Consider the block diagram of Fig. 4.9 which has only one sampling operation. This situation may arise if the error sensor had significant dynamics which precede the sampling action of the A/D converter. $H(s)$ represents the sensor dynamics. Again, we write the equations (all symbols are Laplace transforms)

$$E = R - Y, \tag{4.42}$$

$$U = HE, \tag{4.43}$$

$$Y = U^*G, \tag{4.44}$$

and again we sample

$$E^* = R^* - Y^*, \tag{4.45}$$

$$U^* = (HE)^*, \tag{4.46}$$

$$Y^* = U^*G^*. \tag{4.47}$$

How do we solve? In (4.46) we need E, not E^*. So we must go back to (4.42):

$$U^* = (H(R - Y))^*$$
$$= (HR)^* - (HY)^*.$$

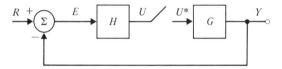

Fig. 4.9 A simple system that does not have a transfer function.

Using (4.44) for Y, we have

$$U^* = (HR)^* - (HU^*G)^*.$$

Taking out the periodic U^* in the second term on the right gives

$$U^* = (HR)^* - U^*(HG)^*.$$

Solving, we get

$$U^* = \frac{(HR)^*}{1 + (HG)^*}. \tag{4.48}$$

From (4.41), we can solve for Y^*:

$$Y^* = \frac{(HR)^*}{1 + (HG)^*} G^*. \tag{4.49}$$

Equation (4.49) displays a curious fact. The transform of the input is bound up with $H(s)$ and *cannot* be divided out to give a transfer function! This system displays an important fact that all our facile manipulations of samples, etc., may cause us to neglect: a sampled-data system is *time varying*. The response depends on the time *relative to the sampling instant* at which the signal is applied. Only when the input samples *alone* are required to generate the output samples can we obtain a transfer function. The time variation occurs on the taking of samples. In general, as in Fig. 4.9, the entire input signal $r(t)$ is involved in the system response and the transfer-function concept fails. Even in the absence of a transfer function, however, the techniques developed here permit study of stability and response to specific inputs such as step and ramp signals.

We need to know the general rules of block-diagram analysis. In solving Fig. 4.9 we found ourselves working with U, the signal which was sampled. This is in fact the key to the problem. Given a block diagram with several samplers, *always select the variables at the inputs to the samplers as the unknowns*. Being sampled, these variables have periodic transforms and will always "come free" after the equation sampling process and give a set of starred variables for which we can solve.

Consider a final example, Fig. 4.10. We select E and M as independent variables, and write

$$E(s) = R - M^*G_2, \tag{4.50}$$

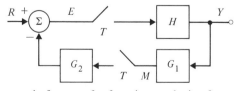

Fig. 4.10 A final example for transfer-function analysis of sampled-data systems.

$$M(s) = E^*HG_1. \tag{4.51}$$

Next we sample these signals, and use the "if periodic, then out" rule from (4.17):

$$E^* = R^* - M^*G_2^*, \tag{4.52}$$

$$M^* = E^*(HG_1)^*. \tag{4.53}$$

We solve these equations by substituting for M^* in (4.52) from (4.53):

$$E^* = R^* - E^*(HG_1)^*G_2^*$$

$$= \frac{R^*}{1 + (HG_1)^*G_2^*}. \tag{4.54}$$

To obtain Y we use the equation

$$Y = E^*H$$

$$= \frac{R^*H}{1 + (HG_1)^*G_2^*}, \tag{4.55}$$

and

$$Y^* = \frac{R^*H^*}{1 + (HG_1)^*G_2^*}. \tag{4.56}$$

In this case we have a transfer function. Why? To obtain the z-transform of the samples of the output, we would let $e^{sT} = z$ in (4.56). From (4.55) we can solve for the continuous output which consists of impulses applied to $H(s)$ in this case.

4.6 SUMMARY

In this chapter we have considered the analysis of mixed systems that are partly discrete and partly continuous, taking the continuous point of view. We used impulse modulation to represent the sampling process, and derived the transfer functions of filters which would represent zero-order and first-order hold action. We show that the transform of a sampled signal is periodic and that sampling introduces aliasing which may be interpreted both in the frequency and the time domains. From the condition of no aliasing we derived the sampling theorem.

Finally, we presented the block-diagram analysis of sampled-data systems showing that proper techniques including the treatment of the sampler inputs as unknowns would lead to solution for the output transforms. However, we also found that not every sampled-data system has a transfer function.

PROBLEMS AND EXERCISES

4.1 Derive Eq. (4.41).

4.2 Sketch a signal which shows hidden oscillations.

4.3 Consider the circuit of Fig. 4.11. By plotting the response to a signal which is zero for all sample instants except $t = 0$ and is 1.0 at $t = 0$, show that this circuit implements a first-order hold.

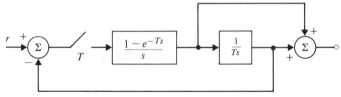

Figure 4.11

4.4 Sketch the step response $y(t)$ of the system shown in Fig. 4.12 for $K = \frac{1}{2}, 1, 2$.

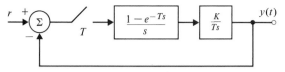

Figure 4.12

4.5 Sketch the response of a *second*-order hold circuit to a step input. What might be the major disadvantage of this data extrapolator? See Fig. 4.13.

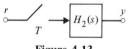

Figure 4.13

4.6 Find the transform of the output, $Y(s)$, and its samples $Y^*(s)$ for the block diagrams shown in Fig. 4.14. Indicate if a transfer function exists in each case.

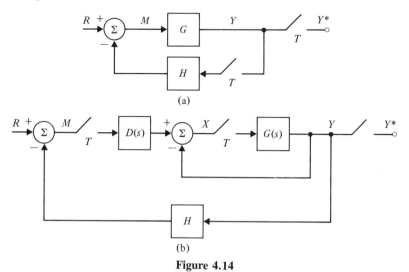

(a)

(b)

Figure 4.14

4.7 In Appendix A are sketched several process transfer functions. Assume they are preceded by a zero-order hold and compute the resulting discrete transfer function for

a) $G_1(s) = 1/s^2$.
b) $G_2(s) = e^{-1.5s}/(s + 1)$; assume $T = 1.0$ in this case.
c) $G_3(s) = 1/s(s + 1)$
d) $G_7(s) = e^{-1.5s}/s(s + 1)$; assume $T = 1.0$ in this case.
e) $G_8(s) = 1/(s^2 - 1)$.

4.8 One technique for examining the response of a sampled-data system between sampling instants is to predict the response a fraction of a sample period and sample the result. The effect is as shown in the block diagram of Fig. 4.15 and described by the equation

$$Y^*(s, m) = R^*(s)\, \mathcal{Z}\{G(s)e^{mTs}\}.$$

The function $\mathcal{Z}\{G(s)e^{mTs}\}$ is called the *modified z-transform* of $G(s)$. Let

$$G(s) = \frac{1}{s + 1}, \quad T = 1, \quad R(s) = \frac{1}{s}$$

a) Compute $y(t)$ by constructing the samples $y(kT)$ from $Y^*(s)$ and observing that with this plant, $y(t)$ is an exponential decay with unit time constant over the intersample interval. Sketch the response for five sample intervals.

b) Let $m = \frac{1}{2}$ and compute the samples corresponding to $Y^*(s; m)$ [or $Y(z; m)$]. Plot on the same sketch of part (a) and verify that the midway points have been found.

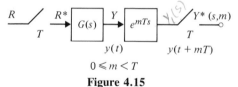

$$0 \leqslant m < T$$

Figure 4.15

5 / Design of Digital Control Systems Using Transform Techniques

5.1 INTRODUCTION

The idea of controlling processes that evolve in time is ubiquitous. Systems from airplanes to the national rate of unemployment, from unmanned space vehicles to human blood pressure, are considered fair targets for control. Over a period of three decades from about 1930 until 1960, a body of control theory was developed based on electronic feedback amplifier design modified for servomechanism problems. This theory was coupled with electronic technology suitable for implementing the required dynamic compensators to give a set of approaches to solve control problems now often called *classical techniques*. The landmark references to this theory are Nyquist (1932), Bode (1945), and Evans (1950). For random inputs, the work of Wiener (1948) should be added. An excellent pedagogical presentation of these methods is given in Truxal, (1955). The unifying theme of these methods is the use of Laplace or Fourier transform representations of the system dynamics and the control specifications; hence, we refer to them here as *transform techniques* after the central role of the frequency domain in the approach.

In this chapter, we discuss briefly the use of transform techniques in the design of digital control systems. First, we describe the feasibility of using discrete equivalents to construct a digital control indirectly from a continuous design. Then we turn to the modifications of the transform techniques necessary to make them directly applicable to digital controls. We find that the root locus can be transferred unchanged to the z-plane, but that for Bode techniques to be useful, we need to return to the bilinear (Tustin) transformation which will also be called the w-transform.

As with any engineering design method, design of control systems by transform techniques requires many computations which are greatly facilitated by a good library of well-documented computer programs. In the presentations in this book we mainly treat elementary cases which can be solved by hand with at most a scientific calculator. However, in designing practical systems, and especially in iterating through the methods many times to meet essential specifications, an interactive computer-aided design package with simple plotting graphics is crucial. At the end of each section, we indicate the most important computations needed for the particular methods.

5.2 z-PLANE SPECIFICATIONS OF CONTROL SYSTEM DESIGN

Before describing how the transform techniques can be applied to digital control designs, we must first develop the specifications toward which the design is aimed. To help fix ideas, the specifications will be discussed in the context of the design of the azimuth control of an antenna designed to pick up signals from a low-altitude communications satellite.[1] The equations of motion are taken to be

$$J\ddot{\theta}_a + B\dot{\theta}_a = T_m + T_w, \qquad (5.1)$$

where θ_a is the pointing angle, T_m is the drive-motor torque, and T_w is the wind torque. The system parameters are the moment of inertia of the moving parts, J, and the viscous coefficient, B, consisting of the mechanical friction component and the back emf effect of the electric motor. We will assume that $J/B = 10$ sec. The aim of the design is to measure θ_a and compute T_m so that the error between the angle of the satellite θ_s and the antenna, namely $(\theta_s - \theta_a)$, is always less than 0.01 rad during tracking so that the radio transmissions of the satellite are within the narrow beam of the antenna. The satellite angle which must be followed may be adequately approximated by a fixed velocity

$$\theta_s(t) = (0.01)t. \qquad (5.2)$$

The worst possible wind torque can be approximated as a gust that comes suddenly and holds constant for several seconds. We will approximate this by a step function and require that the transients must be settled out in less than 10 sec to leave a total steady-state error within the tracking-error specifications. A block diagram of the system is shown in Fig. 5.1 with the normalized variables

$$u = T_m/B, \qquad w = T_w/B, \qquad y = \theta_a.$$

The specifications for this system, which are characteristic of many others, may now be listed. They are:

1. Steady-state tracking accuracy

2. Transient accuracy (dynamic response):
   a) rise time
   b) overshoot
   c) settling time

---

[1] This example is discussed in Appendix A.

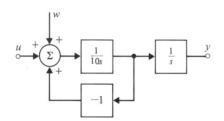

Fig. 5.1 Standardized model of an antenna tracking control.

3. Disturbance rejection:

   a) steady state
   b) transient

4. Control effort required:

   a) maximum magnitude of u
   b) energy $K \int u^2 \, dt$

5. Sensitivity to parameter changes

We will discuss each of these in turn.

   Steady-state accuracy refers to the requirement that after all transients are negligible, the error $r - y$ or, for the antenna, $\theta_s - \theta_a$, must be acceptably small. The two causes of nonzero error are the reference r and the disturbance w. Consider first the reference. Some control systems have a finite nonzero steady-state error when the reference is a constant. Such systems are labeled "Type 0," since there is finite error with a zero-order polynomial input and since the equivalent *unity feedback plant* has finite dc gain. Similarly, a control system that has finite nonzero steady-state error to a first-order polynomial input (a ramp) is called a "Type I" system. And then comes "Type II," *mutatis mutandis*.[2] In each case, the disturbance w is taken to be zero. However, in evaluating the system we must include the effects of w in the final calculations. In general, we add the errors due to reference and disturbance to find a total system error which must be within acceptable (beam width, in this case) limits.

   To compute the steady-state error due to r or w, we assume the system is stable and use the final-value theorem. Suppose the unity feedback system shown in Fig. 5.2 has a reference input which is a step function and that the disturbance is zero. The error will have the transform.

$$E(z) = \frac{R(z)}{1 + D(z)G(z)} \tag{5.3}$$

$$= \frac{z}{z - 1} \frac{1}{1 + D(z)G(z)}. \tag{5.4}$$

---

[2] With necessary changes, i.e., with "second" for "first" and "II" for "I."

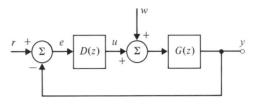

Fig. 5.2 A unity feedback system.

The final value of $e(k)$, if the roots of $1 + DG = 0$ are all inside the unit circle, is, by (2.52),

$$e(\infty) = \lim_{z \to 1} (z - 1) \frac{z}{z - 1} \frac{1}{1 + D(z)G(z)}$$

$$= \frac{1}{1 + D(1)G(1)}$$

$$\triangleq \frac{1}{1 + K_p}. \tag{5.5}$$

Thus, $D(1)G(1)$ is the position constant, K_p, of the Type 0 system. If DG has a pole at $z = 1$, then the error given by (5.5) is zero. Suppose the pole is simple. Then we have a Type I system and we can compute the error to a unit ramp input. Let $r = t1(t)$; then

$$E(z) = \frac{Tz}{(z - 1)^2} \frac{1}{1 + D(z)G(z)}. \tag{5.6}$$

Now the steady-state error is

$$e(\infty) = \lim_{z \to 1} (z - 1) \frac{Tz}{(z - 1)^2} \frac{1}{1 + DG}$$

$$= \lim_{z \to 1} \frac{Tz}{(z - 1)(1 + D(z)G(z))}$$

$$\triangleq \frac{1}{K_v}. \tag{5.7}$$

Thus the velocity constant of the Type I system with unity feedback (as shown in Fig. 5.2) is

$$K_v = \lim_{z \to 1} \frac{(z - 1)(1 + D(z)G(z))}{Tz}. \tag{5.8}$$

Because systems of Type I occur frequently, it is useful to observe that the value of K_v is fixed by the *closed-loop* poles and zeros by a relation given, for the continuous case, by Truxal (1955). Suppose the overall transfer function Y/R is

$H(z)$ and that $H(z)$ has poles p_i and zeros z_i. Then we may write

$$H(z) = K \frac{(z - z_1)(z - z_2) \cdots (z - z_n)}{(z - p_1)(z - p_2) \cdots (z - p_n)}. \tag{5.9}$$

Now suppose that $H(z)$ is the transfer function of a Type I system which implies that the steady-state error of this system to a step is zero and requires that

$$H(1) = 1. \tag{5.10}$$

Furthermore, by definition we can express the error to a ramp as

$$E(z) = R(1 - H(z))$$
$$= \frac{Tz}{(z - 1)^2} (1 - H(z)),$$

and the final value of this error is given by

$$e_\infty = \lim_{z \to 1} (z - 1) \frac{Tz}{(z - 1)^2} (1 - H(z)) = \frac{1}{K_v};$$

therefore (omitting a factor of z in the numerator which makes no difference in the result)

$$\frac{1}{TK_v} = \lim_{z \to 1} \frac{1 - H(z)}{z - 1}. \tag{5.11}$$

Because of (5.10), the limit in (5.11) is indeterminate, and so we may use l'Hospital's rule

$$\frac{1}{TK_v} = \lim_{z \to 1} \frac{(d/dz)(1 - H(z))}{(d/dz)(z - 1)}$$
$$= \lim_{z \to 1} - \frac{dH(z)}{dz}.$$

However, note that by (5.10) again, at $z = 1$,

$$\frac{d}{dz} \ln H(z) = \frac{1}{H} \frac{d}{dz} H(z) = \frac{d}{dz} H(z),$$

so that

$$\frac{1}{TK_v} = \lim_{z \to 1} - \frac{d}{dz} \ln H(z)$$
$$= \lim_{z \to 1} - \frac{d}{dz} \left\{ \ln K \frac{\Pi(z - z_i)}{\Pi(z - p_i)} \right\}$$
$$= \lim_{z \to 1} - \frac{d}{dz} \left\{ \sum \ln (z - z_i) - \sum \ln (z - p_i) + \ln K \right\}$$

$$\frac{1}{TK_v} = \lim_{z \to 1} \left\{ \sum \frac{1}{z - p_i} - \sum \frac{1}{z - z_i} \right\}$$

$$= \sum_{i=1}^{n} \frac{1}{1 - p_i} - \sum_{i=1}^{n} \frac{1}{1 - z_i}. \tag{5.12}$$

We note especially that the farther the poles of the closed loop system from $z = 1$, the larger the velocity constant, but K_v may be increased ($1/K_v$ reduced) by zeros *close* to $z = 1$. From the results of Chapter 2 on dynamic response, we recall that a zero close to $z = 1$ usually signals large overshoot and poor dynamic response. Thus is expressed one of the classic trade-off situations: we must balance low steady-state error against good transient response.

For our antenna problem, we have specified that a ramp of slope 0.01 has steady-state error no more than 0.01 so that $K_v = 1$ is satisfactory for this problem.

Transient accuracy, or dynamic response, refers to the ability of the system to keep the error small as $r(t)$ changes. Specifications of transient performance may be made in the time domain and then translated to the frequency domain either in terms of characteristic pole locations in s or z, or in terms of frequency response features such as bandwidth and resonant peak. We will aim to consider specifications in terms of characteristic root locations in the z-plane and will first obtain the specifications in the s-plane and then use the relation that $z = e^{sT}$ to map the poles and zeros in the s-plane to the z-plane.

First we need to transfer the transient specifications from a time description to an s-plane pole-location requirement. In Fig. 5.3(a) are plotted the step responses of a second-order system with unity direct current (dc) gain, no finite zeros, and various damping ratios. We see immediately that the major influence on percent overshoot is the damping ratio, ζ. In Fig. 5.3(b) is plotted the value of this feature against ζ. A specification on percent overshoot can, for the second-order system, be translated into a specification of ζ. Note also that we should refer to Chapter 2 and Figs. 2.18 and 2.19, where we plotted our finding that an extra zero can also greatly influence overshoot. We must consider both ζ and the zero locations to help meet transient response specifications on percent overshoot. With this proviso, we can say, very roughly, that

$$\% \text{ overshoot} \doteq (1 - \zeta/0.6)100$$

for a second-order system with no finite zeros. Thus, given a requirement on percent overshoot, we require

$$\zeta \geq (0.6)\left(1 - \frac{\% \text{ overshoot}}{100}\right).$$

Another feature of interest is the rise time of the response toward its final value. By inspection of Fig. 5.3(a) we see that the time scale is in terms of ω_n, the distance of the poles from the origin of the s-plane. Thus the rise time will cer-

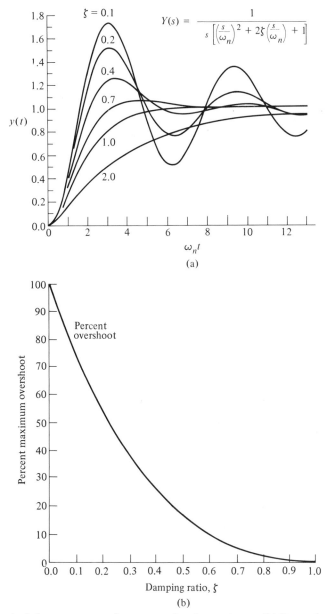

Fig. 5.3 (a) Typical time responses for a second-order system.   (b) Dependence of overshoot on damping ratio for second-order system.   (From R. C. Dorf, *Modern Control Systems,* 3rd ed. Reading, Mass.: Addison-Wesley, 1980, pp. 113, 116.)

tainly be shorter as ω_n is increased. While there is some dependence of rise time on ζ, we may take the curve for $\zeta = 0.5$ to be about the center of the distribution and thus approximate the rise time

$$t_r \doteq 2.5/\omega_n,$$

where we take t_r to be the time necessary for the response to rise from 0.1 to 0.9. A requirement on t_r thus becomes a requirement that ω_n satisfy

$$\omega_n \geq 2.5/t_r.$$

The final time domain feature which is of importance to us is the settling time. This is the time required for the response to settle to within some small fraction of its steady-state value and stay there. For the prototype second-order system, we can return to the mathematics of the solution to conclude that the transient is of the form

$$y(t) = 1 - e^{-\zeta \omega_n t} \cos (\omega_d t + \phi),$$

where $\omega_n = \omega_n \sqrt{1 - \zeta^2}$. The point is that the transient portion of this signal is contained in an envelope of $e^{\zeta \omega_n t}$, where $- \zeta \omega_n$ is the real part of the root location. Thus, we can require that $\zeta \omega_n$ be large enough that the transient will be squeezed into whatever error tolerance band we choose. A typical value of the error tolerance is 1% for which we compute the envelope function to be

$$e^{-\zeta \omega_n t_s} \leq 0.01,$$

and thus

$$\zeta \omega_n t_s \geq 4.6,$$
$$t_s \doteq 4.6/\zeta \omega_n,$$

or

$$- \operatorname{Re} \{s_i\} = \zeta \omega_n \geq 4.6/t_s.$$

We need now convert these specifications into guidelines on the placement of poles and zeros in the z-plane in order to guide the design of digital controls. We do so by pole-and-zero mapping via $z = e^{sT}$. Thus the restriction on percent overshoot has been expressed as a restriction on damping ratio, ζ. In the z-plane, curves of pole locations for constant ζ are logarithmic spirals as sketched in Fig. 5.4(a) for $\zeta = 0.5$. The forbidden region is indicated by the partial hatching. The restriction on rise time is the requirement that the natural frequency be greater than a certain value. In the z-plane the curves of constant ω_n are lines drawn at right angles to the constant ζ spirals. A given value is sketched in Fig. 5.4(b), again with the hatching on the undesirable side of the line. The final time-domain specification was in terms of settling time. In this case, the real parts of the roots, $- \zeta \omega_n$, were restricted. Since the s-z mapping has the z-plane root radius at $r =$

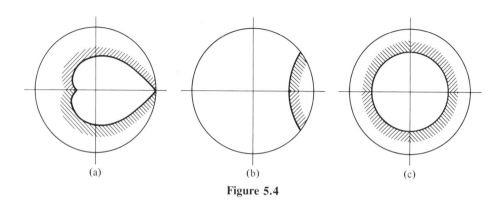

Figure 5.4

$e^{-\zeta\omega_n T}$, we see at once that a settling-time restriction maps into a restriction that the z-plane poles should be inside a circle given by

$$r_0 = e^{-4.6T/t_s}$$

which, when sketched, looks like Fig. 5.4(c)

The final effect is drawn in Fig. 5.5, where we take $T = 1$ sec and

$$\% \text{ overshoot} \leq 15\% \Rightarrow \zeta \geq 0.5,$$

$$t_r \leq 8 \Rightarrow \omega_n \geq \frac{2.5}{8},$$

$$t_s \leq 20 \Rightarrow r \leq 0.8.$$

The region forbidden by these specifications is indicated by the partial hatching along the line for $\omega_n = \pi/10T$ to the circle of radius 0.8 to the spiral of $\zeta = 0.5$ and then on around the spiral. These curves are approximate as befits a design process which is essentially a set of guidelines to the closed-loop system pole locations. The final design must be checked by simulation and/or experiment and modifications made as indicated by the manner in which the first design fails to meet the specifications. For example, if a trial design has a settling time which is too long, then the radius of the poles should be reduced, and so on, for corrections to overshoot and rise time.

The effectiveness of the system in *disturbance rejection* is readily studied with the topology of Fig. 5.2. From the figure, if we take $r = 0$, we find

$$E(z) = \frac{WG}{1 + DG}. \tag{5.13}$$

If the loop gain DG is large compared to 1, then (5.13) reduces to

$$E(z) = W/D(z). \tag{5.14}$$

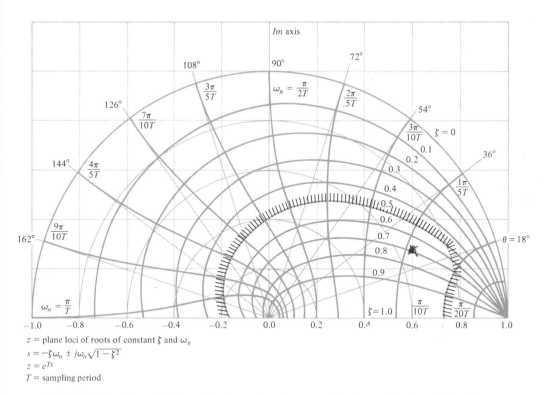

z = plane loci of roots of constant ζ and ω_n

$s = -\zeta\omega_n \pm j\omega_n\sqrt{1-\zeta^2}$

$z = e^{Ts}$

T = sampling period

Fig. 5.5 Plot of acceptable region for poles of a second-order system to satisfy dynamic response specification.

Thus the extent of disturbance reduction is given by the amount of *gain* which precedes the disturbance in the loop. In particular, if $w(k)$ is a constant, then an integrator in D[pole at $z = 1$] will cause the steady-state error due to the disturbance to be zero. From the point of view of frequency response, a disturbance at w in Fig. 5.2 will be rejected over the frequency range where $|DG| \gg 1$ and $|D| \gg 1$, and also over the range where $|G| \ll 1$, $|DG| \ll 1$.

The *control effort* required to perform a control task is important on several counts. Since all physical variables are bounded, the device that provides the control, such as the motor that drives the tracking antenna, can only put out a certain maximum torque even when turned on fully. It is pointless to try to get 100 ft-lb out of a 10 oz-in motor! Conversely, after completion of a design meeting dynamic response specifications, we can simulate a worst-case transient, and from the size of the control signal required, determine the size of the motor necessary to meet these specifications. In addition to peak control, $|u|$, we are sometimes interested in the total heat generated by the drive motor. This, too, will influence the motor size and design (and expense) needed. Usually this number is proportional to $\int_0^\infty u^2 \, dt$. Another measure of control effort arises in gas jets used for attitude control of satellites where the total fuel used is a proper measure of control effort, and the fuel expenditure is proportional to $\int_0^\infty |u| \, dt$. The theory and applications of optimal control are an effort to include these objectives directly in the design. In this chapter, we restrict ourselves to analysis of control effort after the fact, and we suggest that a simulation of the final design will determine whether design is satisfactory from the point of view of control effort. If given a choice, we will prefer that design which gives the smallest value of control effort while meeting the error (dynamic and steady-state) specifications.

Finally, *sensitivity to parameter changes* needs to be studied separately for changes in plant parameters and for changes in controller parameters. As to changes in the parameters of the plant, the situation is very much like the disturbance-signal rejection, and both features are contained in the concept of robustness as discussed briefly in Chapter 2. The larger the gain of the feedback loop around the offending parameter, the lower the sensitivity of the transfer function to changes in that parameter. Since in the most common cases we have very slowly varying parameters, we are led to design for high gain at $z = 1$ which corresponds to constant or dc signals. If this high gain is in front of the disturbance, then we will also achieve good disturbance rejection. The second aspect concerns the effects of changes in the controller, $D(z)$. Here, we have control over the topology and a design choice can be made to minimize the effects of parameter changes in $D(z)$. Furthermore, in a digital control, the effects of round-off errors and truncation in realization of parameters in $D(z)$ are important; in Chapter 7 we discuss selection of canonical realizations to minimize these effects.

5.3 DESIGN BY DISCRETE EQUIVALENT

Let us return to the primary problem at hand, the design of a digital control for systems like the tracking antenna to meet the given specifications. The first technique is to do the design in the s-plane, using root locus or Bode techniques to derive a satisfactory $D(s)$ as the controller. This step totally ignores the fact that a sampler and digital computer will eventually be used. Having $D(s)$, we then convert the design to a digital control by considering $D(s)$ to be a filter transfer function, and we apply one of the techniques from Chapter 3 to obtain an equivalent $D(z)$.

If we ask for $\zeta = 0.5$ (overshoot about 17%) and settling time of 10 sec ($10\zeta\omega_n = 4.6$), we find that[3]

$$D(s) = \frac{10s + 1}{s + 1} \tag{5.15}$$

---

[3] Using root locus and canceling the one pole of $G(s)$.

will do.   This controller, which is a lead network, results in K_v of 1 and thus will satisfy the steady-state error requirement of the job statement.   What should $D(z)$ be?   We could use any of the methods of Chapter 3; we will illustrate the simple pole-zero mapping which is quickest for use with the root-locus technique.   First, we must select the sampling period.   We have designed the system to have $\omega_n = 1$ rad/sec and $\omega_d = \sqrt{3}/2$.   Thus the natural (closed-loop) period will be 7.25 sec. We want to sample as slowly as possible to minimize the task's demand on the computer but the quality of the dynamic response generally goes down as T gets larger.   A good rule of thumb for maximum T is to sample between 6 and 10 times per cycle, which leads to T in the range of 1 sec.   We will select $T = 1$ and simulate the result.[4]   In (5.15) we have two first-order factors, one having $\omega_1 T = 1$ and the other having $\omega_2 T = \frac{1}{10}$.   Using the rules of Section 3.3 on pole-zero mapping, we compute

$$\text{zero at } z = e^{-0.1} = 0.9048,$$
$$\text{pole at } z = e^{-1} = 0.3679,$$
$$\text{gain at } z = 1 = K\frac{1 - 0.9048}{1 - 0.3679}; \tag{5.16}$$

therefore

$$K = 6.6397,$$

and we have the compensation

$$D(z) = 6.6397\frac{z - 0.9048}{z - 0.3679}, \tag{5.17}$$

which can be readily converted to a difference equation and implemented for the digital controller.

To analyze the behavior of this compensation, we must determine the z-transform of the continuous plant (Fig. 5.1) preceded by a zero-order hold (ZOH). The ZOH is necessary to convert the output of the computer to a continuous signal for control of the plant.   The procedure for finding the z-transform of a $G(s)$ preceded by a ZOH was developed in Chapters 3 and 4.   Applying (3.22) to $G(s)$ of Fig. 5.1 we obtain

$$G(z) = \frac{z - 1}{z} \mathcal{Z}\left\{\frac{a}{s^2(s + a)}\right\}, \tag{5.18}$$

which is

$$G(z) = \frac{z - 1}{z} \mathcal{Z}\left\{\frac{1}{s^2} - \frac{1}{as} + \frac{1}{a}\frac{1}{s + a}\right\}.$$

---

[4] Selection of sampling rates is discussed in Chapter 10.   With this large value of T we should expect substantial errors in the approximation.

Using the tables in Appendix B, we find

$$G(z) = \frac{z-1}{z} \left\{ \frac{Tz}{(z-1)^2} - \frac{z}{a(z-1)} + \frac{1}{a} \frac{z}{z - e^{-aT}} \right\}$$

$$= \frac{Az + B}{a(z-1)(z - e^{-aT})},$$

$$A = e^{-aT} + aT - 1, \qquad B = 1 - e^{-aT} - aTe^{-aT}.$$

For $T = 1$, $a = 0.1$ as in this case, this evaluates to

$$G(z) = 0.04837 \frac{z + 0.9672}{(z-1)(z - 0.9048)}. \tag{5.19}$$

Combining (5.17) and (5.19) we find $K_v = 0.998$ and the closed-loop roots are at a radius of 0.82 and an angle of 50°. A plot of the step response of the resulting system design is shown in Fig. 5.6.

The response sketched in Fig. 5.6 does not correspond to a damping ratio of $\zeta = 0.5$ since that would lead us to expect an overshoot of about 17% (see Fig. 5.3), and this response shows nearly 50% overshoot. The roots are actually near $\zeta = 0.2$. Clearly the accuracy of the approximation is not adequate in this case. A likely explanation is the fact that even if $D(z)$ generates nearly the same sample values as $D(s)$, the zero-order hold reconstruction of u is only an approximation to the continuous u assumed in the design of $D(s)$. In fact, Fig. 3.5 shows that $\hat{e}$ is at

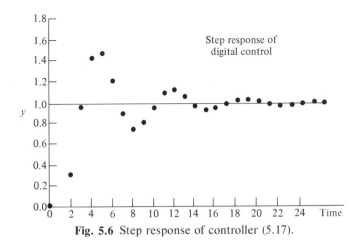

Fig. 5.6 Step response of controller (5.17).

best an approximation to a delayed version of $e(t)$, delayed about $T/2$ sec. A time delay is well known to produce phase lag and generally leads to a less stable design if not taken properly into account.[5] A sample period of $T = 0.2$ will give a response much closer to that of the continuous design.

While the simple second-order system described in this section can be designed by hand calculations, more complex systems designs would be greatly enhanced by the availability of computer-aided design tools. Among the most essential are

1. POLLY: A program to compute the roots of a polynomial in case the plant transfer function is not available in factored form.
2. ROOT LOCUS: A program to compute and, ideally, to plot on a cathode ray tube and/or on paper the root locus of a characteristic polynomial. Some programs are based on repeated calls to POLLY but these methods lead to very nonuniform spacing of the roots in the s-plane (or z-plane). Other programs track and trace a single branch of the locus using the angle criterion that if the locus is defined by $1 + KG(s) = 0$, then on a locus point the angle of $G(s_0) = 180°$. See Ash and Ash (1968).
3. STEP: A program to compute and plot the step response of an open or closed loop system to either reference or disturbance inputs.

5.4 ROOT LOCUS IN THE z-PLANE

The root locus is the locus of points where roots of a characteristic equation may be found as some real parameter varies from zero to large values.[6] From Fig. 5.2 and block-diagram analysis, the characteristic equation of the single loop system is

$$1 + D(z)G(z) = 0. \tag{5.20}$$

The significant thing about (5.20) is that this is exactly the same equation as that found for the s-plane root locus. The implication is that the mechanics of drawing the root loci are the same in the z-plane as in the s-plane except the pole locations mean different things when we come to interpret the system stability and dynamic response. Suppose, for example, we design the antenna system for a sampling

---

[5] At $\omega_d = \sqrt{3}/2$ and $T = 1$ sec, a delay of $T/2$ introduces a phase lag of $\sqrt{3}/4$ rad, or about 25°. Since $\zeta = 0.5$ corresponds to a phase margin (see Section 5.5) of only about 50°, then the delay has taken half of that. In fact, a phase margin of 25° corresponds to $\zeta = 0.2$, which is about the correct value for the step response of Fig. 5.6.

[6] Sometimes we are interested in negative parameter values and look at root loci for the parameter in the entire range $-\infty \leq K < \infty$. The loci for positive gain are the most common.

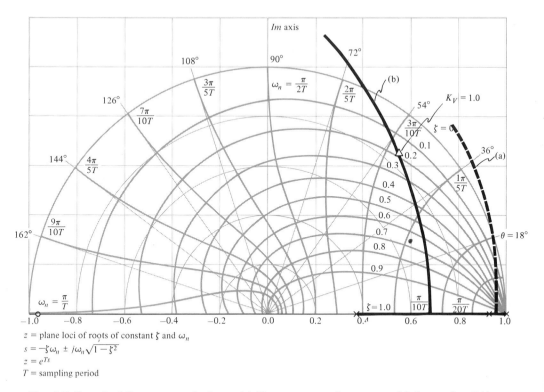

Fig. 5.7 Root loci for antenna design. (a) Uncompensated system. (b) Locus for $D(z)$ given by (5.13).

period of $T = 1$ sec. Then $G(z)$ is given by (5.19) which we can rewrite (approximating the constants) in root locus form:

$$1 + K \frac{z + 0.97}{(z - 1)(z - 0.91)}.$$

From study of the root locus we should remember that this locus, with two poles and one zero, is a circle centered at the zero (-0.97) and breaking away from the real axis between the two real poles at the point where K versus z is a maximum for real z.[7] Here it is almost halfway, a bit closer to 0.91 than 1.0. A plot is given in Fig. 5.7 as a dashed circle (a).

From the root locus of the uncompensated system (Fig. 5.7a) it is clear that some dynamic compensation is required if we are to get satisfactory response from this system. The radius of the roots never gets less than 0.95 and the system goes unstable at $\theta = 25°$ where $K \doteq 0.92$.

---

[7] See Problem 5.3.

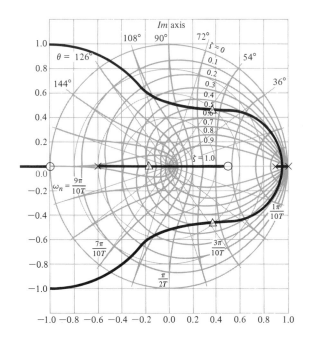

Fig. 5.8 Root locus for antenna design $D(z)$ given by (5.21).

If we cancel the plant pole at 0.91 with a zero, and add a pole at 0.37, we are using the compensation of (5.17) and the locus is also sketched for this case in Fig. 5.7 as the solid curve (b). The point where $K_v = 1$ is marked $\triangle$, and we can see that a damping ratio of about 0.2 is to be expected with $\theta \doteq 50°$ or about seven samples $(= \frac{360}{50})$ per cycle, as we saw in the step response of Fig. 5.6. A better choice of compensation may be expected if we placed the compensation pole on the negative real axis and the compensation zero so that the root locus passed through the point with radius 0.6 (to satisfy the settling time requirement) and on the 0.5 curve for ζ. We place the compensator pole at $z = -0.6$, where it is essentially out of the effective part of the root locus and yet far enough inside the unit circle to afford adequate settling time. The root locus is sketched in Fig. 5.8; the step response corresponding to the closed-loop poles at the triangles ($\triangle$) is shown in Fig. 5.9. The controller for this case is

$$D(z) = 20.674 \frac{z - 0.5}{z + 0.6}. \tag{5.21}$$

The dc gain of $D(z)$ is 6.46, which is equal to K_v since the gain of $(z - 1)G(z)$ is unity. This exceeds the specification, and on the next iteration the design may be

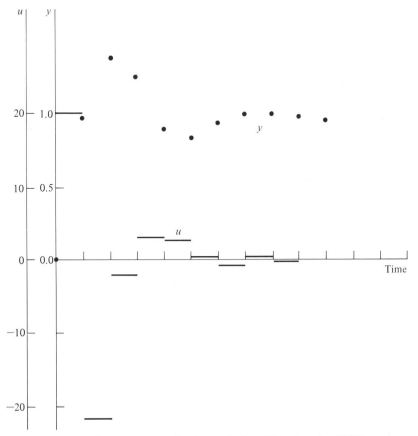

Fig. 5.9 Step response of antenna design $D(z)$ given by (5.21).

made slower by accepting somewhat less rise time and more settling time in exchange for less gain (lower control effort) and less overshoot. Changing the controller to

$$D(z) = 9\frac{(z - 0.8)}{(z + 0.8)} \tag{5.22}$$

gives a K_v of 1.0. The root locus and step responses are shown in Fig. 5.10. Here the output response is slow in settling, but the overshoot and peak control effort are improved over the previous design. The oscillatory nature of the control, which requires significantly more control effort, could be reduced by moving the controller pole from -0.8 to -0.6.

The computer programs POLLY, ROOT LOCUS, and STEP constitute the major aids to design using the root locus technique.

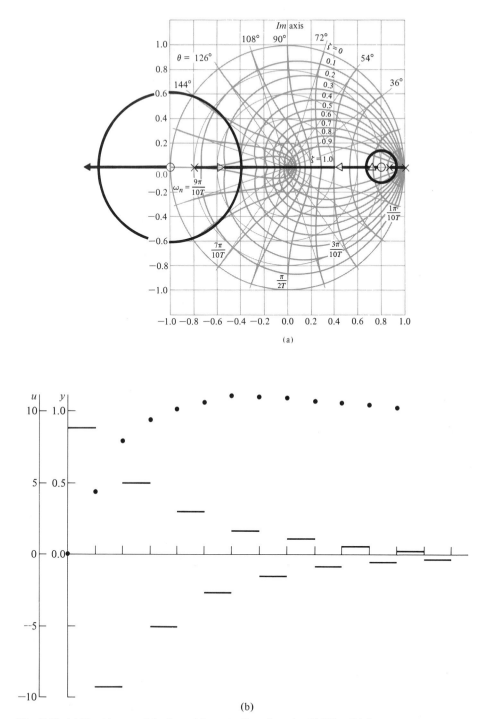

Fig. 5.10 (a) Root locus of design with controller given by (5.22). (b) Step responses corresponding to the controller of (5.22) with the antenna design.

5.5 FREQUENCY RESPONSE METHODS: THE w-TRANSFORM

The frequency response methods for control system design were developed from the original work of Bode (1945) on feedback amplifier techniques. They depend on several facts. These are:

1. The system error constant (mainly K_v) can be read directly from the low-frequency asymptote of the gain plot.
2. Nyquist's stability criterion can be applied and dynamic response specifications can be readily interpreted in terms of gain and phase margins which are easily seen on the plot of log gain and phase versus log frequency.
3. The corrections to the gain and phase curves introduced by a trial pole or zero of a compensator can be quickly and easily computed, *using the gain curve alone*.

It is not possible to review these points adequately in this brief account. The books by James, Nichols, and Phillips (1947), Clark (1962), and Ogata (1970) give better pedagogic treatments than Bode provides. We can, however, illustrate the difficulties of applying this design technique to discrete systems and show a remedy for most of them. The difficulty is almost entirely with point 3. Point 1 is not critical since the error constant is only a single point away, and no great difficulty attaches to computing it. Likewise, with respect to point 2, the essential features of the Nyquist stability criterion remain unchanged, and we can speak of gain and phase margins as in the continuous case. However, there may be some difficulty in calculating the gain and phase curves for a discrete system as z takes on values around the unit circle and the interpretation of phase margins will be modified if sampling very slowly.

The key to Bode's design technique is his proof that with a minimum phase transfer function, the phase is uniquely determined by an integral of the slope of the magnitude curve on a log-log plot. Furthermore, if the function is rational, these slopes are readily and adequately approximated by constants! Thus we have the result that the amplitude curve must cross unity gain (zero log) at a slope (in approximation) of -1 if the phase is to remain above the stability boundary of $-180°$. This is the essence of point 3.

Unfortunately, discrete transfer functions are typically not rational functions, but rather the frequency appears in the form $z = e^{j\omega T}$ and the simplicity of Bode's design technique is altogether lost in the z-plane. The cure is to make a transformation to a different plane (called w) where the simplicity is regained. As a matter of fact, we have already seen the transformation needed in Chapter 3 when we used

$$s = \frac{z-1}{z+1}\frac{2}{T}$$

to convert rational functions of s into approximate but realizable functions of z.

To enable easy use of Bode's design methods, we use the transformation some-what in the reverse order, to convert a given function of z into a rational function of w, so we can use Bode's design techniques.

We thus propose the substitution for z of a new variable w, where

$$w \triangleq \frac{2}{T} \frac{z - 1}{z + 1} \tag{5.23}$$

or

$$w = \frac{2}{T} \frac{e^{sT} - 1}{e^{sT} + 1} = \frac{2}{T} \tanh \frac{sT}{2}.$$

If s is pure imaginary $(=j\omega)$ and therefore $z = e^{j\omega T}$ (on the unit circle), then

$$w \triangleq jv = j\frac{2}{T} \tan \frac{\omega T}{2}, \tag{5.24}$$

and we see that while z goes around the unit circle, the w-plane frequency, v, stays *real* and goes from 0 to ∞. We chose the scale factor[8] of $2/T$ in (5.23) to make sure that the error constant would come out correctly and so that w-plane transfer functions would approach those in the s-plane as T went to zero. The process is best illustrated by an example.

Consider again the antenna model described in Appendix A.2, that is,

$$G(s) = \frac{1}{s(s/0.1 + 1)}, \tag{5.25}$$

which, when preceded by a ZOH ($T = 1$ sec) has the discrete transfer function

$$G(z) = 0.048 \frac{(z + 0.967)}{(z - 1)(z - 0.905)}. \tag{5.26}$$

To transform to the w-plane, we substitute the inverse of (5.23), or

$$z = \frac{1 + wT/2}{1 - wT/2} \tag{5.27}$$

into (5.26) and obtain:[9]

$$G(w) = -\frac{(w/120 + 1)(w/2 - 1)}{w(w/0.0999 + 1)}. \tag{5.28}$$

---

[8] It is more common to perform this transformation without the $2/T$ scale factor, in which case the transformed frequency ($\triangleq v'$) differs from the real frequency, ω, by $v' = \tan \omega T/2$ and is essentially a frequency that is scaled to the sample rate; that is, $v' = 1$ when $\omega = \omega_s/4$.

[9] We use the same symbol, G, for three distinct functions in (5.25), (5.26), and (5.28). The arguments s, z, and w identify the function as well as the variable.

The algebra in computing $G(w)$ from $G(z)$ can be somewhat tedious; therefore Appendix B contains a table of a few common transfer functions and a general formula for making the conversion.

Note that the gain of $G(w)$ is precisely the same as $G(s)$; for example, it is unity in both cases. This will always be true for a $G(w)$ computed using the definition of w given in (5.23), but will not be true for the more common definition discussed in footnote 8 on p. 114. The gain of 1 in (5.28) is the K_v of the uncompensated discrete system, as the reader may verify using (5.8), and also applies to the continuous system as can be verified from (5.25). We also note that in (5.28) the denominator looks very similar to that of $G(s)$ and that the denominators will be the same as T approaches zero. This would also have been true for any zeros of $G(w)$ that corresponded to zeros of $G(s)$, but our example did not have any. Our example also shows the creation of a right-hand plane zero of $G(w)$ at $2/T$ and the creation of a fast left-hand plane zero when compared to the original $G(s)$. The two "created" zeros can be attributed to the sampling-and-hold operations and thus depend on the sample rate. They both become faster and thus less important to the design problem as the sample rate is increased. In general, one or both of these additional zeros usually occur.

In our example, the most important feature added to $G(s)$ in transforming to $G(w)$ is the right-hand plane zero at $w = 2$ rad/sec($=2/T$) which will introduce serious distortion as ν approaches 2. From (5.24) we see that this point corresponds to $\omega T/2 = \tan^{-1} 1 = 45°$ or $\omega T = 90°$. Since the limit of frequency occurs at $\omega = \omega_s/2$ ($\omega T = 180°$), frequencies of interest can approach $\omega = \omega_s/4$ and will be affected by this zero. More significantly, the zero at 2 is in the right half of the plane so that Bode's gain-phase integral does not apply to this one term. We will need to be especially cautious as we get close to $\nu = 2$. The magnitude and phase of $G(j\nu)$ are the magnitude and phase of $G(z)$ as z takes on values around the unit circle, and since $G(j\nu)$ is a rational function of ν we can apply all the standard straight-line approximations to the log magnitude and phase curves. Nyquist's stability criterion applies to $G(j\nu)$ in the w-plane just as it does to $G(j\omega)$ in the s-plane because in both cases we are determining the number of zeros of $(1 + G)$ (unstable system roots) in the right-hand plane. Therefore, the gain and phase margins of classical Bode designs apply directly to $G(j\nu)$. The rules are the same, so let us play the game.

We want a compensation that will keep $K_v = 1$ and will have a damping of about $\zeta = 0.5$. Using the rule of thumb that the phase margin (PM) $\simeq 100\zeta$, Saucedo and Shiring (1968) suggest that the PM should be $\simeq 50°$. A sketch of the asymptotes of the magnitude of (5.28) is shown in Fig. 5.11 by the solid line and is set to give $K_v = 1$. The phase margin for the system without compensation is on the order of 10°. Note that the zero at $\nu = 2$ contributes phase lag because it is in the right-hand plane. Let us try compensating by placing a zero on top of the pole at $\nu = 0.0999$. How about the compensation pole . . . or do we need a pole? In continuous systems, we always include a pole to avoid noise amplification and to

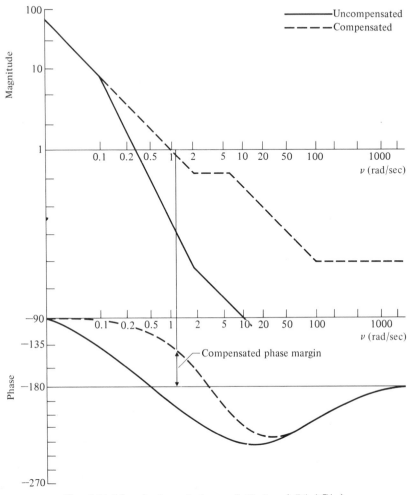

Fig. 5.11 Magnitude and phase of $G(w)$ and $D(w)G(w)$.

make it easier to build. A w-plane compensation with no pole yields a z-plane pole at $z = -1$, which leads to a marginally unstable compensation and an unstable closed-loop system in spite of a positive phase margin[10]; therefore, to stay out of trouble, always use a zero and a pole together.

Where should we place the pole? The crossover frequency will be at approximately $\nu = 1$, where the phase of the uncompensated system was $-21°$ and the compensation zero added $84°$; therefore the pole can cause a phase lag of up to $13°$

---

[10] The resulting system has a higher-order numerator than denominator and one must revert to Nyquist's -1 encirclement criterion to determine stability.

for our desired 50° phase margin.   Placing the pole at $\nu = 6$ meets this requirement, so our compensation is

[handwritten: place @ zero on top of G(w) pole]

$$D(w) = \frac{1 + w/0.0999}{1 + w/6},$$ (5.29)

[handwritten: and place a pole to get needed phase]

and the compensated loop gain and phase are the dashed curves in Fig. 5.11. Note that the phase in the vicinity of $\nu = 10$ corresponds to a magnitude slope of -3 rather than -1; thus compensating nonminimum phase systems (as most w-plane designs are) based on magnitude alone is treacherous.

Reversing the w-transformation, we now use (5.23) and let

$$w = \frac{2}{T}\frac{z - 1}{z + 1}$$

in (5.29) to compute $D(z)$.   The result is

$$D(z) = \frac{15.8(z - 0.905)}{z + 0.5}.$$ (5.30)

A root locus of the system

$$DG = (15.8)\frac{(1 - 0.905z^{-1})(0.048)(z^{-1})(1 + 0.967z^{-1})}{(1 + 0.5z^{-1})(1 - z^{-1})(1 - 0.905z^{-1})}$$

$$= 0.758\frac{z + 0.967}{(z - 1)(z + 0.5)}$$ (5.31)

is the circle centered at -0.967 in Fig. 5.12(a).   The closed-loop pole corresponding to the root locus gain of 0.758 is marked $\triangle$.   The resulting roots do not have the desired damping of $\zeta = 0.5$.   In fact, the damping is $\zeta = 0.37$.   The step response is shown in Fig. 5.12(b).   This breakdown in the phase-margin/damping rule of thumb can sometimes occur in discrete frequency-response design when the sampling is very slow.   In this particular design example, the sampling rate is only 3.4 times faster than the closed-loop roots and is probably slower than the rate one would typically select.

Figure 5.13 compares the damping of the example (5.31) with varying loop gains and thus phase margins.   The numbers shown adjacent to the discrete curve indicate the ratio of the closed-loop root frequencies to the sample rate and show how the damping approaches the rule of thumb as the sample rate multiple increases.   Above sample rate multiples of 5, one can use the "×100" rule with confidence.

In addition to the programs POLLY and STEP, design by frequency-response methods can be greatly enhanced by program

4. UNITCIRCLE: A program to compute and plot on Nyquist (polar) or Bode (log-log) coordinates the magnitude and phase of a transfer function as z varies around the unit circle.   Having such a program permits design of the poles and zeros of $D(z)$ directly in the z-plane without transforming to the w-plane.

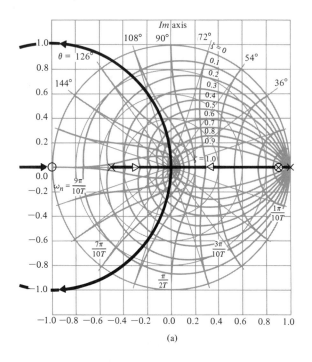

(a)

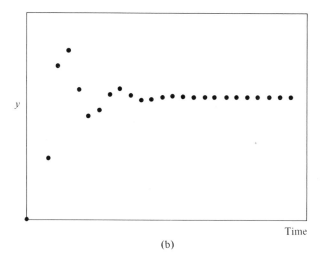

(b)

Fig. 5.12 (a) Root locus of system designed in the w-plane.   (b) Step response of closed-loop antenna control designed in the w-plane.

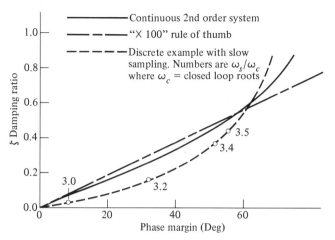

Fig. 5.13 System damping versus phase margin.

5.6 DIRECT DESIGN
METHOD OF RAGAZZINI
[Ragazzini and Franklin (1958)]

Much of the style of the transform design techniques we have been discussing in this chapter grew out of the limitations of technology which was available for realization of the compensators with pneumatic components or electric networks and amplifiers. In particular, many constraints were imposed in order to assure the realization of electric compensator networks $D(s)$ as networks consisting only of resistors and capacitors.[11] In the digital computer, such limitations on realization are, of course, not relevant, and one can ignore these particular constraints. One design method which eliminates these constraints begins from the very direct point of view that we are given a plant (plus hold) discrete transfer function $G(z)$, that we want to construct a desired transfer function, $H(z)$, between R and Y, and that we have the computer transfer function, $D(z)$, to do the job. The overall transfer function is given by the formula

$$H(z) = \frac{DG}{1 + DG},$$

from which we get the design formula

$$D(z) = \frac{1}{G(z)} \frac{H(z)}{1 - H(z)}. \tag{5.32}$$

---

[11] In the book by Truxal (1955), where much of this theory is collected at about the height of its first stage of development, a chapter is devoted to RC network synthesis.

From formula (5.32) we can see that this design calls for a $D(z)$ which will cancel the plant effects and add whatever is necessary to give the desired result. The problem is to discover and implement constraints on $H(z)$ so that we do not ask for the impossible.

First, let us consider the constraint of causality. From z-transform theory we know that if $D(z)$ is causal, then as $z \to \infty$, its transfer function is well behaved; it does not have a pole at infinity. Looking at (5.32), we see that if $G(z)$ were to have a zero at infinity, then $D(z)$ would have a pole there unless we request an $H(z)$ which is such as to cancel it. Thus we have the constraint that for $D(z)$ to be causal

$H(z)$ must have a zero at infinity of the same order
as the zero of $G(z)$ at infinity.   (5.33)

This requirement has an elementary interpretation in the time domain: $G(z)$ has a zero at infinity because the pulse response of the plant has a delay of at least one sample time. If there is a transportation lag in the plant, then the delay may be several samples and $G(z)$ may start with $z^{-\ell}$. The causality requirement on $H(z)$ is that the closed-loop system must have at least as long a delay as the plant has.

Considerations of stability add a second constraint. The roots of the characteristic equation of the closed-loop system are the roots of the equation

$$1 + D(z)G(z) = 0. \tag{5.34}$$

We can express (5.34) as a polynomial if we identify $D = c(z)/d(z)$ and G as $b(z)/a(z)$ where a, b, c, and d are polynomials. Then the characteristic polynomial is

$$ad + bc = 0. \tag{5.35}$$

Now suppose there is a common factor in DG, as would result if $D(z)$ were called upon to cancel a pole or zero of $G(z)$. Let this factor be $z - \alpha$ and suppose it is a pole of $G(z)$, so we can write $a(z) = (z - \alpha)\bar{a}(z)$, and to cancel it we have $c(z) = (z - \alpha)\bar{c}(z)$. Then (5.35) becomes

$$(z - \alpha)\bar{a}(z)d(z) + b(z)(z - \alpha)\bar{c}(z) = 0,$$
$$(z - \alpha)[\bar{a}d + b\bar{c}] = 0. \tag{5.36}$$

In other words—perhaps it was obvious from the start—a common factor *remains a factor of the characteristic polynomial*. If this factor is outside the unit circle, the system is unstable! How do we avoid such cancellation? Considering again (5.32), we see that if $D(z)$ is not to cancel a pole of $G(z)$, then that factor of $a(z)$ must also be a factor of $1 - H(z)$. Likewise, if $D(z)$ is not to cancel a zero of $G(z)$, such zeros must be factors of $H(z)$. Thus we write the constraints:[12]

---

[12] Roots on the unit circle are also unstable by some definitions and good practice indicates that we should not cancel singularities outside the radius of desired settling time. See Fig. 5.5 and the discussion associated with it.

$1 - H(z)$ must contain as zeros all the poles of $G(z)$ that are
outside the unit circle.   (5.37)

$H(z)$ must contain as zeros all the zeros of $G(z)$ that are
outside the unit circle.   (5.38)

Consider finally the constraint of steady-state accuracy. Since $H(z)$ is the
overall transfer function, the error transform is given by

$$E(z) = R(z)(1 - H(z)).$$ (5.39)

Thus if the system is to be Type I with velocity constant K_v, we must have zero
steady-state error to a step and $1/K_v$ error to a unit ramp. The first requirement is

$$e(\infty) = \lim_{z \to 1} (z - 1) \frac{1}{z - 1} [1 - H(z)] = 0,$$

which implies

$$H(1) = 1.$$ (5.40)

The velocity constant requirement is that

$$e(\infty) = \lim_{z \to 1} (z - 1) \frac{Tz}{(z - 1)^2} [1 - H(z)] = \frac{1}{K_v}.$$ (5.41)

From (5.39) we know that $1 - H(z)$ is zero at $z = 1$, so that to evaluate the limit in
(5.41), it is necessary to use l'Hospital's rule with the result [see (5.11) and fol-
lowing]

$$-T \frac{dH}{dz} \bigg|_{z=1} = \frac{1}{K_v}.$$ (5.42)

An example will best illustrate the application of these constraints. Consider
again the plant described by the transfer function (5.19) and suppose we ask for
the same design that led to (5.15) as a continuous controller. The continuous
closed-loop system has a characteristic equation

$$s^2 + s + 1 = 0.$$

With a sampling period $T = 1$ sec, this maps to the discrete characteristic equation

$$z^2 - 0.7859z + 0.36788 = 0.$$ (5.43)

Let us therefore ask for a design which is stable, has $K_v = 1$, and has poles at the
roots of (5.43) plus, if necessary, additional poles at $z = 0$, where the transient is
as short as possible. The form of $H(z)$ is thus

$$H(z) = \frac{b_0 + b_1 z^{-1} + b_2 z^{-2} + b_3 z^{-3} + \cdots}{1 - 0.7859z^{-1} + 0.36788z^{-2}}.$$ (5.44)

The causality design constraint, using (5.33) and (5.18), requires that

$$H(z)\big|_{z=\infty} = 0$$

or

$$b_0 = 0. \tag{5.45}$$

Equations (5.37) and (5.38) add no constraints since $G(z)$ has all poles and zeros inside the unit circle except for the single zero at ∞, which is taken care of by (5.45). The steady-state error requirement leads to

$$
\begin{aligned}
H(1) &= 1 \\
&= \frac{b_1 + b_2 + b_3 + \cdots}{1 - 0.7859 + 0.36788} = 1.
\end{aligned} \tag{5.46}
$$

Therefore

$$b_1 + b_2 + b_3 + \cdots = 0.58198$$

and

$$-T\frac{dH}{dz}\bigg|_{z=1} = \frac{1}{K_v}.$$

Since in this case, both T and K_v are 1, we use (5.46) and the derivative with respect to z^{-1} to obtain

$$
\begin{aligned}
1 = \frac{1}{K_v} &= \frac{dH}{dz^{-1}}\bigg|_{z=1} \\
&= \frac{(0.58198)[b_1 + 2b_2 + 3b_3 + \cdots] - [0.58198][-0.7859 + 0.36788(2)]}{(0.58198)(0.58198)}
\end{aligned}
$$

or

$$\frac{b_1 + 2b_2 + 3b_3 + \cdots - [-0.05014]}{0.58198} = 1. \tag{5.47}$$

Since we have only two equations to satisfy, we need only two unknowns and we can truncate $H(z)$ at b_2. The resulting equations are

$$b_1 + b_2 = 0.58198, \qquad b_1 + 2b_2 = 0.53184,$$

which have the solution

$$b_1 = 0.63212, \qquad b_2 = -0.05014. \tag{5.48}$$

Thus the final design gives an overall transfer function

$$H(z) = \frac{0.63212z - 0.05014}{z^2 - 0.7859z + 0.36788}. \tag{5.49}$$

We shall also need

$$1 - H(z) = \frac{(z - 1)(z - 0.41802)}{z^2 - 0.7859z + 0.36788}. \tag{5.50}$$

We know that $H(1) = 1$ so that $1 - H(z)$ must have a zero at $z = 1$. Now, turning to the basic design formula, (5.32), we compute

$$D(z) = \frac{(z - 1)(z - 0.9048)(0.63212)}{(0.04837)(z + 0.9672)} \frac{(z - 0.07932)}{(z - 1)(z - 0.41802)}$$

$$= 13.068 \frac{(z - 0.9048)}{(z + 0.9672)} \frac{(z - 0.07932)}{(z - 0.41802)}.$$

A plot of the step response of the resulting design is provided in Fig. 5.14, which also shows the control effort. We can see the oscillation of $u(k)$ which is asso-

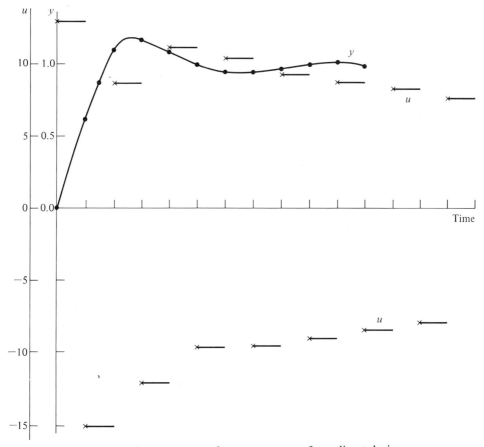

Fig. 5.14 Step response of antenna system from direct design.

ciated with the pole of $D(z)$ at -0.967, which is quite near the unit circle. To avoid this effect, we could introduce another term in $H(z)$, $b_3 z^{-3}$, and require that $H(z)$ be zero at $z = -0.9672$, so this zero of $G(z)$ is not canceled by $D(z)$. The result will be a simpler $D(z)$ with a slightly more complicated $H(z)$. However, rather than pursue this method further, we will wait until the more powerful method of pole assignment by state variable analysis is developed in the next chapter, where computer algorithms are more readily provided.

5.7 A SECOND EXAMPLE: TEMPERATURE CONTROL VIA MIXING

In Appendix A we describe a process of temperature control via mixing and derive the transfer function

$$G_3(s) = \frac{e^{-s\tau_d}}{s/a + 1}.$$

As a second example of digital design by transform techniques, we consider this case with a zero-order hold, unity sampling period, and system time constant ($T = a = 1$) and a $1\frac{1}{2}$ period delay ($\tau_d = 1.5$). First, we need to compute the discrete transfer function as

$$G_3(z) = \mathcal{Z}\left\{\frac{1 - e^{-Ts}}{s}\frac{e^{-1.5s}}{s + 1}\right\}, \tag{5.51}$$

which is given in the appendix as

$$G_3(z) = 0.3935\frac{z + 0.6065}{z^2(z - 0.3679)}. \tag{5.52}$$

As it stands, this transfer function has unity gain to a constant control and will have a steady-state error to a constant command or disturbance. If we assume that such behavior in the steady state is unacceptable, we can correct the problem by including integral control via the transfer function

$$D_I(z) = \frac{z}{z - 1}.$$

The effective plant transfer function is now

$$D_I G_3 = 0.3935\frac{z + 0.6065}{z(z - 1)(z - 0.3679)}. \tag{5.53}$$

The unity feedback root locus of this transfer function is sketched in Fig. 5.15 with the roots corresponding to $K_v = 1$ marked $\square$; at $\zeta = 0.5$, the locations are marked $\triangle$. Since the roots are outside the unit circle for $K_v = 1$, the design cannot achieve such high gain without further compensation.

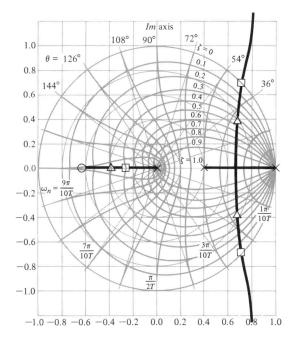

Fig. 5.15 Root locus of mixing flow plant with discrete integral control.

A lead network compensation which cancels the plant pole at $z = 0.3679$ and the plant zero at $z = -0.6065$ is

$$D_L = K \frac{z - 0.3679}{z + 0.6065}.$$

With this compensation, the plant is

$$D_L D_I G_3(z) = K_v \frac{1}{z(z - 1)};$$

the root locus is sketched in Fig. 5.16. The system bandwidth has been raised from $\theta = 21°$ (0.39 rad) to $\theta = 41°$ (0.733 rad), and K_v has been raised from 0.3 to 0.45. To raise K_v further, we would consider a lag compensation. Suppose we wish to raise K_v to 1.0. Then we must have

$$D_\ell = \frac{z - z_i}{z - p_i},$$

where $D_\ell(1) = 1/0.45 = 2.22$. If we take $z_i = 0.9$, then we compute $p_i = 0.9545$. The lag pole-zero pair are very close to each other and do not change the root

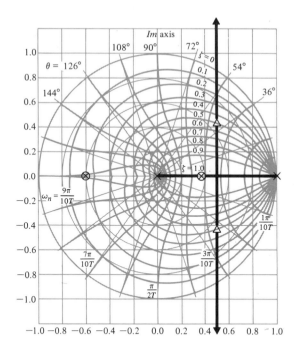

Fig. 5.16 Root locus of mixing flow plant with lead compensation and discrete integral control.

locus near the dominant poles significantly. However, the lag network does introduce a small but very slow transient whose effect on dynamic response needs to be evaluated, especially in terms of the response to disturbances.

5.8 SUMMARY

In this chapter we have reviewed the philosophy and specifications of control system design by transform techniques and discussed four such methods. First we developed the relations between the time-domain specifications of overshoot, rise time, and settling time and poles in the z-plane. Using the theory and techniques of discrete equivalents, we then showed how a continuous design can be converted into a discrete design. With a sampling period giving only eight samples per cycle, we found that the approximation was quite coarse and would require substantial adjustment to meet the design specifications. As a second design approach we discussed the root locus in the z-plane. We saw that the root locus is the same as for s-plane designs but the relations to time-domain response must refer to the z-plane. Our third design method based on transform techniques used the Bode plots. For this we found it necessary to use the bilinear transformation to the w-plane. Our final method was a direct transfer-function calculation wherein we found causality and stability constraints on an overall transfer function so that an acceptable compensator can be derived. Here we found that canceling poles near the unit circle may have undesirable effects. In the final section we presented a design by root locus methods for a plant which required the introduction of discrete integral control.

PROBLEMS AND EXERCISES

5.1 Use the $z = e^{sT}$ mapping function and prove that the curve of constant ζ in s is a logarithmic spiral in z.

5.2 Sketch the acceptable region in the s-plane for the specification on the antenna given before (5.15) and sketch the s-plane root locus corresponding to the controller of (5.15).

5.3 *Root-locus review.* The following root loci illustrate important features of the root-locus technique.

i) This locus is typical of the behavior near $s = 0$ of a double integrator with lead compensation or a single integration with a lag network and one additional real pole. Sketch the locus for values of P_1 of 5, 9, and 20. Pay close attention to the real axis break-in and break-away points

$$1 + K \frac{s + 1}{s^2(s + P_1)}.$$

ii) This locus illustrates the possibility of complex multiple roots and shows the value of departure angles. Plot the locus for $a = 0, +.5$, and $+2$. Be sure to note the departure angles from the complex poles in each case:

$$1 + K \frac{1}{s(s + 1)((s + a)^2 + 4)}.$$

iii) This locus illustrates the use of complex zeros to compensate for the presence of complex poles due to vibration modes. Be sure to compute (estimate) the angles of departure and arrival. Sketch the loci for $\omega = 1$ and $\omega = 3$. Which case is unconditionally stable (stable for all positive K less than the design value)?

$$1 + K \frac{(s + 1)^2 + \omega^2}{s(s^2 + 4)} = 0$$

iv) For the following root locus, show that the locus is a circle of radius $\sqrt{P_1 P_2}$ centered at the origin (location of the zero). Can this result be translated to the case of two poles and a zero on the negative real axis?

$$1 + K \frac{s}{(s - P_1)(s - P_2)} = 0$$

5.4 Appendix A gives the transfer function of a satellite attitude control as

$$G_1(z) = K \frac{z + 1}{(z - 1)^2}.$$

a) Sketch the root locus of this system as a function of K with unity feedback. What is the type of the uncompensated system?
b) Add a lead network so that the dominant poles are at $\zeta = 0.5$ and $\theta = 45°$. Plot the closed-loop step response.

5.5 a) Use Ragazzini's direct design method to find a compensation for the satellite transfer function of Appendix A such that all the closed loop poles are at $z = 0$.
b) Sketch the root locus of the resulting design.
c) Sketch the step response of the resulting design. Let $T = 1.0$.

5.6 Repeat the design of the antenna by pole-zero mapping of the $D(s)$ of (5.15) but use sample period $T = 0.2$.

5.7 It is possible to suspend a mass of magnetic material by means of an electromagnet whose current is controlled by the position of the mass [Woodson and Melcher (1968)]. A schematic of a possible setup is shown in Fig. 5.17. The equations of motion are

$$m\ddot{x} = -mg + f(x, I),$$

where the force on the ball due to the electromagnet is given by $f(x, I)$. At equilibrium, the magnet force balances the gravity force, and suppose we call the current there I_0. If we write $I = I_0 + i$ and expand f about $x = 0$ and $I = I_0$, and neglect higher-order terms, we obtain

$$m\ddot{x} = k_1 x + k_2 i.$$

Reasonable values are $m = 0.02$ kg, $k_1 = 20$ N/m, $k_2 = 0.4$ N/A.

a) Compute the transfer function from i to x and draw the (continuous) root locus for simple feedback $i = -Kx$.
b) Let the sample period be 0.02 sec and compute the plant discrete transfer function when used with a zero-order hold.

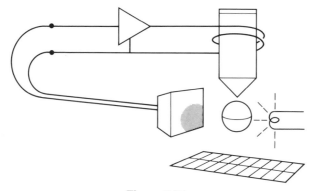

Figure 5.17

c) Design a digital control for the magnetic levitation to meet the specifications $t_r \leq$ 0.1 sec, $t_s \leq 0.4$ sec, % overshoot $\leq 20\%$.

d) Plot a root locus of your design versus k_1 and discuss the possibility of balancing balls of various masses.

e) Plot a step response of your design to an initial disturbance displacement on the ball and show both x and the control current i. If the sensor can measure x over a range of only $\pm\frac{1}{4}$ cm, and the amplifier can provide a current of only 1 A, what is the *maximum* displacement possible for control, neglecting the nonlinear terms in $f(x, I)$?

5.8 Design compensation using frequency response methods for a $1/s^2$ plant (Appendix A.1) that yields a bandwidth of approximately 10 rad/sec. Pick two candidate sample rates, one where $\nu T/2$ at crossover is approximately 1.3, and one fairly fast sample rate yielding a $\nu T/2$ of approximately 0.2. Design each case to have the same phase margin (approximately 30°), then compare the damping of the equivalent s-plane roots resulting from the two designs.

6 / Design of Digital Control Systems Using State-Space Methods

6.1 INTRODUCTION

In Chapter 5, we discussed how to design digital controllers using transform techniques, methods now commonly designated as "classical design." The goal of this chapter is to solve the identical problem using different techniques which are based on the state-space or modern control formulation. The difference in the two approaches is entirely in the design method since the end result, a set of difference equations providing control, is identical. Advantages of modern control are especially apparent when engineers design controllers for systems with more than one control input or sensed output; however, to illustrate the ideas of state-space design, we will devote our efforts in this chapter to single input/output systems. Techniques for the multi-input–multi-output design are discussed in Chapter 9.

6.2 SYSTEM REPRESENTATION

A continuous, linear, constant-coefficient system of differential equations can always be expressed as a set of first-order matrix differential equations[1]:

$$\dot{\mathbf{x}} = \mathbf{F}\mathbf{x} + \mathbf{G}u + \mathbf{G}_1 w, \tag{6.1}$$

where u is the control input to the system, and w is a disturbance input. The output can be expressed as a linear combination of the state, $\mathbf{x}$, and the input as

$$y = \mathbf{H}\mathbf{x} + Ju. \tag{6.2}$$

The representations (6.1) and (6.2) are not unique. Given one state representation, any nonsingular linear transformation of that state $\boldsymbol{\xi} = \mathbf{T}\mathbf{x}$ is also an allow-

---

[1] We assume the reader has some knowledge of matrices. The results we require and references to study material are given in Appendix C. To distinguish vectors and matrices, we will use bold-face type.

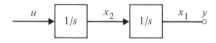

Fig. 6.1 Satellite attitude control in classical representation.

able alternative realization of the same system. Most commonly for control problems, the parameter J in (6.2) is zero and will often be omitted.

A simple application of state representation to the equations of the satellite attitude-control example shown in Fig. 6.1, and described in Appendix A yields

$$\begin{bmatrix} \dot{x}_1 \\ \dot{x}_2 \end{bmatrix} = \underbrace{\begin{bmatrix} 0 & 1 \\ 0 & 0 \end{bmatrix}}_{F} \begin{bmatrix} x_1 \\ x_2 \end{bmatrix} + \underbrace{\begin{bmatrix} 0 \\ 1 \end{bmatrix}}_{G} u,$$

$$\theta = y = \underbrace{\begin{bmatrix} 1 & 0 \end{bmatrix}}_{H} \begin{bmatrix} x_1 \\ x_2 \end{bmatrix},$$

(6.3)

which, in this case, turns out to be a rather involved way of writing

$$\ddot{\theta} = u.$$

If we consider an alternative state and let $\boldsymbol{\xi} = \mathbf{T}\mathbf{x}$ in (6.1) and (6.2), we find

$$\begin{aligned} \dot{\boldsymbol{\xi}} = \mathbf{T}\dot{\mathbf{x}} &= \mathbf{T}(\mathbf{F}\mathbf{x} + \mathbf{G}u + \mathbf{G}_1 w) \\ &= \mathbf{T}\mathbf{F}\mathbf{x} + \mathbf{T}\mathbf{G}u + \mathbf{T}\mathbf{G}_1 w, \\ \dot{\boldsymbol{\xi}} &= \mathbf{T}\mathbf{F}\mathbf{T}^{-1}\boldsymbol{\xi} + \mathbf{T}\mathbf{G}u + \mathbf{T}\mathbf{G}_1 w, \\ y &= \mathbf{H}\mathbf{T}^{-1}\boldsymbol{\xi} + Ju. \end{aligned}$$

If we designate the system matrices for the new state $\boldsymbol{\xi}$ as $\mathbf{A}, \mathbf{B}, \mathbf{C}, \mathbf{D}$, then

$$\dot{\boldsymbol{\xi}} = \mathbf{A}\boldsymbol{\xi} + \mathbf{B}u + \mathbf{B}_1 w, \qquad y = \mathbf{C}\boldsymbol{\xi} + Du,$$

where

$$\mathbf{A} = \mathbf{T}\mathbf{F}\mathbf{T}^{-1}, \qquad \mathbf{B} = \mathbf{T}\mathbf{G}, \qquad \mathbf{B}_1 = \mathbf{T}\mathbf{G}_1, \qquad \mathbf{C} = \mathbf{H}\mathbf{T}^{-1}, \qquad D = J.$$

As an illustration, we can let $\xi_1 = x_2$ and $\xi_2 = x_1$ in (6.3); or, in matrix notation, the transformation to interchange the states is

$$\mathbf{T} = \begin{bmatrix} 0 & 1 \\ 1 & 0 \end{bmatrix}.$$

In this case $\mathbf{T}^{-1} = \mathbf{T}$, and application of the transformation equations to the system matrices of (6.3) gives

$$\mathbf{A} = \begin{bmatrix} 0 & 0 \\ 1 & 0 \end{bmatrix}, \qquad \mathbf{B} = \begin{bmatrix} 1 \\ 0 \end{bmatrix}, \qquad \mathbf{C} = \begin{bmatrix} 0 & 1 \end{bmatrix}.$$

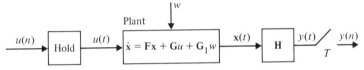

Fig. 6.2 System definition with sampling operations shown.

In general, the state method of representing a dynamic system is very useful because it standardizes the information required into three matrices, **F, G**, and **H**, no matter how complicated the system is.

For the digital control problem, we wish to establish a general method for obtaining the difference equations which represent the behavior of the continuous plant. Figure 6.2 depicts the portion of our system under consideration. Ultimately, the digital controller will take the samples $y(n)$, operate on that sequence by means of a difference equation, and put out a sequence of numbers, $u(n)$, which are the inputs to the plant. The loop will, therefore, be closed. To analyze the result, we must be able to relate the samples of the output $y(n)$ to the samples of the control $u(n)$. To do this, we must solve (6.1).

The homogeneous (no input) solution to (6.1) is

$$\mathbf{x}_h(t) = e^{\mathbf{F}(t-t_0)}\mathbf{x}(t_0), \tag{6.4}$$

where, by definition, the matrix exponential is

$$e^{\mathbf{F}(t-t_0)} = \mathbf{I} + \mathbf{F}(t - t_0) + \mathbf{F}^2 \frac{(t - t_0)^2}{2!} + \mathbf{F}^3 \frac{(t - t_0)^3}{3!} + \cdots$$

$$= \sum_{k=0}^{\infty} \mathbf{F}^k \frac{(t - t_0)^k}{k!}. \tag{6.5}$$

This is proved by differentiating (6.4) with respect to t,[2] yielding

$$\dot{\mathbf{x}}_h(t) = \mathbf{F}\left[\mathbf{I} + \mathbf{F}(t - t_0) + \mathbf{F}^2 \frac{(t - t_0)^2}{2!} + \cdots \right]\mathbf{x}(t_0)$$

or

$$\dot{\mathbf{x}}_h(t) = \mathbf{F}\mathbf{x}_h(t),$$

which is the unforced portion of (6.1).

It can be shown that the solution given by (6.4) is unique, which leads to very interesting properties of the matrix exponential. For example, consider two val-

---

[2] Which can be done by differentiating (6.5) term by term. We can also obtain the solution by assuming a series solution $\mathbf{x}_h(t) = \Sigma \mathbf{A}_i t^i$ and solving for $\mathbf{A}_i$ to satisfy (6.1) when $u = w = 0$.

ues of t: t_1 and t_2. We have

$$\mathbf{x}(t_1) = e^{\mathbf{F}(t_1-t_0)}\mathbf{x}(t_0)$$

and

$$\mathbf{x}(t_2) = e^{\mathbf{F}(t_2-t_0)}\mathbf{x}(t_0).$$

Since t_0 is arbitrary also, we can express $\mathbf{x}(t_2)$ as if the equation solution began at t_1, for which

$$\mathbf{x}(t_2) = e^{\mathbf{F}(t_2-t_1)}x(t_1).$$

Substituting for $\mathbf{x}(t_1)$ gives

$$\mathbf{x}(t_2) = e^{\mathbf{F}(t_2-t_1)}e^{\mathbf{F}(t_1-t_0)}\mathbf{x}(t_0).$$

We now have two separate expressions for $\mathbf{x}(t_2)$, and, if the solution is unique, these must be the same. Hence we conclude that

$$e^{\mathbf{F}(t_2-t_0)} = e^{\mathbf{F}(t_2-t_1)}e^{\mathbf{F}(t_1-t_0)} \tag{6.6}$$

for all t_2, t_1, t_0. Note especially that if $t_2 = t_0$, then

$$\mathbf{I} = e^{-\mathbf{F}(t_1-t_0)}e^{\mathbf{F}(t_1-t_0)}.$$

Thus we can obtain the inverse of $e^{\mathbf{F}t}$ by merely changing the sign of t. We will use this result in computing the particular solution to (6.1).

The particular solution when u is not zero is obtained by using the method of *variation of parameters*.[3] We guess the solution to be in the form

$$\mathbf{x}_p(t) = e^{\mathbf{F}(t-t_0)}\mathbf{v}(t), \tag{6.7}$$

where $\mathbf{v}(t)$ is a vector of variable parameters to be determined [as contrasted to the constant parameters $\mathbf{x}(t_o)$ in (6.7)]. Substituting (6.7) into (6.1), we obtain

$$\mathbf{F}e^{\mathbf{F}(t-t_0)}\mathbf{v} + e^{\mathbf{F}(t-t_0)}\dot{\mathbf{v}} = \mathbf{F}e^{\mathbf{F}(t-t_0)}\mathbf{v} + \mathbf{G}u,$$

and, using the fact that the inverse is found by changing the sign of the exponent, we can solve for $\dot{\mathbf{v}}$ as

$$\dot{\mathbf{v}}(t) = e^{-\mathbf{F}(t-t_0)}\mathbf{G}u(t).$$

Assuming that the control $u(t)$ is zero for $t < t_0$, we can integrate $\dot{\mathbf{v}}$ from t_0 to t to obtain

$$\mathbf{v}(t) = \int_{t_0}^{t} e^{-\mathbf{F}(\tau-t_0)}\mathbf{G}u(\tau)\, d\tau.$$

---

[3] Due to Joseph Louis Lagrange, French mathematician (1736–1813). We assume $w = 0$, but since the equations are linear, the effect of w can be added later.

Hence, from (6.6),

$$\mathbf{x}_p(t) = e^{\mathbf{F}(t-t_0)} \int_{t_0}^{t} e^{-\mathbf{F}(\tau-t_0)} \mathbf{G}u(\tau) \, d\tau,$$

and simplifying, using the results of (6.6), we obtain the particular solution (convolution)

$$\mathbf{x}_p(t) = \int_{t_0}^{t} e^{\mathbf{F}(t-\tau)} \mathbf{G}u(\tau) \, d\tau. \tag{6.8}$$

The total solution for $w = 0$ and $u \neq 0$ is the sum of (6.4) and (6.8):

$$\mathbf{x}(t) = e^{\mathbf{F}(t-t_0)} \mathbf{x}(t_0) + \int_{t_0}^{t} e^{\mathbf{F}(t-\tau)} \mathbf{G}u(\tau) \, d\tau. \tag{6.9}$$

We wish to use this solution over one sample period to obtain a difference equation; hence we juggle the notation a bit (let $t = nT + T$ and t_0 equal nT) and arrive at a particular version of (6.9):

$$\mathbf{x}(nT + T) = e^{\mathbf{F}T} \mathbf{x}(nT) + \int_{nT}^{nT+T} e^{\mathbf{F}(nT+T-\tau)} \mathbf{G}u(\tau) \, d\tau. \tag{6.10}$$

This result is not dependent on the type of hold since u is specified in terms of its continuous time history, $u(\tau)$, over the sample interval. A common and typically valid assumption is that of a zero-order hold (ZOH) with no delay, that is,

$$u(\tau) = u(nT), \qquad nT \leq \tau < nT + T.$$

If some other hold is implemented or if there is a delay between the application of the control from the ZOH and the sample point, this fact can be accounted for in the evaluation of the integral in (6.10). The equations for a delayed ZOH are given in Appendix A at the end of this chapter.

To facilitate the solution of (6.10) for a ZOH with no delay, let

$$\eta = nT + T - \tau.$$

Then we have

$$\mathbf{x}(nT + T) = e^{\mathbf{F}T} \mathbf{x}(nT) + \int_{0}^{T} e^{\mathbf{F}\eta} \, d\eta \, \mathbf{G}u(nT). \tag{6.11}$$

If we define

$$\mathbf{\Phi} = e^{\mathbf{F}T}, \tag{6.12a}$$

$$\mathbf{\Gamma} = \int_{0}^{T} e^{\mathbf{F}\eta} \, d\eta \, \mathbf{G}, \tag{6.12b}$$

Eqs. (6.11) and (6.2) reduce to difference equations in standard form:

$$\mathbf{x}(n + 1) = \mathbf{\Phi}\mathbf{x}(n) + \mathbf{\Gamma}u(n) + \mathbf{\Gamma}_1 w(n),$$
$$y(n) = \mathbf{H}\mathbf{x}(n), \tag{6.13}$$

where we include the effect of a disturbance, w. If w is a constant, then $\mathbf{\Gamma}_1$ is given by (6.12b) with $\mathbf{G}$ replaced by $\mathbf{G}_1$. If w is an impulse, then $\mathbf{\Gamma}_1 = \mathbf{G}_1$ and $w(n) = w_0\delta_0(n)$.[4] The $\mathbf{\Phi}$ series expansion,

$$\mathbf{\Phi} = e^{\mathbf{F}T} = \mathbf{I} + \mathbf{F}T + \frac{\mathbf{F}^2T^2}{2!} + \frac{\mathbf{F}^3T^3}{3!} + \cdots,$$

can also be written

$$\mathbf{\Phi} = \mathbf{I} + \mathbf{F}T\mathbf{\Psi}, \tag{6.14}$$

where

$$\mathbf{\Psi} = \mathbf{I} + \frac{\mathbf{F}T}{2!} + \frac{\mathbf{F}^2T^2}{3!} + \cdots$$

The $\mathbf{\Gamma}$ integral in (6.12) can be evaluated term by term to give

$$\mathbf{\Gamma} = \sum_{k=0}^{\infty} \frac{\mathbf{F}^k T^{k+1}}{(k + 1)!} \mathbf{G}$$

$$= \sum_{k=0}^{\infty} \frac{\mathbf{F}^k T^k}{(k + 1)!} T\mathbf{G}$$

$$\mathbf{\Gamma} = \mathbf{\Psi}T\mathbf{G}. \tag{6.15}$$

We evaluate $\mathbf{\Psi}$ by a series in the form

$$\mathbf{\Psi} \approx \mathbf{I} + \frac{\mathbf{F}T}{2}\left(\mathbf{I} + \frac{\mathbf{F}T}{3}\left(\cdots \frac{\mathbf{F}T}{N - 1}\left(\mathbf{I} + \frac{\mathbf{F}T}{N}\right)\right)\cdots\right), \tag{6.16}$$

which has better numerical properties than the direct series of powers. We then find $\mathbf{\Gamma}$ from (6.15) and $\mathbf{\Phi}$ from (6.14). A discussion of the selection of N and a technique to compute $\mathbf{\Psi}$ for comparatively large T is given by Källström (1973), and a review of various methods is found in Moler and Van Loan (1978). Some of these results are described in Appendix A to Chapter 6. The program logic for computation of $\mathbf{\Phi}$ and $\mathbf{\Gamma}$ for simple cases is given in Fig. 6.3.

To compare this method of representing the plant with the discrete transfer functions obtained in Chapter 5, we can take the z-transform of (6.13) with $w = 0$

---

[4] δ_0 is the discrete or Kronecker delta, which is zero except at $n = 0$, where it is 1.0.

1. Select sampling period T and description matrices **F** and **G**.
2. Matrix **I** ← Identity
3. Matrix **Ψ** ← **I**
4. k ← 11; Comment: we are using $N = 11$ in (6.16).
5. If $k = 1$, go to step 9.
6. Matrix **Ψ** ← **I** + $\dfrac{FT}{k}$ **Ψ**
7. k ← $k - 1$
8. Go to step 5.
9. Matrix **Γ** ← T**ΨG**
10. Matrix **Φ** ← **I** + FT**Ψ**

To compute
Φ & Γ

Fig. 6.3 Program logic to compute **Φ** and **Γ** from **F**, **G**, and T for simple cases. (The left arrow, ←, is to be read "is replaced by.")

and obtain

$$[z\mathbf{I} - \mathbf{\Phi}]\mathbf{X}(z) = \mathbf{\Gamma}U(z) \tag{6.17a}$$

$$Y(z) = \mathbf{H}\mathbf{X}(z); \tag{6.17b}$$

therefore

$$\frac{Y(z)}{U(z)} = \mathbf{H}[z\mathbf{I} - \mathbf{\Phi}]^{-1}\mathbf{\Gamma}. \tag{6.18}$$

For the satellite attitude-control example, the **Φ** and **Γ** matrices are easy to calculate using (6.14) and (6.15) and the values for **F** and **G** defined in (6.3) since $\mathbf{F}^2 = \mathbf{0}$ in this case:

$$\mathbf{\Phi} = \mathbf{I} + \mathbf{F}T + \frac{\mathbf{F}^2 T^2}{2!} + \cdots = \begin{bmatrix} 1 & 0 \\ 0 & 1 \end{bmatrix} + \begin{bmatrix} 0 & 1 \\ 0 & 0 \end{bmatrix} T + \cdots = \begin{bmatrix} 1 & T \\ 0 & 1 \end{bmatrix},$$

$$\mathbf{\Gamma} = \left[\mathbf{I}T + \mathbf{F}\frac{T^2}{2!} + \frac{\mathbf{F}^2 T^3}{3!} \right] \mathbf{G} = \left\{ \begin{bmatrix} T & 0 \\ 0 & T \end{bmatrix} + \begin{bmatrix} 0 & 1 \\ 0 & 0 \end{bmatrix} \frac{T^2}{2} \right\} \begin{bmatrix} 0 \\ 1 \end{bmatrix} = \begin{bmatrix} \frac{T^2}{2} \\ T \end{bmatrix};$$

hence using (6.18), we obtain

$$\frac{Y(z)}{U(z)} = \begin{bmatrix} 1 & 0 \end{bmatrix} \left\{ z \begin{bmatrix} 1 & 0 \\ 0 & 1 \end{bmatrix} - \begin{bmatrix} 1 & T \\ 0 & 1 \end{bmatrix} \right\}^{-1} \begin{bmatrix} \frac{T^2}{2} \\ T \end{bmatrix}$$

$$= \frac{T^2}{2} \frac{(z + 1)}{(z - 1)^2}, \qquad \rightarrow \det(z\mathbf{I} - \mathbf{\Phi})$$

which is the same result that would be obtained using (3.20) and the z-transform tables. Note that to compute Y/U we find that the denominator is the determinant $\det(z\mathbf{I} - \boldsymbol{\Phi})$, which comes from the matrix inverse in (6.18). This determinant is the characteristic polynomial of the transfer function, and the zeros of the determinant are the poles of the plant. We have two poles at $z = 1$ in this case, corresponding to the two integrations in this plant's equations of motion.

We can explore further the question of poles and zeros and the state-space description by considering again the transform equations (6.17). An interpretation of transfer-function poles from the perspective of the corresponding difference equation is that a pole is a value of z such that the equation has a nontrivial solution when the forcing input is zero. From Eq. (6.17a), this implies that the linear equations

$$[z\mathbf{I} - \boldsymbol{\Phi}]\mathbf{X}(z) = [0]$$

have a nontrivial solution. From matrix algebra the well-known requirement for this is that $\det(z\mathbf{I} - \boldsymbol{\Phi}) = 0$. In the present case,

$$\det[z\mathbf{I} - \boldsymbol{\Phi}] = \det\left[\begin{bmatrix} z & 0 \\ 0 & z \end{bmatrix} - \begin{bmatrix} 1 & T \\ 0 & 1 \end{bmatrix}\right]$$

$$= \det\begin{bmatrix} z - 1 & T \\ 0 & z - 1 \end{bmatrix}$$

$$= (z - 1)^2 = 0,$$

which is the characteristic equation, as we have seen.

Along the same line of reasoning, a system zero is a value of z such that the system output is zero even with a nonzero state and input combination. Thus we are able to find a nontrivial solution for $\mathbf{X}(z_0)$ and $U(z_0)$ such that if $Y(z_0)$ is zero, then z_0 is a zero of the system. Combining the two parts of (6.17), we must satisfy the requirement

$$\begin{bmatrix} z\mathbf{I} - \boldsymbol{\Phi} & -\boldsymbol{\Gamma} \\ \mathbf{H} & 0 \end{bmatrix} \begin{bmatrix} \mathbf{X}(z) \\ U(z) \end{bmatrix} = [0]. \tag{6.19}$$

Once more the condition for the existence of nontrivial solutions is that the determinant of the square coefficient system matrix be zero.[5] For the satellite example,

$$\det\begin{bmatrix} z - 1 & -T & -T^2/2 \\ 0 & z - 1 & -T \\ 1 & 0 & 0 \end{bmatrix} = 1 \cdot \det\begin{bmatrix} -T & -T^2/2 \\ z - 1 & -T \end{bmatrix}$$

$$= +T^2 + \left(\frac{T^2}{2}\right)(z - 1)$$

$$= +\frac{T^2}{2}z + \frac{T^2}{2}$$

$$= +\frac{T^2}{2}(z + 1).$$

---

[5] We do not consider here the case of different numbers of inputs and outputs.

Thus we have a single zero at $z = -1$, as we have seen from the transfer function.

For computer aids, it is obvious that we need a program:

SAMPLE0: A program to compute $\mathbf{\Phi}$ and $\mathbf{\Gamma}$ from $\mathbf{F}$, $\mathbf{G}$, and sample period T. The program assumes a zero-order hold and can use the logic of Fig. 6.3 or Fig. 6.17 of Appendix A to this chapter.

and a matrix step response program:

STEP: A program to solve (6.13) and plot the results.

6.3 CONTROL-LAW DESIGN

One of the attractive features of state-space design methods is that the procedure consists of two independent steps. One step *assumes* that we have all the states at our disposal for feedback purposes. In general, of course, this would be a ridiculous assumption since a practical engineer would not, as a rule, find it necessary to purchase this large number of sensors, especially since he knows that he would not need them, using classical design methods. The assumption that all states are available merely allows us to proceed with the first design step, namely, the control law. The remaining step is to design an "estimator" (or "observer"[6]) which estimates the entire state vector, given measurements of the portion of the state provided by (6.2). The final control algorithm will consist of the control law and the estimator combined where the control-law calculations are based on the estimated states rather than the actual states. In Sections 6.5 and 6.6 we show that this substitution is reasonable and that the combined control law and estimator can give closed-loop dynamic characteristics which are unchanged from those assumed in designing the control law and estimator separately. The dynamic system we obtain from the combined control law and estimator is called the *controller*. The first step is to get a good control law.

The control law is simply the feedback of a linear combination of all the states, that is,

$$u = -\mathbf{Kx} = -[K_1 K_2 \; . \; . \; .] \begin{bmatrix} x_1 \\ x_2 \\ \vdots \end{bmatrix}. \qquad (6.20)$$

Substituting the above result in the difference equation (6.13), we have

$$\mathbf{x}(n + 1) = \mathbf{\Phi x}(n) - \mathbf{\Gamma Kx}(n) + \mathbf{\Gamma}_1 w. \qquad (6.21)$$

If we assume $w = w_0 \delta_0$, a pulse at zero, the z-transform of (6.21) is

$$(z\mathbf{I} - \mathbf{\Phi} + \mathbf{\Gamma K})\mathbf{X}(z) = \mathbf{\Gamma}_1 w_0 .$$

---

[6] The literature [Luenberger (1960)] commonly refers to these devices as "observers"; however, we feel that "estimator" is much more descriptive of their function since "observe" implies a direct measurement. In this book the terms are used interchangeably.

Thus the characteristic equation of the controlled (closed-loop) system is

$$\det |z\mathbf{I} - \mathbf{\Phi} + \mathbf{\Gamma K}| = 0. \tag{6.22}$$

The control-law design then consists of picking the elements of $\mathbf{K}$ so that the roots of (6.22) are in desirable locations.

Given desired root locations,[7] say

$$z_i = \beta_1, \beta_2, \beta_3, \ldots ,$$

the desired control-characteristic equation is

$$\alpha_c(z) = (z - \beta_1)(z - \beta_2)(z - \beta_3) \cdots = 0. \tag{6.23}$$

Hence the required elements of $\mathbf{K}$ are obtained by matching coefficients in (6.22) and (6.23), thus forcing the system characteristic polynomial to be identical to (6.23).

Suppose we want to design a control law for the satellite attitude-control example so that the closed-loop characteristic equation has s-plane roots at a damping ratio of $\zeta = 0.5$ and negative real part $a = +1.8$. Using Fig. 6.4 with a sample period of $T = 0.1$ sec, we obtain $\theta = \sqrt{3}aT = 17.9°$ and $r = e^{-aT} = 0.837$. The desired characteristic equation is then (approximately)

$$(z - 0.837e^{i17.9°})(z - 0.837e^{-i17.9°}) = 0,$$
$$z^2 - 1.6z + 0.70 = 0,$$

and the evaluation of (6.22) for any control law $\mathbf{K}$ leads to

$$\det \left| z \begin{bmatrix} 1 & 0 \\ 0 & 1 \end{bmatrix} - \begin{bmatrix} 1 & T \\ 0 & 1 \end{bmatrix} + \begin{bmatrix} T^2/2 \\ T \end{bmatrix} [K_1 \quad K_2] \right| = 0$$

or

$$z^2 + (TK_2 + (T^2/2)K_1 - 2)z + (T^2/2)K_1 - TK_2 + 1 = 0,$$

and equating coefficients, we obtain two simultaneous equations in the two unknown elements of $\mathbf{K}$:

$$TK_2 + (T^2/2)K_1 - 2 = -1.6,$$
$$(T^2/2)K_1 - TK_2 + 1 = 0.70,$$

which are easily solved for the coefficients and evaluated for $T = 0.1$ sec:

$$K_1 = \frac{0.10}{T^2} = 10, \qquad K_2 = \frac{0.35}{T} = 3.5.$$

---

[7] Section 6.5 discusses the selection of root locations. We can also use the results of the specification discussion in Chapter 5 and the Butterworth or ITAE pole patterns discussed in Chapter 3.

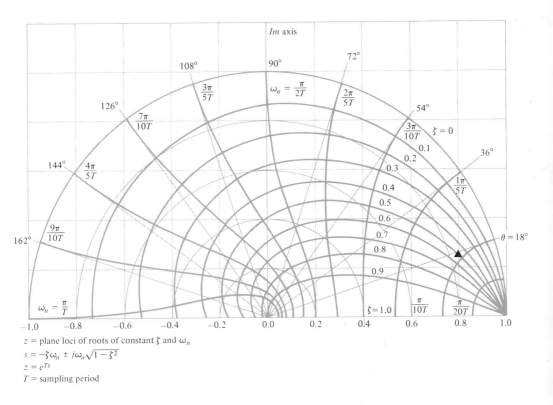

Im axis

$\omega_n = \dfrac{\pi}{2T}$

$\zeta = 0$

$\theta = 18°$

$\omega_n = \dfrac{\pi}{T}$

$\zeta = 1.0$

z = plane loci of roots of constant ζ and ω_n

$s = -\zeta\omega_n \pm j\omega_n\sqrt{1 - \zeta^2}$

$z = e^{Ts}$

T = sampling period

Fig. 6.4 Desired root locations for satellite attitude-control example.

The calculation of the gains using the method illustrated in the previous example becomes rather tedious when the order of the system (and therefore the order of the determinant to be evaluated) is greater than 3. A computer does not solve the tedium unless the computer is used to perform the algebraic manipulations necessary in expanding the determinant in (6.22) to obtain the characteristic equation.

The algebra for finding the specific value of **K** is especially simple if the system matrices happen to be in the form associated with the block diagram of Fig. 2.12. This structure is called "control canonical form" because it is so useful in control-law design. Referring to that figure and taking the states as the outputs of the delay elements, numbered from the left, we get (we assume $b_0 = 0$ for this case)

$$\mathbf{\Phi}_c = \begin{bmatrix} a_1 & a_2 & a_3 \\ 1 & 0 & 0 \\ 0 & 1 & 0 \end{bmatrix}, \qquad \mathbf{\Gamma}_c = \begin{bmatrix} 1 \\ 0 \\ 0 \end{bmatrix}, \qquad \mathbf{H}_c = [b_1 \quad b_2 \quad b_3]. \qquad (6.24)$$

Note that from (2.15), the characteristic polynomial of this system is $a(z) = z^3 - a_1 z^2 - a_2 z - a_3$. The key idea here is that the elements of the first row of Φ_c are exactly the coefficients of the characteristic polynomial of the system. If we now form the closed-loop system matrix $\Phi_c - \Gamma_c K$, we find

$$\Phi_c - \Gamma_c K = \begin{bmatrix} a_1 - K_1 & a_2 - K_2 & a_3 - K_3 \\ 1 & 0 & 0 \\ 0 & 1 & 0 \end{bmatrix}. \tag{6.25}$$

By inspection, the characteristic equation of (6.25) is

$$z^3 - (a_1 - K_1)z^2 - (a_2 - K_2)z - (a_3 - K_3) = 0.$$

Thus, if the desired root locations result in the characteristic equation

$$z^3 - \alpha_1 z^2 - \alpha_2 z - \alpha_3 = 0,$$

then the necessary values for control gains are

$$K_1 = a_1 - \alpha_1, \qquad K_2 = a_2 - \alpha_2, \qquad K_3 = a_3 - \alpha_3. \tag{6.26}$$

Conceptually, then, we have a design method: Given an arbitrary (Φ, Γ) and a desired characteristic equation $\alpha(z) = 0$, we convert (by redefinition of the states) (Φ, Γ) to control form (Φ_c, Γ_c) and solve for the gain by (6.26). Since this gain is for states in the control form, we must, finally, express the result back in terms of the original states.

The first question this process raises is existence: Is it always possible to find an equivalent (Φ_c, Γ_c) for arbitrary (Φ, Γ)? The answer is almost always "yes." The exception occurs in certain pathological systems dubbed "uncontrollable" for which no control will give arbitrary root locations. These systems have certain modes or subsystems which are unaffected by the control. Uncontrollability is best exhibited by a realization (selection of states) where the system states are defined so that each state represents a natural mode of the system. If all the roots of (6.22) are distinct, then (6.13) written in this way (normal mode or "Jordan canonical form") becomes

$$\mathbf{x}(n+1) = \begin{bmatrix} \lambda_1 & & & \bigcirc \\ & \lambda_2 & & \\ & & \ddots & \\ \bigcirc & & & \lambda_n \end{bmatrix} \mathbf{x}(n) + \begin{bmatrix} \Gamma_1 \\ \Gamma_2 \\ \vdots \\ \Gamma_n \end{bmatrix} u(n) \tag{6.27}$$

and explicitly exhibits the criterion for controllability: when in this form, no element in Γ can be zero. If any Γ element were zero, no control would influence that normal mode directly and the associated state would remain uncontrolled. A good physical understanding of the system being controlled usually prevents any attempt to design a controller for an uncontrollable system; however, there is a mathematical test for controllability applicable to any system description which may be an additional aid in discovering this condition, and a discussion of this test is contained in Section 6.7.

The second question, if the system is found to be controllable and a gain is known to exist, is that of computational complexity. The process described above of converting to (Φ_c, Γ_c) needs to be organized to make the design easy to use. A very convenient formula has been derived by Ackermann (1972), and the proof for it is repeated in Appendix B to Chapter 6. The relation is[8]:

$$\mathbf{K} = [0 \ . \ . \ . \ 0 \ 1][\Gamma \ \ \Phi\Gamma \ \ \Phi^2\Gamma \ . \ . \ . \ \Phi^{n-1}\Gamma]^{-1}\alpha_c(\Phi), \qquad (6.28)$$

where $\mathscr{C} = [\Gamma \ \Phi\Gamma \ . \ . \ .]$ is called the controllability matrix, n is the order of system or number of states, and we substitute Φ for z in $\alpha_c(z)$ to form

$$\alpha_c(\Phi) = \Phi^n - \alpha_1\Phi^{n-1} - \alpha_2\Phi^{n-2} \ . \ . \ . - \alpha_n\mathbf{I}, \qquad (6.29)$$

where the α_i's are the coefficients of the desired characteristic equation, that is,

$$\alpha_c(z) = |z\mathbf{I} - \Phi + \Gamma\mathbf{K}| = z^n - \alpha_1 z^{n-1} - \cdot \cdot \cdot - \alpha_n. \qquad (6.30)$$

Applying this formula to the satellite attitude-control example, we find that

$$\alpha_1 = 1.6, \qquad \alpha_2 = -0.70,$$

and therefore

$$\alpha_c(\Phi) = \begin{bmatrix} 1 & 2T \\ 0 & 1 \end{bmatrix} - 1.6 \begin{bmatrix} 1 & T \\ 0 & 1 \end{bmatrix} + 0.70 \begin{bmatrix} 1 & 0 \\ 0 & 1 \end{bmatrix} = \begin{bmatrix} 0.1 & 0.4T \\ 0 & 0.1 \end{bmatrix}.$$

Furthermore

$$[\Gamma \ \ \Phi\Gamma] = \begin{bmatrix} T^2/2 & 3T^2/2 \\ T & T \end{bmatrix}$$

and

$$[\Gamma \ \ \Phi\Gamma]^{-1} = 1/T^2 \begin{bmatrix} -1 & +3T/2 \\ 1 & -T/2 \end{bmatrix},$$

and finally

$$\mathbf{K} = [K_1 \ \ K_2] = (1/T^2)[0 \ \ 1] \begin{bmatrix} -1 & 3T/2 \\ 1 & -T/2 \end{bmatrix} \begin{bmatrix} 0.1 & 0.4T \\ 0 & 0.1 \end{bmatrix};$$

---

[8] We note that the matrix $\mathscr{C}$ in (6.28) may be poorly conditioned and should not be inverted, but rather the equations $b^T\mathscr{C} = e^T$ should be solved by a stable method such as gaussian elimination with pivoting. Also we note that careful selection of states and their amplitude scaling will help avoid trouble in computing K.

therefore

$$[K_1 \quad K_2] = \frac{1}{T^2}[0.1 \quad 0.35T] = [10 \quad 3.5],$$

which is the same result as that obtained earlier.

A program logic for application of Ackermann's formula to compute the control law is given in Fig. 6.5.

---

1. Read in $\mathbf{\Phi}$, $\mathbf{\Gamma}$, T, and N_s, the number of states.
2. Comment: first we will read in the desired pole locations in the s-plane, convert them to z-plane polynomial coefficients, and construct $\alpha(\mathbf{\Phi})$.
3. $\mathbf{I} \leftarrow$ identity matrix, $N_s \times N_s$
4. $\mathbf{ALPHA} \leftarrow \mathbf{I}$
5. $k \leftarrow 1$
6. If $k > N_s$, go to step 18.
7. Read in pole location k as $a + jb$.
8. If $b = 0$, go to step 14.
9. $A_1 \leftarrow -2 \exp(aT) \cos bT$
10. $A_2 \leftarrow \exp(2aT)$
11. $\mathbf{ALPHA} \leftarrow \mathbf{ALPHA} \times (\mathbf{\Phi} \times \mathbf{\Phi} + A_1\mathbf{\Phi} + A_2\mathbf{I})$
12. $k \leftarrow k + 2$
13. Go to step 6.
14. $A_1 \leftarrow \exp(aT)$
15. $\mathbf{ALPHA} \leftarrow \mathbf{ALPHA} \times (\mathbf{\Phi} - A_1 \times \mathbf{I})$
16. $k \leftarrow k + 1$
17. Go to step 6.
18. Comment: now we construct the controllability matrix.
19. $\mathbf{C} \leftarrow \mathbf{I}$
20. $\mathbf{E} \leftarrow \mathbf{\Gamma}$
21. $k \leftarrow 1$
22. If $k > N_s$, go to step 28.
23. Comment: replace column k of $\mathbf{C}$ by $\mathbf{E}$.
24. $\mathbf{C}[\quad ; k] \leftarrow \mathbf{E}$
25. $k \leftarrow k + 1$
26. $\mathbf{E} \leftarrow \mathbf{\Phi} \times \mathbf{E}$
27. Go to step 22.
28. Comment: now solve for the control law, first form e_n^T as the last row of $\mathbf{I}$.
29. $\mathbf{E} \leftarrow \mathbf{I}[N_s; \quad]$
30. Solve $\mathbf{BC} = \mathbf{E}$ for $\mathbf{B}$.
31. $\mathbf{K} = \mathbf{B} \times \mathbf{ALPHA}$
32. END

Fig. 6.5 Program logic for computing control law $\mathbf{K}$ via Ackermann's formula.

6.4 ESTIMATOR DESIGN

The control law designed in the last section assumed that all states were available for feedback. Since, typically, not all states are measured, the purpose of this section is to show how to determine algorithms which will reconstruct all the states, given measurements of a portion of them. If the state is $\mathbf{x}$, then the estimate is $\hat{\mathbf{x}}$, and the idea is to let $u = -\mathbf{K}\hat{\mathbf{x}}$, replacing the true states by their estimates in the control law.

6.4.1 Prediction Estimators

One method of estimating the states which may come to mind is to construct a model of the plant dynamics,

$$\hat{\mathbf{x}}(n + 1) = \mathbf{\Phi}\hat{\mathbf{x}}(n) + \mathbf{\Gamma}u(n), \tag{6.31}$$

where $\hat{\mathbf{x}}$ denotes an estimate of the actual state, $\mathbf{x}$. We know $\mathbf{\Phi}$, $\mathbf{\Gamma}$, and $u(n)$, and hence this estimator should work if we can obtain the correct $\mathbf{x}(0)$ and set $\hat{\mathbf{x}}(0)$ equal to it. Figure 6.6 depicts this "open-loop" estimator. If we define the error in the estimate as

$$\tilde{\mathbf{x}} \overset{\Delta}{=} \mathbf{x} - \hat{\mathbf{x}}, \tag{6.32}$$

then this system is described by

$$\tilde{\mathbf{x}}(n + 1) = \mathbf{\Phi}\tilde{\mathbf{x}}(n). \tag{6.33}$$

Thus, if the initial value of $\hat{\mathbf{x}}$ is off, then the error dynamics are those of the uncompensated plant, $\mathbf{\Phi}$. Since one of the main objectives of control is to fix up the unsatisfactory dynamics of $\mathbf{\Phi}$, it is not likely that (6.33) represents adequate performance for the error! However, if we feed back the difference between the

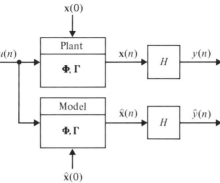

Fig. 6.6 Open-loop estimator.

measured output and the estimated output and constantly correct the model with this error signal, this divergence should be minimized. This scheme is shown in Fig. 6.7, and the equation for it is

$$\hat{x}(n + 1) = \Phi\hat{x}(n) + \Gamma u(n) + L[y(n) - H\hat{x}(n)]. \tag{6.34}$$

We will call this a *prediction estimator* because the estimate, $\hat{x}(n + 1)$, is one cycle ahead of the measurement, $y(n)$.

A difference equation describing the behavior of the error is again obtained by subtracting the $\hat{x}$ equation (6.34) from the actual equation (6.13),

$$\bar{x}(n + 1) = [\Phi - LH]\bar{x}(n). \tag{6.35}$$

This is a homogeneous equation but the dynamics are given by $[\Phi - LH]$, and if this system matrix represents a fast stable system, $\bar{x}$ will converge to zero in a satisfactory way for any value of $\bar{x}(0)$. In other words, $\hat{x}(n)$ will converge to $x(n)$ regardless of the value of $\hat{x}(0)$ and can do so faster than the normal (open-loop) motion of $x(n)$. If the values used for Φ and Γ in the estimator are not exactly those existing in the actual plant, then the dynamics of the error will not be given exactly by (6.35). However, typically L can be chosen so that the system is stable and the error is acceptably small.

To select L, we take the same approach as we did when designing the control law. If we specify the desired estimator root locations in the z-plane, L is uniquely determined, provided y is a scalar and the system is "observable." We may have an unobservable system if some of its modes do not appear at the given measurements. For example, if only derivatives of certain states are measured, and these states do not affect the dynamics, a constant of integration is obscured. This situation occurs with a $1/s^2$ plant if only velocity is measured, for then it is impossible to deduce the initial condition of the position. For an oscillator, a velocity measurement is sufficient to estimate position because the acceleration and consequently the velocity observed are affected by position. A mathematical test for observability is given in Section 6.7.

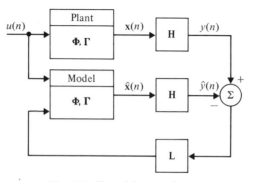

Fig. 6.7 Closed-loop estimator.

As in the controller design, two methods are available for the computation of L. The first is to expand the determinant and match the coefficients in like powers of z on the two sides of

$$|z\mathbf{I} - \mathbf{\Phi} + \mathbf{LH}| = (z - \beta_1)(z - \beta_2) \ldots (z - \beta_n), \tag{6.36}$$

where the β's are the desired estimator root locations[9] and represent how fast the estimator state converges to the correct value of the plant state.

The second is to use Ackermann's[10] estimator formula

$$\mathbf{L} = \alpha_e(\mathbf{\Phi}) \begin{bmatrix} \mathbf{H} \\ \mathbf{H\Phi} \\ \mathbf{H\Phi}^2 \\ \cdot \\ \cdot \\ \cdot \\ \mathbf{H\Phi}^{n-1} \end{bmatrix}^{-1} \begin{bmatrix} 0 \\ 0 \\ \cdot \\ \cdot \\ \cdot \\ 0 \\ 1 \end{bmatrix}, \tag{6.37}$$

which is derived in Appendix B to this chapter along with the control version.[11] The coefficient matrix with rows $\mathbf{H\Phi}^j$ is called the *observability matrix*.

6.4.2 Current Estimators

As was already noted, the previous form of the estimator equation (6.34) arrives at the state estimate $\hat{\mathbf{x}}(n)$ after receiving measurements up through $y(n - 1)$. This means that the current value of *control*[12] does not depend on the current value of error in the observations and thus may not be as accurate as it could be. For high-order systems controlled with a slow computer or any time the sample rates are fast compared to the computation time, this delay between making an observation and using it in a control may be a blessing. In many systems, however, the computation time required to evaluate (6.34) is quite short compared to the sample period, and practical engineers become restless waiting for the end of the sample period and watching the sampled signal becoming old and unused. Therefore, an alternative estimator formulation which bases the current estimate $\hat{\mathbf{x}}(n)$ on the current measurement $y(n)$ is to separate the prediction of the next state from the correction of the estimate based on the observation of $y(n)$. If we have an estimate at

---

[9] The following section discusses how one should select these roots in relation to the control roots and how both sets of roots appear in the combined system.

[10] See reference on p. 143.

[11] Note that if we take the transpose of $\mathbf{\Phi} - \mathbf{LH}$, we get $\mathbf{\Phi}^T - \mathbf{H}^T\mathbf{L}^T$, which is the same form as the system matrix $\mathbf{\Phi} - \mathbf{\Gamma K}$ of the control problem. Therefore if we substitute $\mathbf{\Phi}^T$ for $\mathbf{\Phi}$, $\mathbf{H}^T$ for $\mathbf{\Gamma}$ and $\mathbf{L}^T$ for $\mathbf{K}$, we can use the control design results. Equation (6.37) is the consequence.

[12] We plan to use $u_n = -\mathbf{K}\hat{\mathbf{x}}_n$, in place of $u_n = -\mathbf{K}\mathbf{x}_n$.

n, $\hat{x}(n)$, then we would predict the next state by a model of the plant as[13]

$$\bar{x}(n + 1) = \Phi\hat{x}(n) + \Gamma u(n). \tag{6.38a}$$

Now, at the time $n + 1$, we observe $y(n + 1)$ and correct $\bar{x}$ by using the observation to compute

$$\hat{x}(n + 1) = \bar{x}(n + 1) + L(y(n + 1) - H\bar{x}(n + 1)). \tag{6.38b}$$

We shall call $\hat{x}(n)$ the *current estimate* of $x(n)$. In practice, this estimator cannot be implemented exactly because it is impossible to sample, perform calculations, and output with absolutely no time elapsed. However, as we saw when we discussed implementations in Chapter 2, the calculation of u based on (6.38) can be arranged to minimize computational delays. We will assume that this procedure introduces a negligible error here, just as we did in Chapter 5, when we had a direct feed term, b_0, in the controller.

The error equation for the current estimator is similar to the error equation for the prediction estimator that was given in (6.35). The current-estimator error equation is obtained by subtracting (6.38) from (6.13), taking $\tilde{x} = x - \hat{x}$, and is

$$\tilde{x}(n + 1) = [\Phi - LH\Phi]\tilde{x}(n). \tag{6.39}$$

Therefore, the gain matrix L is obtained exactly as before, except that H is replaced by $H\Phi$. The two methods, (6.36) and (6.37), become:

a) Match coefficients of:

$$\|[zI - \Phi + LH\Phi]\| = \alpha_e(z) \tag{6.40}$$

where α_e is the desired characteristic polynomial of the estimator error, or

b) use Ackermann's formula:[14]

$$L = \alpha_e(\Phi) \begin{bmatrix} H\Phi \\ H\Phi^2 \\ H\Phi^3 \\ \cdot \\ \cdot \\ \cdot \\ H\Phi^n \end{bmatrix}^{-1} \begin{bmatrix} 0 \\ 0 \\ \cdot \\ \cdot \\ \cdot \\ 0 \\ 1 \end{bmatrix}. \tag{6.41}$$

---

[13] These equations are the form of the Kalman filter. See Chapter 9.

[14] If Φ is singular, as can happen with systems having time delay, (6.41) cannot be used. In that case, the current-estimator structure forces some of the estimator poles to be at the origin of the z-plane. Since some components of $\tilde{x}(n + 1)$ will be forced to zero by the Φ matrix, there is no need to use correction via L on these.

6.4.3 Reduced-Order Estimators[15]

The estimators discussed so far are designed to reconstruct the entire state vector, given measurements of some of the states. One might therefore ask: Why bother to reconstruct the states that are measured directly? The answer is: You don't have to, although, when there is significant noise on the measurements, the estimator for the full state vector yields superior results.

To pursue an estimator for only the unmeasured states, let us partition the state vector into two parts: $\mathbf{x}_a$ is the portion directly measured which is y, and $\mathbf{x}_b$ is the remaining portion to be estimated. The complete system description, like (6.13), becomes

$$\begin{bmatrix} \mathbf{x}_a(n+1) \\ \mathbf{x}_b(n+1) \end{bmatrix} = \begin{bmatrix} \mathbf{\Phi}_{aa} & \mathbf{\Phi}_{ab} \\ \mathbf{\Phi}_{ba} & \mathbf{\Phi}_{bb} \end{bmatrix} \begin{bmatrix} \mathbf{x}_a(n) \\ \mathbf{x}_b(n) \end{bmatrix} + \begin{bmatrix} \mathbf{\Gamma}_a \\ \mathbf{\Gamma}_b \end{bmatrix} u(n), \tag{6.42}$$

$$y(n) = \begin{bmatrix} I & 0 \end{bmatrix} \begin{bmatrix} \mathbf{x}_a(n) \\ \mathbf{x}_b(n) \end{bmatrix}, \tag{6.43}$$

and the portion describing the dynamics of the unmeasured states is

$$\mathbf{x}_b(n+1) = \mathbf{\Phi}_{bb}\mathbf{x}_b(n) + \underbrace{\mathbf{\Phi}_{ba}\mathbf{x}_a(n) + \mathbf{\Gamma}_b u(n)}_{\text{known ''input''}}, \tag{6.44}$$

where the right-hand two terms are known and can be considered as an input into the $\mathbf{x}_b$ dynamics. If we reorder the $\mathbf{x}_a$ portion of (6.42), we obtain

$$\underbrace{\mathbf{x}_a(n+1) - \mathbf{\Phi}_{aa}\mathbf{x}_a(n) - \mathbf{\Gamma}_a u(n)}_{\text{known ''measurement''}} = \mathbf{\Phi}_{ab}\mathbf{x}_b(n). \tag{6.45}$$

Note that this is a relationship between a measured quantity on the left and the unknown state on the right. Therefore, Eqs. (6.44) and (6.45) have the same relationship to the state $\mathbf{x}_b$ that the original equation, (6.13), had to the entire state $\mathbf{x}$. Following this reasoning, we arrive at the desired estimator by making the following substitutions into the estimator equations:

$$\mathbf{x} \leftarrow \mathbf{x}_b,$$
$$\mathbf{\Phi} \leftarrow \mathbf{\Phi}_{bb},$$
$$\mathbf{\Gamma}u(n) \leftarrow \mathbf{\Phi}_{ba}x_a(n) + \mathbf{\Gamma}_b u(n),$$
$$y(n) \leftarrow x_a(n+1) - \mathbf{\Phi}_{aa}x_a(n) - \mathbf{\Gamma}_a u(n),$$
$$\mathbf{H} \leftarrow \mathbf{\Phi}_{ab}.$$

Thus the reduced-order estimator equations are

$$\hat{\mathbf{x}}_b(n+1) = \mathbf{\Phi}_{bb}\hat{\mathbf{x}}_b(n) + \mathbf{\Phi}_{ba}\mathbf{x}_a(n) + \mathbf{\Gamma}_b u(n)$$
$$+ \mathbf{L}[x_a(n+1) - \mathbf{\Phi}_{aa}x_a(n) - \mathbf{\Gamma}_a u(n) - \mathbf{\Phi}_{ab}\hat{\mathbf{x}}_b(n)]. \tag{6.46}$$

---

[15] Reduced-order estimators (or observers) were originally proposed by Luenberger (1964). This development follows Gopinath (1971).

Subtracting (6.46) from (6.44) yields the error equation

$$\tilde{\mathbf{x}}_b(n + 1) = [\boldsymbol{\Phi}_{bb} - \mathbf{L}\boldsymbol{\Phi}_{ab}]\tilde{\mathbf{x}}_b(n), \tag{6.47}$$

and therefore $\mathbf{L}$ is selected exactly as before, that is, (a) by picking roots of

$$|z\mathbf{I} - \boldsymbol{\Phi}_{bb} + \mathbf{L}\boldsymbol{\Phi}_{ab}| = \alpha_e(z) \tag{6.48}$$

to be in desirable locations, or (b) using Ackermann's formula

$$\mathbf{L} = \alpha_e(\boldsymbol{\Phi})_{bb} \begin{bmatrix} \boldsymbol{\Phi}_{ab} \\ \boldsymbol{\Phi}_{ab}\boldsymbol{\Phi}_{bb} \\ \boldsymbol{\Phi}_{ab}\boldsymbol{\Phi}_{bb}^2 \\ \cdot \\ \cdot \\ \cdot \\ \boldsymbol{\Phi}_{ab}\boldsymbol{\Phi}_{bb}^{n-2} \end{bmatrix}^{-1} \begin{bmatrix} 0 \\ 0 \\ \cdot \\ \cdot \\ \cdot \\ 0 \\ 1 \end{bmatrix}. \tag{6.49}$$

We note here that Gopinath (1971) proved that if a full estimator as given by (6.34) exists, then the reduced-order estimator given by (6.46) also exists; that is, we can place the roots of (6.48) anywhere we choose by choice of $\mathbf{L}$.

6.5 REGULATOR DESIGN: COMBINED CONTROL LAW AND ESTIMATOR

If we take the control law (Section 6.3) and implement it, using an estimated state vector (Section 6.4), the control system can be completed. A schematic of such a system is shown in Fig. 6.8. However, since we designed the control law assuming that the true state, $\mathbf{x}$, was fed back instead of $\hat{\mathbf{x}}$, it is of interest to examine what effect this has on the system dynamics. The controlled plant equation (6.21) becomes

$$\mathbf{x}(n + 1) = \boldsymbol{\Phi}\mathbf{x}(n) - \boldsymbol{\Gamma}\mathbf{K}\hat{\mathbf{x}}(n), \tag{6.50}$$

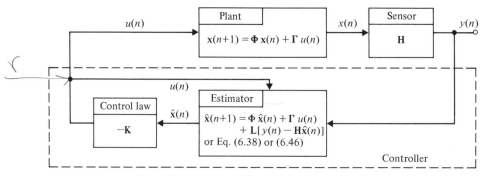

Fig. 6.8 Estimator and controller mechanization.

which can also be written in terms of the state error using Eq. (6.32):

$$\mathbf{x}(n + 1) = \mathbf{\Phi}\mathbf{x}(n) - \mathbf{\Gamma}\mathbf{K}(\mathbf{x}(n) - \tilde{\mathbf{x}}(n)). \tag{6.51}$$

Combining this with the estimator error equation (6.35),[16] we obtain two coupled equations which describe the behavior of the complete system:

$$\begin{bmatrix} \tilde{\mathbf{x}}(n + 1) \\ \mathbf{x}(n + 1) \end{bmatrix} = \begin{bmatrix} \mathbf{\Phi} - \mathbf{LH} & 0 \\ \mathbf{\Gamma K} & \mathbf{\Phi} - \mathbf{\Gamma K} \end{bmatrix} \begin{bmatrix} \tilde{\mathbf{x}}(n) \\ \mathbf{x}(n) \end{bmatrix}. \tag{6.52}$$

The characteristic equation is

$$\begin{vmatrix} z\mathbf{I} - \mathbf{\Phi} + \mathbf{LH} & 0 \\ \mathbf{\Gamma K} & z\mathbf{I} - \mathbf{\Phi} + \mathbf{\Gamma K} \end{vmatrix} = 0, \tag{6.53}$$

which, because of the zero matrix in the upper right, can be written as

$$|z\mathbf{I} - \mathbf{\Phi} + \mathbf{LH}|\, |z\mathbf{I} - \mathbf{\Phi} + \mathbf{\Gamma K}| = \alpha_e(z)\alpha_c(z) = 0. \tag{6.54}$$

In other words, the characteristic-equation roots of the combined system consist of the sum of the estimator roots and the control roots which are unchanged from those obtained assuming actual state feedback. The fact that the combined control-estimator system has the same poles as those of the control alone and the estimator alone is a special case of a separation principle by which control and estimation can be designed separately yet used together.

To compare this method of design to the methods discussed in Chapter 5, we note from Fig. 6.8 that the portion within the dashed line corresponds to a classical compensation. The difference equation for this controller or "state-space designed compensator" is obtained by including the control feedback (since it is part of the controller) in the estimator equation, yielding

$$\hat{\mathbf{x}}(n + 1) = [\mathbf{\Phi} - \mathbf{\Gamma K} - \mathbf{LH}]\hat{\mathbf{x}}(n) + \mathbf{L}y(n),$$
$$u(n) = -\mathbf{K}\hat{\mathbf{x}}(n). \tag{6.55}$$

The poles of the controller alone are obtained from

$$|z\mathbf{I} - \mathbf{\Phi} + \mathbf{\Gamma K} + \mathbf{LH}| = 0 \tag{6.56}$$

and need not be determined during a state-space design effort.

In the previous sections, we developed techniques to compute $\mathbf{K}$ and $\mathbf{L}$ (which completely define the controller), given the desired locations of the roots of the characteristic equations of the control and the estimator. We now know that these desired root locations will be the closed-loop system poles. The same meter sticks which applied to the classical design and were discussed in Section 5.2 also apply to picking these roots. In practice, it is convenient to pick the control roots to satisfy the performance specifications and actuator limitations, and then to pick the estimator roots somewhat faster (approximately by a factor of 4)

---

[16] We show only the predictive estimator case. The other estimators give very similar results.

so that the total response is dominated by the response due to the slower control poles. It does not cost anything in terms of actuator hardware to increase the estimator gains (and hence speed of response) since they only appear in the computer. A fast estimator root merely implies that it converges to the correct values quickly. The upper limit to estimator speed of response is based on noise rejection characteristics and sensitivity to modeling errors. The limit can be determined by simulation, or a balance between noise and speed of response can be obtained by the use of optimal estimation techniques [see Bryson and Ho (1968)].[17]

As an example of the complete design, we will add a state estimator to the satellite attitude-control considered in Section 6.3. The system equations of motion are described by (6.3) in continuous time and by (6.13) in discrete time with

$$\mathbf{\Phi} = \begin{bmatrix} 1 & T \\ 0 & 1 \end{bmatrix}, \qquad \mathbf{\Gamma} = \begin{bmatrix} T^2/2 \\ T \end{bmatrix}, \qquad \mathbf{H} = \begin{bmatrix} 1 & 0 \end{bmatrix}. \tag{6.57}$$

We will also consider an impulse disturbance torque for which

$$\mathbf{\Gamma}_1 = \begin{bmatrix} 0 \\ 1 \end{bmatrix}. \tag{6.58}$$

In Section 6.3 we selected a sampling rate of 0.1 sec and control poles at $z = 0.837\angle 18°$ which resulted in the control law

$$\mathbf{K} = \begin{bmatrix} 10 & 3.5 \end{bmatrix}. \tag{6.59}$$

First we will design a reduced-order estimator for this system. The partitioned matrices look like

$$\begin{bmatrix} \Phi_{aa} & \Phi_{ab} \\ \hline \Phi_{ba} & \Phi_{bb} \end{bmatrix} = \begin{bmatrix} 1 & T \\ \hline 0 & 1 \end{bmatrix} \quad \begin{bmatrix} \Gamma_a \\ \Gamma_b \end{bmatrix} = \begin{bmatrix} T^2/2 \\ T \end{bmatrix} = \begin{bmatrix} 0.005 \\ 0.1 \end{bmatrix},$$

$$\begin{bmatrix} x_1 \\ x_2 \end{bmatrix} = \begin{bmatrix} \text{the measured position state } y \\ \text{the velocity to be estimated} \end{bmatrix},$$

where Φ_{aa}, etc., are all scalars. Therefore L is a scalar also, and there is only one estimator root to pick, the root corresponding to the speed at which the estimate of scalar velocity converges. From (6.48) we pick L from

$$z - 1 + LT = 0.$$

For the estimator to be about four times faster than the control, we pick z at $+0.5$ (≈ 0.8374); therefore $LT - 1 = -0.5$ and $L = 5$. The estimator equation, (6.46), is

$$\hat{x}_b(n + 1) = \hat{x}_b(n) + (0.1)u(n) + 5.0 \left[y(n + 1) - y(n) - \frac{0.01}{2} u(n) - (0.1)\hat{x}_b(n) \right].$$

---

[17] See also Chapter 9.

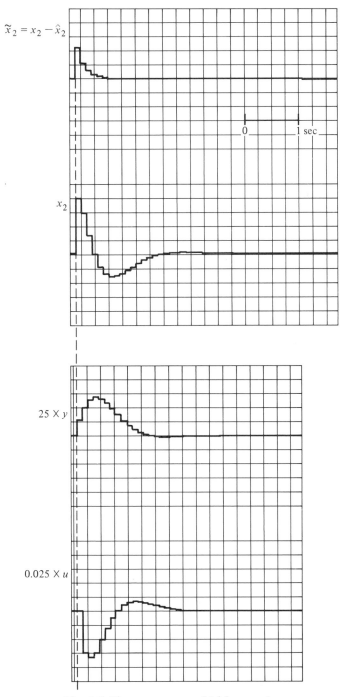

Fig. 6.9 Time responses of $1/s^2$ example.

Substituting in $u(n)$ from the control law,

$$u(n) = -10y(n) - 3.5\hat{x}_b(n), \qquad (6.60)$$

we obtain a simplified estimator

$$\hat{x}_b(n + 1) = 0.238\hat{x}_b(n) + 5.0y(n + 1) - 5.75y(n). \qquad (6.61)$$

These last two difference equations complete the design and can be used to control the plant to the desired specifications. Figure 6.9 shows the time responses of the controlled-system states and the $\hat{x}_b$-estimate to a large initial velocity error as would be the result of an input from the disturbance, and demonstrates the different root characteristics.

To relate this controller design to a classical design, we compute the z-transform of (6.61) and (6.60), obtaining

$$\frac{U(z)}{Y(z)} = -27.5\frac{z - 0.819}{z - 0.238}. \qquad (6.62)$$

This compensation now looks very much like the classic lead compensation that would be used for a $1/s^2$ system. A sketch of the root locus is given in Fig. 6.10 with the root locations corresponding to a gain of 27.5 indicated by the $\triangle$'s.

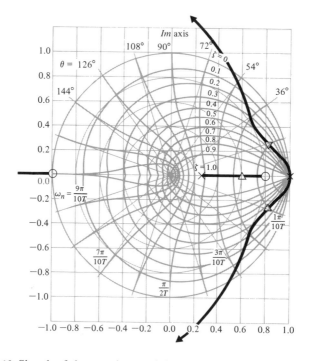

Fig. 6.10 Sketch of the root locus of the satellite attitude-control system.

6.6 SERVODESIGN: INTRODUCTION OF
THE REFERENCE INPUT BY FEEDFORWARD CONTROL

The controller obtained by combining the control law studied in Section 6.3 with any of the estimators of Section 6.4 is essentially a regulator design. We design the characteristic equations of the control and the estimator to give satisfactory natural mode transients to disturbances such as $w(k)$, but no mention is made of a reference input or of the considerations necessary to obtain good transient response to command changes. To study these matters we will consider first the full-order predictive estimator and introduce later the modifications necessary or desirable to apply the ideas to the current and reduced-order estimators.

Let us begin by repeating the plant, control, and estimator equations

$$\left.\begin{array}{l} \mathbf{x}(k + 1) = \boldsymbol{\Phi}\mathbf{x}(k) + \boldsymbol{\Gamma}u(k) \\ y(k) = \mathbf{H}\mathbf{x}(k) \end{array}\right\} \text{ plant;} \tag{6.63}$$

$$\left.\begin{array}{l} \hat{\mathbf{x}}(k + 1) = (\boldsymbol{\Phi} - \boldsymbol{\Gamma}\mathbf{K} - \mathbf{L}\mathbf{H})\hat{\mathbf{x}}(k) + \mathbf{L}y(k) \\ u(k) = -\mathbf{K}\hat{\mathbf{x}}(k) \end{array}\right\} \text{ controller.} \tag{6.64}$$

If we have a reference input, $r(k)$, the most general linear manner to introduce r to these equations is to add a term $\mathbf{M}r$ to $\hat{\mathbf{x}}(k + 1)$ and a term Nr to the control equation, u. Here $\mathbf{M}$ is an $n \times 1$ matrix, and N is a scalar. The controller, with these additions, becomes

$$\begin{array}{l} \hat{\mathbf{x}}(k + 1) = (\boldsymbol{\Phi} - \boldsymbol{\Gamma}\mathbf{K} - \mathbf{L}\mathbf{H})\hat{\mathbf{x}}(k) + \mathbf{L}y(k) + \mathbf{M}r(k), \\ u(k) = -\mathbf{K}\hat{\mathbf{x}}(k) + Nr(k). \end{array} \tag{6.65}$$

Since $r(k)$ is an external signal, introduction of these terms will have no influence on the characteristic equation of the combined control-estimator plant system. In transfer function terms, the selection of $\mathbf{M}$ and N will affect only the zeros of the transfer function from r to y. Three reasonable choices present themselves. We will list them and then discuss each in turn.

1. Select $\mathbf{M}$ and N so that the state estimator error is independent of r.
2. Select $\mathbf{M}$ and N so that only the output error, $\epsilon = (r - y)$, is used in the control.
3. Select $\mathbf{M}$ and N to give the designer maximum flexibility in setting the dynamic and static error responses of the system.

From the point of view of the estimator performance, the first of these methods seems highly reasonable. If $\hat{\mathbf{x}}$ is to generate a good approximation to $\mathbf{x}$, then surely $\tilde{\mathbf{x}}$ should be free of external excitation. Computation of the necessary form for $\mathbf{M}$ and N is readily found. The estimator-error equation is given by (6.63) and (6.65), with the plant output substituted into the estimator equations

① State estimate error
indep of r

and the control law substituted into the plant equation

$$
\begin{aligned}
\mathbf{x}(k + 1) - \hat{\mathbf{x}}(k + 1) &= \mathbf{\Phi x} + \mathbf{\Gamma}(-\mathbf{K}\hat{\mathbf{x}} + Nr) \\
&\quad - (\mathbf{\Phi} - \mathbf{\Gamma K} - \mathbf{LH})\hat{\mathbf{x}} - \mathbf{LHx} - \mathbf{M}r, \\
\tilde{\mathbf{x}}(k + 1) &= (\mathbf{\Phi} - \mathbf{LH})\tilde{\mathbf{x}}(k) + \mathbf{\Gamma}Nr - \mathbf{M}r.
\end{aligned} \tag{6.66}
$$

The sufficient condition that r not appear in (6.66) is that

$$
\mathbf{M} = \mathbf{\Gamma}N. \tag{6.67}
$$

Since $\mathbf{\Gamma}$ is an $n \times 1$ matrix and $\mathbf{M}$ is an $n \times 1$ matrix, (6.67) fixes $\mathbf{M}$ to within the constant N. We will return to the selection of N, the gain factor on the reference input, after discussing the alternative methods of selecting $\mathbf{M}$.[18]

The second approach mentioned above was that only the output error would be used. This solution is sometimes forced on the control designer because the sensor only measures the error. For example, many thermostats have an output which is the difference between the temperature to be controlled and the reference or set-point temperature. No absolute indication of the reference temperature is available to the controller. Likewise, some radar tracking systems have a reading which is proportional to the pointing error, and this signal alone must be used for control. In these cases we must select $\mathbf{M}$ and N so that r appears in (6.65) only in terms of the error. The requirement is easy to meet:

$$
N = 0, \qquad \mathbf{M} = -\mathbf{L}. \tag{6.68}
$$

② only output error $E = (r - y)$

Then the estimator equation is

$$
\hat{\mathbf{x}}(k + 1) = (\mathbf{\Phi} - \mathbf{\Gamma K} - \mathbf{LH})\hat{\mathbf{x}} + \mathbf{L}(y - r).
$$

Our final suggestion is that $\mathbf{M}$ and N be selected so that the designer has maximum flexibility in satisfying response constraints. This will be accomplished if $\mathbf{M}$ and N can be selected to give arbitrary locations to the zeros of the transfer function from r to y. First, we will discuss briefly the relation between the matrices of a state variable design and the zeros of the corresponding closed-loop transfer function. We consider first the situation when control is computed from state feedback (no estimator is involved). Suppose we wish to use state feedback but include a reference input in the control by the equation $u = -\mathbf{K}\mathbf{x} + Nr$. Then the equations of motion of the closed-loop system are

$$
\begin{aligned}
\mathbf{x}(k + 1) &= \mathbf{\Phi x}(k) + \mathbf{\Gamma}(-\mathbf{K}x + Nr), \\
y &= \mathbf{Hx},
\end{aligned}
$$

---

[18] Note that with this choice of M we can write the controller equations as

$$
\begin{aligned}
\hat{\mathbf{x}}_{k+1} &= (\mathbf{\Phi} - \mathbf{LH})\hat{\mathbf{x}}_k + \mathbf{\Gamma}u_k + \mathbf{L}y_k, \\
u_k &= -\mathbf{K}\hat{\mathbf{x}}_k + Nr.
\end{aligned}
$$

In this form, if the true control is subject to saturation, the same control limits apply to $\hat{x}$ and the nonlinearity does not influence the $\tilde{x}$-equation.

which may be written

$$\mathbf{x}(k + 1) = (\mathbf{\Phi} - \mathbf{\Gamma K})\mathbf{x} + \mathbf{\Gamma}Nr,$$
$$y = \mathbf{Hx}. \qquad (6.69)$$

The condition for a zero of this system is given by application of Eq. (6.19) to the matrices of (6.69). The result is

$$\begin{bmatrix} z\mathbf{I} - \mathbf{\Phi} + \mathbf{\Gamma K} & -\mathbf{\Gamma} \\ \mathbf{H} & 0 \end{bmatrix} \begin{bmatrix} \mathbf{x}_0 \\ Nr_0 \end{bmatrix} = \mathbf{0}.$$

The values of z for which these equations have a solution will not be changed if we change variables, so that the last column multiplied by $+\mathbf{K}$ is added to the first column. The equations become

$$\begin{bmatrix} z\mathbf{I} - \mathbf{\Phi} & -\mathbf{\Gamma} \\ \mathbf{H} & 0 \end{bmatrix} \begin{bmatrix} \mathbf{x}_0 \\ Nr_0 - \mathbf{Kx}_0 \end{bmatrix} = \mathbf{0}. \qquad (6.70)$$

But now the coefficient matrix in (6.70) is independent of the feedback gains $\mathbf{K}$. We have proved that the zeros of the system are *not* changed by state feedback of the type given in (6.69). The plant zeros are invariant under state variable feedback.

Now we consider the controller of (6.65). If we have a zero in transmission from r to u, then of necessity we have a zero in transmission from r to y unless there is a coinciding point of infinite transmission, which is to say a pole at the same value of z. It is therefore sufficient to treat the controller alone. The equations for a zero from r to u (we let $y = 0$ since we only care about the effects of r) in (6.65) are, again using (6.19),

$$\begin{bmatrix} z\mathbf{I} - \mathbf{\Phi} + \mathbf{\Gamma K} + \mathbf{LH} & -\dfrac{\mathbf{M}}{N} \\ -\mathbf{K} & 1 \end{bmatrix} \begin{bmatrix} \mathbf{x}_0 \\ Nr_0 \end{bmatrix} = [\mathbf{0}]. \qquad (6.71)$$

Since the coefficient matrix in (6.71) is square, the condition for a nontrivial solution is that the determinant of this matrix be zero. Thus we have

$$\det \begin{bmatrix} z\mathbf{I} - \mathbf{\Phi} + \mathbf{\Gamma K} + \mathbf{LH} & -\dfrac{\mathbf{M}}{N} \\ -\mathbf{K} & 1 \end{bmatrix} = 0.$$

If we add K times the last column to the first column, we find

$$\det \begin{bmatrix} z\mathbf{I} - \mathbf{\Phi} + \mathbf{\Gamma K} + \mathbf{LH} - \dfrac{\mathbf{M}}{N}\mathbf{K} & -\dfrac{\mathbf{M}}{N} \\ 0 & 1 \end{bmatrix} = 0,$$

$$\det \begin{bmatrix} z\mathbf{I} - \mathbf{\Phi} + \mathbf{\Gamma K} + \mathbf{LH} - \dfrac{\mathbf{M}}{N}\mathbf{K} \end{bmatrix} = \gamma(z) = 0. \qquad (6.72)$$

Now (6.72) is exactly like the equation we found in the estimator design for selection of **L** for a specified estimator-error characteristic equation, with the exception that here we are selecting **M**/N for a specified polynomial of zeros, $\gamma(z)$, in the reference input to the control transfer function. Thus we do have substantial freedom to influence the dynamic response by the choice of **M**—we can select an arbitrary nth-order polynomial to be the numerator of the transfer function from r to u and thus from r to y. If the roots of this polynomial are not canceled by poles of the system, then they will be zeros of transmission of the system.

It is interesting now to return to examine the first rule for the selection of **M** and N in light of the definition of zeros of transmission. Under the first rule, we let **M** = $\mathbf{\Gamma}N$ as given in (6.67). If we substitute this value in (6.72), we find

$$\det [z\mathbf{I} - \mathbf{\Phi} + \mathbf{LH}] = 0. \tag{6.73}$$

But (6.73) is exactly the equation from which **L** was selected to make the characteristic polynomial of the estimator error equal $\alpha_e(z)$. Thus we see that the selection of **M** and N to remove r from the error equation is the same thing as a selection of input feed-forward gains which place zeros on top of the poles of the estimator in the overall transfer function.

We can also consider the second rule from the point of view of zero locations. If we take **M** = $-L$ and $N = 0$, the zeros are such that a solution exists for the matrix equation

$$\begin{bmatrix} z\mathbf{I} - \mathbf{\Phi} + \mathbf{\Gamma K} + \mathbf{LH} & \mathbf{L} \\ -\mathbf{K} & 0 \end{bmatrix} \begin{bmatrix} \mathbf{x}_0 \\ r_0 \end{bmatrix} = [\mathbf{0}]. \tag{6.74}$$

If we multiply the last column on the right by **H** and subtract from the first n columns, we remove the **LH** term in the upper left corner. If we multiply the last row by $\mathbf{\Gamma}$ on the left and add these new rows to the top n rows, we eliminate the $\mathbf{\Gamma K}$ term. Thus the condition for zero is

$$\det \begin{bmatrix} z\mathbf{I} - \mathbf{\Phi} & \mathbf{L} \\ -\mathbf{K} & 0 \end{bmatrix} = 0. \tag{6.75}$$

If we multiply the first rows by $\mathbf{K}(z\mathbf{I} - \mathbf{\Phi})^{-1}$ and add to the last row, we obtain the condition

$$\det \begin{bmatrix} z\mathbf{I} - \mathbf{\Phi} & \mathbf{L} \\ 0 & \mathbf{K}(z\mathbf{I} - \mathbf{\Phi})^{-1}\mathbf{L} \end{bmatrix} = 0, \tag{6.76}$$

which is

$$\det (z\mathbf{I} - \mathbf{\Phi})\mathbf{K}(z\mathbf{I} - \mathbf{\Phi})^{-1}\mathbf{L} = 0. \tag{6.77}$$

Thus the zeros are those which result if we replace the plant input matrix by **L** and its output matrix by **K**. If we wish to use error-only control, we must accept the zeros that come with **K** and **L**. In the context of the example at the end of Section 6.5, this is the zero at $z = 0.819$ given in Eq. (6.62). We shall see shortly how the arbitrary zeros from **M** selected as in (6.72) appear in the controller realization.

We may summarize our findings with respect to the introduction of the reference input as follows. When the reference signal is included in the controller equations, we obtain a transfer function of the form

$$\frac{Y(z)}{R(z)} = \frac{\gamma(z)b(z)}{\alpha_e(z)\alpha_c(z)}. \tag{6.78}$$

The polynomial $\alpha_c(z)$ is selected by the designer and results in a control gain $\mathbf{K}$ so that $\det(z\mathbf{I} - \boldsymbol{\Phi} + \boldsymbol{\Gamma}\mathbf{K}) = \alpha_c(z)$. The polynomial α_e is selected by the designer and results in an estimator gain $\mathbf{L}$, so that $\det[z\mathbf{I} - \boldsymbol{\Phi} + \mathbf{LH}] = \alpha_e(z)$. The polynomial $\gamma(z)$ may be selected by the designer so that $\gamma(z) = \alpha_e(z)$, from which $\mathbf{M}/N$ is given by (6.67), or else $\gamma(z)$ may be accepted as given by (6.75), so that error-only control is used, or finally $\gamma(z)$ may be given arbitrary coefficients by selecting $\mathbf{M}/N$ from (6.72).

We now turn to the selection of the gain, N, in case we select either alternative 1 or 3 for $\mathbf{M}$ (for error control, $N = 0$). For these cases, the zeros of the polynomial $\gamma(z)$ are determined by the ratio $\mathbf{M}/N$. How then should we select N? One selection which seems reasonable is to pick N such that when r and y are both unchanging, the (dc) gain from r to u is the *negative* of the dc gain from y to u.[19] The consequences of this choice are that our controller can be structured as error control plus generalized derivative feedback, and if the system is capable of Type I behavior, that capability will be realized. We need first to derive the formula for N and then to show that the claimed properties are realized. To compute N, then, we need to compute the dc gains from r to u and from y to u for the controller of (6.65). We first set $\hat{\mathbf{x}}(k + 1) = \hat{\mathbf{x}}(k) = \hat{\mathbf{x}}_0$ and $r = 0$ and $y = y_0$. The equations become, with $\boldsymbol{\Phi} - \boldsymbol{\Gamma}\mathbf{K} - \mathbf{LH} = \boldsymbol{\theta}$,

$$\mathbf{x}_0 = \boldsymbol{\theta}\mathbf{x}_0 + \mathbf{L}y_0, \qquad u_0 = -\mathbf{K}\mathbf{x}_0.$$

Eliminating $\mathbf{x}_0$, we get

$$\mathbf{x}_0 = (\mathbf{I} - \boldsymbol{\theta})^{-1}\mathbf{L}y_0, \qquad u_0 = -\mathbf{K}(\mathbf{I} - \boldsymbol{\theta})^{-1}\mathbf{L}y_0. \tag{6.79}$$

To get the dc controller gain from r, we repeat this process but with $y = 0$ and $r = r_0$. Again,

$$\mathbf{x}_0 = \boldsymbol{\theta}\mathbf{x}_0 + \mathbf{M}r_0, \qquad u_0 = -\mathbf{K}\mathbf{x}_0 + Nr_0,$$

and, as above,

$$\mathbf{x}_0 = (\mathbf{I} - \boldsymbol{\theta})^{-1}\mathbf{M}r_0, \qquad u_0 = (-\mathbf{K}(\mathbf{I} - \boldsymbol{\theta})^{-1}\mathbf{M} + N)r_0. \tag{6.80}$$

---

[19] A reasonable alternative is to select N so that the overall closed-loop gain at $z = 1$ is 1. The two choices are the same if the plant has a pole at $z = 1$, and the method proposed above requires inversion of matrices of order n, whereas the overall method would require inversion of a matrix of order $2n$; the difference may be significant.

If we equate the negative of the gain in (6.79) to that in (6.80), we can solve for N:

$$N - \mathbf{K}(\mathbf{I} - \boldsymbol{\theta})^{-1}\mathbf{M} = \mathbf{K}(\mathbf{I} - \boldsymbol{\theta})^{-1}\mathbf{L},$$

$$N\left(1 - \mathbf{K}(\mathbf{I} - \boldsymbol{\theta})^{-1}\frac{\mathbf{M}}{N}\right) = \mathbf{K}(\mathbf{I} - \boldsymbol{\theta})^{-1}\mathbf{L},$$

$$N = \frac{\mathbf{K}(\mathbf{I} - \boldsymbol{\theta})^{-1}\mathbf{L}}{1 - \mathbf{K}(\mathbf{I} - \boldsymbol{\theta})^{-1}\overline{\mathbf{M}}}. \qquad (6.81)$$

In (6.81) we have written $\overline{\mathbf{M}}$ for $\mathbf{M}/N$ to indicate that it is $\overline{\mathbf{M}}$ which is the outcome of selection of zero locations via either (6.67) or (6.72).

Our equations thus far have been developed in terms of the predictor estimator, but they can be generalized to include the current estimator or the reduced-order estimator quite readily. If we assume a general form for the equations of the controller, then each version of the estimator becomes a special case. For example, general controller equations are

$$\mathbf{x}_c(k + 1) = \mathbf{A}\mathbf{x}_c(k) + \mathbf{B}y(k) + \mathbf{M}r(k),$$
$$u(k) = \mathbf{C}\mathbf{x}_c(k) + Dy(k) + Nr(k). \qquad (6.82)$$

In the case of the predictor estimator, the specific values for the description matrices in (6.82) are found by comparison with (6.65) to be

$$\mathbf{A} = \boldsymbol{\Phi} - \boldsymbol{\Gamma}\mathbf{K} - \mathbf{L}\mathbf{H}, \qquad \mathbf{B} = \mathbf{L}, \qquad \mathbf{C} = -\mathbf{K}, \qquad D = 0, \qquad (6.83)$$

and the state is the estimate so that $\mathbf{x}_c = \hat{\mathbf{x}}$. For the current estimator it is necessary to return to the basic equations (6.38) and manipulate them into the form of (6.82). Since this process involves at least one step which may not be obvious, it will be instructive to go through the steps in some detail. We repeat (6.38) for ease of reference:

$$\bar{\mathbf{x}}(k + 1) = \boldsymbol{\Phi}\hat{\mathbf{x}}(k) + \boldsymbol{\Gamma}u(k),$$
$$\hat{\mathbf{x}}(k + 1) = \bar{\mathbf{x}}(k + 1) + \mathbf{L}(y(k + 1) - \mathbf{H}\mathbf{x}(k + 1)),$$
$$u(k) = -\mathbf{K}\hat{\mathbf{x}}(k). \qquad (6.38)$$

We could write the final equations in terms of either $\bar{\mathbf{x}}$ or $\hat{\mathbf{x}}$. Since $\hat{\mathbf{x}}$ is the best estimate, we will eliminate $\bar{\mathbf{x}}$ in (6.38). First we write the second equation as

$$\hat{\mathbf{x}}(k + 1) = (\mathbf{I} - \mathbf{L}\mathbf{H})\bar{\mathbf{x}}(k + 1) + \mathbf{L}y(k + 1),$$

and then we substitute the first equation into this form to obtain

$$\hat{\mathbf{x}}(k + 1) = (\mathbf{I} - \mathbf{L}\mathbf{H})\{\boldsymbol{\Phi}\hat{\mathbf{x}}(k) + \boldsymbol{\Gamma}\mathbf{u}(k)\} + \mathbf{L}y(k + 1).$$

Now, if $u = -\mathbf{K}\hat{\mathbf{x}}$, then

$$\hat{\mathbf{x}}(k + 1) = (\mathbf{I} - \mathbf{L}\mathbf{H})(\boldsymbol{\Phi} - \boldsymbol{\Gamma}\mathbf{K})\hat{\mathbf{x}}(k) + \mathbf{L}y(k + 1)$$
$$u(k) = -\mathbf{K}\hat{\mathbf{x}}.$$

We find $y(k + 1)$ on the right-hand side, and this term must be eliminated if we are to have the form of (6.82). The key is to redefine the state by collecting *all* terms occurring at $(k + 1)$th time on the left of the equation, and use the resulting collection as the new state, $\mathbf{x}_c(k + 1)$. In this case, we let

$$\mathbf{x}_c(k + 1) = \hat{\mathbf{x}}(k + 1) - \mathbf{L}y(k + 1)$$

or

$$\hat{\mathbf{x}}(k + 1) = \mathbf{x}_c(k + 1) + \mathbf{L}y(k + 1).$$

Now we can write

$$\mathbf{x}_c(k + 1) = \{(\mathbf{I} - \mathbf{LH})(\mathbf{\Phi} - \mathbf{\Gamma K})\{\mathbf{x}_c(k) + \mathbf{L}y(k)\} + \mathbf{L}y(k + 1)\} - \mathbf{L}y(k + 1),$$
$$\mathbf{x}_c(k + 1) = (\mathbf{I} - \mathbf{LH})(\mathbf{\Phi} - \mathbf{\Gamma K})\mathbf{x}_c(k) + (\mathbf{I} - \mathbf{LH})(\mathbf{\Phi} - \mathbf{\Gamma K})\mathbf{L}y(k),$$
$$u(k) = -\mathbf{K}\{\mathbf{x}_c(k) + \mathbf{L}y(k)\}.$$

As a final step we introduce the reference input terms

$$\mathbf{x}_c(k + 1) = (\mathbf{I} - \mathbf{LH})(\mathbf{\Phi} - \mathbf{\Gamma K})\mathbf{x}_c(k) + (\mathbf{I} - \mathbf{LH})(\mathbf{\Phi} - \mathbf{\Gamma K})\mathbf{L}y(k) + \mathbf{M}r(k),$$
$$u(k) = -\mathbf{K}\mathbf{x}_c(k) - \mathbf{KL}y(k) + Nr(k). \tag{6.84}$$

Comparing (6.84) to (6.82) the controller matrices are

$$\mathbf{A} = (\mathbf{I} - \mathbf{LH})(\mathbf{\Phi} - \mathbf{\Gamma K}), \qquad \mathbf{B} = (\mathbf{I} - \mathbf{LH})(\mathbf{\Phi} - \mathbf{\Gamma K})\mathbf{L},$$
$$\mathbf{C} = -\mathbf{K}, \qquad D = -\mathbf{KL}. \tag{6.85}$$

The state estimate is

$$\hat{\mathbf{x}}(k) = \mathbf{x}_c(k) + \mathbf{L}y(k). \qquad P\ 149$$

For the reduced-order estimator, we must return to (6.46) and collect all the terms. The algebra is perhaps tedious but very direct and will not be repeated here. The results are

$$\mathbf{A} = \mathbf{\Phi}_{bb} - \mathbf{\Gamma}_b\mathbf{K}_b + \mathbf{L}\mathbf{\Gamma}_a\mathbf{K}_b - \mathbf{L}\mathbf{\Phi}_{ab},$$
$$\mathbf{B} = \mathbf{\Phi}_{ba} - \mathbf{\Gamma}_b\mathbf{K}_a - \mathbf{L}\mathbf{\Phi}_{aa} + \mathbf{L}\mathbf{\Gamma}_a\mathbf{K}_a + \mathbf{AL},$$
$$\mathbf{C} = -\mathbf{K}_b, \qquad D = -\mathbf{K}_b\mathbf{L} - \mathbf{K}_a, \tag{6.86}$$
$$\hat{\mathbf{x}}_b = \mathbf{x}_c + \mathbf{L}y.$$

In every case, if $N \neq 0$, we define $\overline{\mathbf{M}} = \mathbf{M}/N$, and the zeros of the controller are given by the roots of $\gamma(z)$ defined as

$$\det[z\mathbf{I} - \mathbf{A} + \overline{\mathbf{M}}\mathbf{C}] = \gamma(z). \tag{6.87}$$

The gain for the reference input is N which, if $\overline{\mathbf{M}}$ is selected from (6.87), can be computed from the generalization of (6.81) for the general matrices given in (6.82), namely

$$N = \overline{\left[\frac{D + \mathbf{C}(\mathbf{I} - \mathbf{A})^{-1}\mathbf{B}}{1 + \mathbf{C}(\mathbf{I} - \mathbf{A})^{-1}\overline{\mathbf{M}}}\right]}. \tag{6.88}$$

The two special cases for zero selection are to restrict control to output-error information only and to eliminate r from the state error equation.

To allow only output error, we take [referring to (6.82)],

$$\mathbf{M} = -\mathbf{B}, \qquad N = -D. \tag{6.89}$$

To eliminate r from the $\bar{\mathbf{x}}$-equations, we must force $\gamma(z)$ in (6.87) to equal $\alpha_e(z)$. For this we have

$$\text{Predictor estimator:} \quad \overline{\mathbf{M}} = \mathbf{\Gamma}. \tag{6.90a}$$

$$\text{Current estimator:}[20] \quad \overline{\mathbf{M}} = (\mathbf{I} - \mathbf{LH})\mathbf{\Gamma}. \tag{6.90b}$$

$$\text{Reduced-order estimator:} \quad \overline{\mathbf{M}} = \mathbf{\Gamma}_b - \mathbf{L}\mathbf{\Gamma}_a. \tag{6.90c}$$

We will illustrate the introduction of a reference input by again considering the satellite attitude control for which control and estimator designs have already been provided in Sections 6.3 and 6.4. Since the use of error control is trivial [we would select $\mathbf{M}$ and N by (6.89)], we will here treat selection of system zeros to eliminate the forcing term r from the $\bar{\mathbf{x}}$-equations. First, we need to identify the element values for this example. The plant matrices are

$$\mathbf{\Phi} = \begin{bmatrix} 1 & 0.1 \\ 0 & 1 \end{bmatrix} = \begin{bmatrix} \Phi_{aa} & \Phi_{ab} \\ \Phi_{ba} & \Phi_{bb} \end{bmatrix},$$

$$\mathbf{\Gamma} = \begin{bmatrix} 0.005 \\ 0.1 \end{bmatrix} = \begin{bmatrix} \Gamma_a \\ \Gamma_b \end{bmatrix},$$

$$\mathbf{K} = [10 \quad 3.5] = [K_a \quad K_b],$$

$$L = 5.$$

Substituting these values in (6.86), we compute

$$\begin{aligned} A &= 1 - (0.1)(3.5) + 5(0.005)(3.5) - 5(0.1) \\ &= 0.2375, \\ B &= 0 - (0.1)(10) - 5(1) + 5(0.005)(10) + (0.2375)(5) \\ &= -4.5625, \\ C &= -3.5, \\ D &= -(3.5)(5) - 10 \\ &= -27.5. \end{aligned}$$

Substituting in (6.90c) yields

$$\overline{M} = (0.1) - 5(0.005) = 0.075,$$

---

[20] We have not derived this result. It follows by direct computation of $\bar{\mathbf{x}}$ using (6.84) and also by noting that if we let $u = -\mathbf{K}\hat{\mathbf{x}} + Nr$ in (6.38) *before* substituting the control in the estimator equations, we can be sure the Nr-term will cancel in the $\bar{\mathbf{x}}$-equations, and thus r will be appropriately introduced into the estimator equations. This technique can be extended to let the control variable introduced into the estimator equation be subject to the same saturation limits as are in effect in the plant.

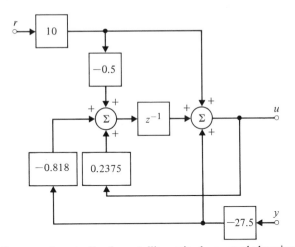

Fig. 6.11 Block diagram of controller for satellite attitude-control showing reference input.

and, substituting in (6.88) gives

$$N = \left(- \frac{-27.5 + (-3.5)(1 - 0.2375)^{-1}(-4.5625)}{1 + (-3.5)(1 - 0.2375)^{-1}(0.075)} \right)$$

$$= 10;$$

therefore

$$M = \overline{M}N = 0.75.$$

The controller transfer functions are

$$\frac{U}{Y} = D + C(zI - A)^{-1}B = -27.5 \frac{z - 0.818}{z - 0.2375} \tag{6.91}$$

and

$$\frac{U}{R} = N + C(zI - A)^{-1}M = 10 \frac{z - 0.5}{z - 0.2375}. \tag{6.92}$$

A block diagram of these transfer functions which shows clearly the feedforward structure with respect to the reference input appears in Fig. 6.11. An alternative structure which shows more clearly the consequences of our choice of dc gain (N) is found by writing the transfer functions (6.91) and (6.92) in terms of $z - 1$ and eliminating the reference input in favor of the error, $e = r - y$. Thus if, in terms of transfer functions,

$$U = H_1 R + H_2 Y,$$

then

$$U = H_1 E + (H_2 + H_1) Y,$$

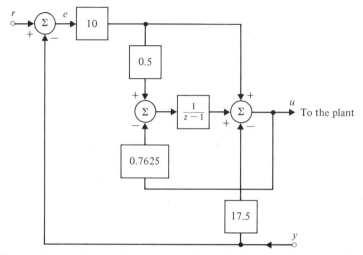

Fig. 6.12 Block diagram of controller showing derivative feedback and error feedforward.

and thus

$$H_2 + H_1 = 10\,\frac{z - 0.5}{z - 0.2375} - 27.5\,\frac{z - 0.818}{z - 0.2375}$$

$$= -17.5\,\frac{z - 1}{z - 1 + 0.7625}. \tag{6.93}$$

This equation shows that the output, y, is fed back through the system error by the transfer function H_1 and, in addition, that there is "derivative" feedback via (6.93). The structure is illustrated in Fig. 6.12.

6.7[21] CONTROLLABILITY AND OBSERVABILITY

Controllability and observability are properties which describe structural features of a dynamic system. These concepts were explicitly identified and studied by Kalman (1960) and Kalman, Ho, and Narendra (1961). We will discuss only a few of the known results for linear constant systems which have one input and one output.

We have encountered these concepts already in connection with design of control laws and estimator gains. We suggested in Section 6.3 that if the matrix $\mathscr{C}$ given by

$$\mathscr{C} = [\boldsymbol{\Gamma} \vdots \boldsymbol{\Phi\Gamma} \vdots \cdots \vdots \boldsymbol{\Phi}^{n-1}\boldsymbol{\Gamma}]$$

[21] This section contains material which may be omitted without loss of continuity.

is nonsingular, then by a transformation of the state we can convert the given description into the control canonical form and construct a control law such that the closed-loop characteristic equation can be given arbitrary (real) coefficients. A proof of this result and the development of Ackermann's formula for computing the gain are given in Appendix B of this chapter. We can begin our discussion of controllability, therefore, by making the definition (the first of three):

I. The system $(\boldsymbol{\Phi}, \boldsymbol{\Gamma})$ is controllable if for every nth-order polynomial $\alpha_c(z)$, there exists a control law $u = -\mathbf{Kx}$ such that the characteristic polynomial of $\boldsymbol{\Phi} - \boldsymbol{\Gamma K}$ is $\alpha_c(z)$.

And, from the results of that same Appendix B, we have the test:

$(\boldsymbol{\Phi}, \boldsymbol{\Gamma})$ are controllable if and only if the rank of $\mathscr{C} = [\boldsymbol{\Gamma} \vdots \boldsymbol{\Phi}\boldsymbol{\Gamma} \vdots \cdots \vdots \boldsymbol{\Phi}^{n-1}\boldsymbol{\Gamma}]$ is n.

The idea of pole placement which is used above to define controllability is essentially a z-transform concept. A time-domain definition is the following:

II. The system $(\boldsymbol{\Phi}, \boldsymbol{\Gamma})$ is controllable if for every $\mathbf{x}_0$ and $\mathbf{x}_1$ there is a finite N and a sequence of controls $u(0), u(1), u(N)$ such that if the system has state x_0 at $k = 0$, it is forced to state $\mathbf{x}_1$ at $k = N$.

In this definition we are considering the direct action of the control u on the state $\mathbf{x}$ and are not concerned at all with modes or characteristic equations. Let us develop a test for controllability for definition II. The system equations are

$$\mathbf{x}(k + 1) = \boldsymbol{\Phi}\mathbf{x}(k) + \boldsymbol{\Gamma}u(k),$$

and, solving for a few steps, we find that if $\mathbf{x}(0) = \mathbf{x}_0$, then

$$\mathbf{x}(1) = \boldsymbol{\Phi}\mathbf{x}_0 + \boldsymbol{\Gamma}u(0),$$
$$\mathbf{x}(2) = \boldsymbol{\Phi}\mathbf{x}(1) + \boldsymbol{\Gamma}u(1)$$
$$= \boldsymbol{\Phi}^2 x_0 + \boldsymbol{\Phi}\boldsymbol{\Gamma}u(0) + \boldsymbol{\Gamma}u(1),$$

$$\vdots$$

$$\mathbf{x}(N) = \boldsymbol{\Phi}^N\mathbf{x}_0 + \sum_{j=0}^{N-1} \boldsymbol{\Phi}^{N-1-j}\boldsymbol{\Gamma}u(j)$$

$$= \boldsymbol{\Phi}^N x_0 + [\boldsymbol{\Gamma} \vdots \boldsymbol{\Phi}\boldsymbol{\Gamma} \vdots \cdots \vdots \boldsymbol{\Phi}^{N-1}\boldsymbol{\Gamma}] \begin{bmatrix} u(N-1) \\ \vdots \\ \vdots \\ u(0) \end{bmatrix}.$$

If $\mathbf{x}(N)$ is to equal $\mathbf{x}_1$, then we must be able to solve the equations

$$[\boldsymbol{\Gamma} \vdots \boldsymbol{\Phi\Gamma} \vdots \cdots \vdots \boldsymbol{\Phi}^{N-1}\boldsymbol{\Gamma}] \begin{bmatrix} u(N) \\ u(N-1) \\ \cdot \\ \cdot \\ \cdot \\ u(0) \end{bmatrix} = \mathbf{x}_1 - \boldsymbol{\Phi}^N x_0.$$

We have assumed that the number of states and hence the number of rows of the coefficient matrix of these equations is n; the number of columns is N. If N is less than n, we cannot possibly find a solution for every $\mathbf{x}_1$. If, on the other hand, N is greater than n, we will add a column $\boldsymbol{\Phi}^n\boldsymbol{\Gamma}$, and so on. But, by the Cayley Hamilton theorem,[22] $\boldsymbol{\Phi}^n$ is a linear combination of lower powers of $\boldsymbol{\Phi}$, and the new columns add no new rank. Therefore we have a solution, and our system is controllable by definition II if and only if the rank of $\mathscr{C}$ is n, exactly the same condition as we found for pole assignment!

Our final definition is closest to the structural character of controllability.

III. The system $(\boldsymbol{\Phi}, \boldsymbol{\Gamma})$ is controllable if every mode in $\boldsymbol{\Phi}$ is connected to the control input.

Because of the generality of modes, we will treat only the case of systems for which $\boldsymbol{\Phi}$ can be transformed to diagonal form. (The double integrator model for the satellite does *not* qualify.) Suppose we have a diagonal $\boldsymbol{\Phi}_\lambda$ matrix and corresponding input matrix $\boldsymbol{\Gamma}_\lambda$ with elements γ_i. Then the structure is as shown in Fig. 6.13. By the definition, the input must be connected to each mode so that no γ_i is

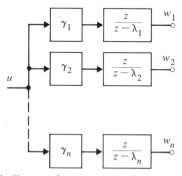

Fig. 6.13 Block diagram for a system with a diagonal $\boldsymbol{\Phi}$-matrix.

---

[22] See Appendix C.

zero. However, this is not enough if the roots λ_i are not distinct. Suppose, for instance, $\lambda_1 = \lambda_2$. Then the equations in the first two states are

$$w_1(k + 1) = \lambda_1 w_1(k) + \gamma_1 u,$$
$$w_2(k + 1) = \lambda_1 w_2(k) + \gamma_2 u.$$

If we now define $\xi = \gamma_2 w_1 - \gamma_1 w_2$, the equation in ξ is

$$\gamma_2 w_1(k + 1) - \gamma_1 w_2(k + 1) = \lambda_1 \gamma_2 w_1(k) - \lambda_1 \gamma_1 w_2(k) + \gamma_1 \gamma_2 u - \gamma_1 \gamma_2 u,$$

which is the same as

$$\xi(k + 1) = \lambda_1 \xi(k).$$

The point is that if any two characteristic roots are equal *in a diagonal Φ_λ system with only one input,* we effectively have a hidden mode which is not connected to the control, and the system is not controllable. Therefore, even in this simple case, we have two conditions for controllability:

i) all characteristic values of Φ_λ are distinct,

ii) no element of Γ_λ is zero.

Now let us consider the controllability matrix of this diagonal system. By direct computation,

$$
\mathcal{C} = \begin{bmatrix}
\gamma_1 & \gamma_1 \lambda_1 & & \gamma_1 \lambda_1^{n-1} \\
\cdot & \gamma_2 \lambda_2 & \cdots & \cdot \\
\cdot & & & \cdot \\
\cdot & & & \cdot \\
\gamma_n & \gamma_n \lambda_n & & \gamma_n \lambda_n^{n-1}
\end{bmatrix}
$$

$$
= \begin{bmatrix}
\gamma_1 & & & 0 \\
 & \gamma_2 & & \\
 & & \cdot & \\
 & & & \cdot \\
0 & & &
\end{bmatrix}
\begin{bmatrix}
1 & \lambda_1 & \lambda_1^2 & & \lambda_1^{n-1} \\
1 & \lambda_2 & & & \\
\cdot & & & & \\
\cdot & & \cdots & & \\
\cdot & & & & \\
1 & \lambda_n & \lambda_n^2 & & \lambda_n^{n-1}
\end{bmatrix}
$$

The controllability matrix is a product of two terms and $\mathcal{C}$ is nonsingular if and only if each factor is nonsingular. The first term has a determinant which is the product of the γ_i, and the second term is nonsingular if and only if the λ_i are distinct! So once again we find that our definition of controllability leads to the same test: the matrix $\mathcal{C}$ must be nonsingular. If $\mathcal{C}$ is nonsingular, then we can assign the system poles by state feedback, we can drive the state to any part of the space in finite time, and we know that every mode is connected to the input.[23]

---

[23] Of course, we showed this latter only for Φ which can be made diagonal. The result is true for general Φ.

As our final remark on the topic of controllability we present a test which is an alternative to testing the rank (or determinant) of $\mathscr{C}$. This is the Rosenbrock-Hautus-Popov (RHP) test [see Rosenbrock (1970), Kailath (1979)]. The system $(\boldsymbol{\Phi}, \boldsymbol{\Gamma})$ is controllable if the system of equations

$$\mathbf{v}'[z\mathbf{I} - \boldsymbol{\Phi} : \boldsymbol{\Gamma}] = \mathbf{0}'$$

has only the trivial solution $\mathbf{v}' = \mathbf{0}'$, or, equivalently,

$$\text{rank}[z\mathbf{I} - \boldsymbol{\Phi} : \boldsymbol{\Gamma}] = n,$$

or there is *no* nonzero $\mathbf{v}'$ such that

$$\text{(i)} \quad \mathbf{v}'\boldsymbol{\Phi} = z\mathbf{v}', \qquad \text{(ii)} \quad \mathbf{v}'\boldsymbol{\Gamma} = 0.$$

This test is equivalent to the rank-of-$\mathscr{C}$ test. It is easy to show that if such a $\mathbf{v}$ exists, then $\mathscr{C}$ is singular. For if a nonzero $\mathbf{v}$ exists such that $\mathbf{v}'\boldsymbol{\Gamma} = 0$ by (ii), then, multiplying (i) by $\boldsymbol{\Gamma}$ on the right, we find

$$\mathbf{v}'\boldsymbol{\Phi}\boldsymbol{\Gamma} = z\mathbf{v}'\boldsymbol{\Gamma} = 0.$$

Then, multiplying by $\boldsymbol{\Phi}\boldsymbol{\Gamma}$, we find

$$\mathbf{v}'\boldsymbol{\Phi}^2\boldsymbol{\Gamma} = z\mathbf{v}'\boldsymbol{\Phi}\boldsymbol{\Gamma} = 0,$$

and so on. Thus we derive $\mathbf{v}'\mathscr{C} = \mathbf{0}'$ has a nontrivial solution, $\mathscr{C}$ is singular, and the system is not controllable. To show that a nontrivial $\mathbf{v}'$ exists if $\mathscr{C}$ is singular, requires a bit more work and is omitted. See Kailath (1979).

We have given two pictures of a noncontrollable system. Either the input is not connected to a dynamic part physically or else two parallel parts have identical characteristic roots. The engineer should be aware of the existence of a third simple situation illustrated in Fig. 6.14. Here the problem is that the mode at $z = \frac{1}{2}$ appears to be connected to the input but is masked by the zero in the preceding member; the result is an uncontrollable system. First we will confirm this allegation by computing the determinant of the controllability matrix. The system

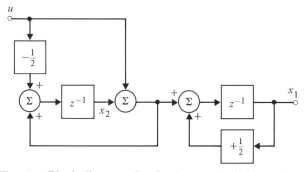

Fig. 6.14 Block diagram of a simple uncontrollable system.

matrices are

$$\Phi = \begin{bmatrix} +\frac{1}{2} & 1 \\ 0 & 1 \end{bmatrix}, \qquad \Gamma = \begin{bmatrix} 1 \\ \frac{1}{2} \end{bmatrix},$$

and

$$\mathcal{C} = [\Gamma \quad \Phi\Gamma] = \begin{bmatrix} 1 & 1 \\ \frac{1}{2} & \frac{1}{2} \end{bmatrix},$$

which is clearly singular. If we compute the transfer function from u to x_1, we find

$$H(z) = \frac{z - \frac{1}{2}}{z - 1} \frac{1}{z - \frac{1}{2}}$$

$$= \frac{1}{z - 1}.$$

Since the natural mode at $z = \frac{1}{2}$ disappears, it is not connected to the input. Finally, if we consider the RHP test,

$$[z\mathbf{I} - \Phi \quad \Gamma] = \begin{bmatrix} z - \frac{1}{2} & -1 & 1 \\ 0 & z - 1 & \frac{1}{2} \end{bmatrix},$$

and let $z = \frac{1}{2}$, then we must test the rank of

$$\begin{bmatrix} 0 & -1 & 1 \\ 0 & -\frac{1}{2} & \frac{1}{2} \end{bmatrix},$$

which is clearly less than two, which means, again, uncontrollable. In conclusion, we have three definitions of controllability: pole assignment, state reachability, and mode coupling to the input. The definitions are equivalent, and the tests for any of these properties are found in the rank of the controllability matrix or in the rank of the input system matrix $[z\mathbf{I} - \Phi \quad \Gamma]$.

We have thus far only discussed controllability. The concept of observability is very parallel to that of controllability and most of the results thus far discussed may be transferred to statements about observability by the simple expedient of substituting the transpose Φ^T for Φ, and $\mathbf{H}^T$ for Γ. The result of these substitutions is a "dual" system. We have already seen an application of duality when we noticed that the conditions for the ability to select an observer gain $\mathbf{L}$ to give the state-error dynamics an arbitrary characteristic equation were that $(\Phi^T, \mathbf{H}^T)$ must be controllable—and we were able to use the same Ackermann formula for estimator gain that we used for control gain. The other properties follow as well. The definitions dual to controllability are:

OI. The system $(\Phi, \mathbf{H})$ is observable if for any nth-order polynomial $\alpha_e(z)$, there exists an estimator gain $\mathbf{L}$ such that the characteristic equation of the state error of the estimator is $\alpha_e(z)$.

OII. The system $(\mathbf{\Phi}, \mathbf{H})$ is observable if for any $\mathbf{x}(0)$, there is a finite N such that $\mathbf{x}(0)$ can be computed from observation of $y(0)$, $y(1)$, . . . , $y(N - 1)$.

OIII. The system $(\mathbf{\Phi}, \mathbf{H})$ is observable if every dynamic mode in $\mathbf{\Phi}$ is connected to the output y via $\mathbf{H}$.

We will consider the development of a test for observability according to definition OII. The system is described by[24]

$$\mathbf{x}(k + 1) = \mathbf{\Phi x}(k), \qquad \mathbf{x}(0) = \mathbf{x}_0$$
$$y(k) = \mathbf{H x}(k),$$

and successive outputs from $k = 0$ are

$$y(0) = \mathbf{H x}_0,$$
$$y(1) = \mathbf{H x}(1) = \mathbf{H \Phi x}_0,$$
$$y(2) = \mathbf{H x}(2) = \mathbf{H \Phi x}(1) = \mathbf{H \Phi}^2 \mathbf{x}_0,$$

$$\cdot$$
$$\cdot$$
$$\cdot$$

$$y(N - 1) = \mathbf{H \Phi}^{N-1} \mathbf{x}_0.$$

In matrix form, these equations are

$$\begin{bmatrix} y(0) \\ \cdot \\ \cdot \\ \cdot \\ y(N - 1) \end{bmatrix} = \begin{bmatrix} \mathbf{H} \\ \mathbf{H \Phi} \\ \cdot \\ \cdot \\ \cdot \\ \mathbf{H \Phi}^{N-1} \end{bmatrix} \mathbf{x}_0.$$

As we saw in the discussion of state controllability, new rows in these equations cannot be independent of previous rows if $N > n$ because of the Cayley-Hamilton theorem. Thus the test for observability is that the matrix

$$\mathcal{O} \triangleq \begin{bmatrix} \mathbf{H} \\ \mathbf{H \Phi} \\ \cdot \\ \cdot \\ \cdot \\ \mathbf{H \Phi}^{n-1} \end{bmatrix}$$

must be nonsingular. If we take the transpose of $\mathcal{O}$ and let $\mathbf{H}^T = \mathbf{\Gamma}$ and $\mathbf{\Phi}^T = \mathbf{\Phi}$, then we find the controllability matrix of $(\mathbf{\Phi}, \mathbf{\Gamma})$, another manifestation of duality.

---

[24] Clearly the input is irrelevant here *if* we assume that all values of $u(k)$ are available in the computation of $x(0)$. If some inputs, such as a disturbance w, are not available, we have a very different problem.

6.8 SUMMARY

In this chapter we have presented methods for the analysis of discrete and sampled data systems in state variable form and methods to design compensators for improved dynamic response using this representation. To obtain the discrete state variable equations from a continuous plant description we solved the matrix differential equations and gave a numerical routine for computing the discrete matrices from the continuous matrices plus sampling time, including the possibility of time delay in the appendix to this chapter.

In Section 6.3 we showed that if the system is controllable, a control gain $\mathbf{K}$ can be found such that the closed-loop characteristic polynomial $\alpha_c(z)$ of $\mathbf{\Phi} - \mathbf{\Gamma K}$ can be given arbitrary real coefficients and thus the closed-loop system poles may be placed anywhere in the z-plane (complex roots must appear in conjugate pairs). We provided Ackermann's formula to compute this gain. Choice of the desired pole locations can be made by comparison to standard forms such as the Butterworth or ITAE polynomials described in Chapter 3. In Section 6.4 we showed how an estimator may be designed such that the characteristic polynomial of the state error-system matrix ($\mathbf{\Phi} - \mathbf{LH}$) may be arbitrarily selected as $\alpha_e(z)$. In Section 6.5 the control law and estimator designs were combined, and we showed that the total system characteristic polynomial is $\alpha_c(z)\alpha_e(z)$, an example of the principle of separation in design.

In Section 6.6 the idea of tracking design was introduced and techniques for the introduction of a reference input were described. We showed that the designer may, by choice of the zeros between the reference input and the control, select to use only error feedback, leave the estimator error unexcited by the reference input, or select the zeros arbitrarily. In the final section, we returned to the concepts of controllability and observability, gave three definitions of each, and showed that the definitions lead to the same test and that the concepts themselves are duals in the sense that a statement about controllability has an equivalent expression in terms of the observability of the dual system.

Appendix A to Chapter 6

State Space Models for Systems with Delay

In Section 6.2 we discussed the calculation of discrete state models from continuous equations of motion. In this appendix we present the formulas for including a time delay in the model and also a time prediction up to one period which corresponds to the modified z-transform.

We begin with a state variable model which includes a delay in control action. The state equations are

$$\dot{\mathbf{x}}(t) = \mathbf{Fx}(t) + \mathbf{G}u(t - \lambda),$$
$$y = \mathbf{Hx}. \tag{6.94}$$

The general solution to (6.94) is given by (6.9),

$$\mathbf{x}(t) = e^{\mathbf{F}(t-t_0)}\mathbf{x}(t_0) + \int_{t_0}^{t} e^{\mathbf{F}(t-\tau)}\mathbf{G}u(\tau - \lambda) \, d\tau.$$

If we let $t_0 = kT$ and $t = kT + T$, then

$$\mathbf{x}(kT + T) = e^{\mathbf{F}T}\mathbf{x}(kT) + \int_{kT}^{kT+T} e^{\mathbf{F}(kT+T-\tau)}\mathbf{G}u(\tau - \lambda) \, d\tau.$$

If we substitute $\eta = kT + t - \tau$ for τ in the integral, we find a modification of (6.11):

$$\mathbf{x}(kT + T) = e^{\mathbf{F}T}\mathbf{x}(kT) + \int_{T}^{0} e^{\mathbf{F}\eta}\mathbf{G}u(kT + T - \lambda - \eta)(-d\eta)$$

$$= e^{\mathbf{F}T}\mathbf{x}(kT) + \int_{0}^{T} e^{\mathbf{F}\eta}\mathbf{G}u(kT + T - \lambda - \eta) \, d\eta.$$

If we now separate the system delay λ into an integral number of sampling periods plus a fraction, we can define an integer ℓ and a number m such that

$$\lambda = \ell T - m, \tag{6.95}$$

and

$$\ell \geq 0,$$
$$0 \leq m < T.$$

With this substitution, we find that the discrete system is described by

$$\mathbf{x}(kT + T) = e^{\mathbf{F}T}\mathbf{x}(kT) + \int_{0}^{T} e^{\mathbf{F}\eta}\mathbf{G}u(kT + T - \ell T + m - \eta) \, d\eta. \tag{6.96}$$

If we sketch a segment of the time axis near $t = kT - \ell T$ (Fig. 6.15), the nature of the integral in (6.96) with respect to the variable η will become clear. The integral runs for η from 0 to T, which corresponds to t from $kT - \ell T + T + m$ backward to $kT - \ell T + m$. Over this period, the control, which we assume is piecewise constant, takes on first the value $u(kT - \ell T + T)$ and then the value $u(kT - \ell T)$.

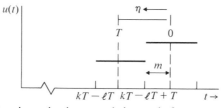

Fig. 6.15 Sketch of a piecewise input and time axis for a system with time delay.

Therefore, we can break the integral in (6.96) into two parts as follows:

$$\mathbf{x}(kT + T) = e^{\mathbf{F}T}\mathbf{x}(kT) + \int_0^m e^{\mathbf{F}\eta}\mathbf{G}\, d\eta\, u(kT - \ell T + T)$$

$$+ \int_m^T e^{\mathbf{F}\eta}\mathbf{G}\, d\eta\, u(kT - \ell T).$$

$$= \mathbf{\Phi}\mathbf{x}(kT) + \mathbf{\Gamma}_1 u(kT - \ell T) + \mathbf{\Gamma}_2 u(kT - \ell T + T). \qquad (6.97)$$

In (6.97) we defined

$$\mathbf{\Phi} = e^{\mathbf{F}T}, \qquad \mathbf{\Gamma}_1 = \int_m^T e^{\mathbf{F}\eta}\mathbf{G}\, d\eta, \qquad \text{and} \qquad \mathbf{\Gamma}_2 = \int_0^m e^{\mathbf{F}\eta}\mathbf{G}\, d\eta. \qquad (6.98)$$

To complete our analysis it is necessary to express (6.97) in standard state-space form. To do this we must consider separately the cases of $\ell = 0$, $\ell = 1$, and $\ell > 1$.

For $\ell = 0$, $\lambda = -m$ according to (6.95), which implies not delay but prediction. Since m is restricted to be less than T, however, the output will not show a sample before $k = 0$ and the discrete system will be causal. The result is that the discrete system computed with $\ell = 0$, $m \neq 0$ will show the response at $t = 0$ which the same system with $\ell = 0$, $m = 0$ would show at $t = m$. In other words, by taking $\ell = 0$ and $m \neq 0$ we pick up the response values *between* the normal sampling instants. In z-transform theory, the transform of the system with $\ell = 0$, $m \neq 0$ is called the *modified-z-transform*.[25] The state-variable form requires that we evaluate the integrals in (6.98). To do so we first convert $\mathbf{\Gamma}_1$ to a form similar to the integral for $\mathbf{\Gamma}_2$. From (6.98) we factor out the constant matrix G to obtain

$$\mathbf{\Gamma}_1 = \int_m^T e^{\mathbf{F}\eta}\, d\eta\, \mathbf{G}.$$

If we set $\sigma = \eta - m$ in this integral,

$$\mathbf{\Gamma}_1 = \int_0^{T-m} e^{\mathbf{F}(m+\sigma)}\, d\sigma\, \mathbf{G}$$

$$= e^{\mathbf{F}m} \int_0^{T-m} e^{\mathbf{F}\sigma}\, d\sigma\, \mathbf{G}. \qquad (6.99)$$

For notational purposes we will define, for any positive nonzero scalar number, a, the two matrices

$$\mathbf{\Phi}(a) = e^{\mathbf{F}a}, \qquad \mathbf{\Psi}(a) = \frac{1}{a}\int_0^a e^{\mathbf{F}\sigma}\, d\sigma. \qquad (6.100)$$

---

[25] See Jury (1964).

In terms of these matrices, we have

$$\begin{aligned}
\boldsymbol{\Gamma}_1 &= \boldsymbol{\Phi}(m)\boldsymbol{\Psi}(T - m)(T - m)\mathbf{G}, \\
\boldsymbol{\Gamma}_2 &= \boldsymbol{\Psi}(m)m\mathbf{G}.
\end{aligned} \tag{6.101}$$

The definition (6.100) is also useful from a computational point of view. If we recall the series definition of the matrix exponential,

$$\boldsymbol{\Phi}(a) = e^{\mathbf{F}a} = \sum_{k=0}^{\infty} \frac{\mathbf{F}^k a^k}{k!},$$

then

$$\begin{aligned}
\boldsymbol{\Psi}(a) &= \frac{1}{a} \int_0^a \sum_{k=0}^{\infty} \frac{\mathbf{F}^k \sigma^k}{k!} \, d\sigma \\
&= \frac{1}{a} \sum_{k=0}^{\infty} \frac{\mathbf{F}^k}{k!} \frac{a^{k+1}}{k + 1} \\
&= \sum_{k=0}^{\infty} \frac{\mathbf{F}^k a^k}{(k + 1)!}.
\end{aligned} \tag{6.102}$$

But now we note that the series for $\boldsymbol{\Phi}(a)$ may be written as

$$\boldsymbol{\Phi}(a) = \mathbf{I} + \sum_{k=1}^{\infty} \frac{\mathbf{F}^k a^k}{k!}.$$

If we let $k = j + 1$ in the sum, then, as in (6.14),

$$\begin{aligned}
\boldsymbol{\Phi}(a) &= I + \sum_{j=0}^{\infty} \frac{\mathbf{F}^{j+1} a^{j+1}}{(j + 1)!} \\
&= I + \sum_{j=0}^{\infty} \frac{\mathbf{F}^j a^j}{(j + 1)!} \, aF \\
&= I + \boldsymbol{\Psi}(a)a\mathbf{F}.
\end{aligned} \tag{6.103}$$

The point of (6.103) is that only the series for $\boldsymbol{\Psi}$ needs to be computed, and from this single sum we can compute $\boldsymbol{\Phi}$ and $\boldsymbol{\Gamma}$.

If we return to the case $\ell = 0$, $m \neq 0$, the discrete state equations are

$$\mathbf{x}(k + 1) = \boldsymbol{\Phi}\mathbf{x}(k) + \boldsymbol{\Gamma}_1 u(k) + \boldsymbol{\Gamma}_2 u(k + 1),$$

where $\boldsymbol{\Gamma}_1$ and $\boldsymbol{\Gamma}_2$ are given by (6.101). In order to put these equations in state-variable form, we must eliminate the term in $u(k + 1)$. To do this, we define a new state $\boldsymbol{\xi}(k) = \mathbf{x}(k) - \boldsymbol{\Gamma}_2 u(k)$. Then the equations are

$$\begin{aligned}
\boldsymbol{\xi}(k + 1) &= \mathbf{x}(k + 1) - \boldsymbol{\Gamma}_2 u(k + 1) \\
&= \boldsymbol{\Phi}\mathbf{x}(k) + \boldsymbol{\Gamma}_1 u(k) + \boldsymbol{\Gamma}_2 u(k + 1) - \boldsymbol{\Gamma}_2 u(k + 1)
\end{aligned}$$

$$\begin{aligned}
\boldsymbol{\xi}(k + 1) &= \boldsymbol{\Phi}[\boldsymbol{\xi}(k) + \boldsymbol{\Gamma}_2 u(k)] + \boldsymbol{\Gamma}_1 u(k) \\
&= \boldsymbol{\Phi}\boldsymbol{\xi}(k) + (\boldsymbol{\Phi}\boldsymbol{\Gamma}_2 + \boldsymbol{\Gamma}_1)u(k) \\
&= \boldsymbol{\Phi}\boldsymbol{\xi}(k) + \boldsymbol{\Gamma}u(k).
\end{aligned} \tag{6.104}$$

The output equation is

$$\begin{aligned}
y(k) &= \mathbf{H}\mathbf{x}(k) \\
&= \mathbf{H}[\boldsymbol{\xi}(k) + \boldsymbol{\Gamma}_2 u(k)] \\
&= \mathbf{H}\boldsymbol{\xi}(k) + \mathbf{H}\boldsymbol{\Gamma}_2 u(k) \\
&= \mathbf{H}_d\boldsymbol{\xi}(k) + J_d u(k).
\end{aligned} \tag{6.105}$$

Thus for $\ell = 0$, the state equations are given by (6.101), (6.104), and (6.105). Note especially that if $m = 0$, then $\boldsymbol{\Gamma}_2 = 0$, and these equations reduce to the previous model with no delay.

Our next case is $\ell = 1$. From (6.97), the equations are given by

$$\mathbf{x}(k + 1) = \boldsymbol{\Phi}\mathbf{x}(k) + \boldsymbol{\Gamma}_1 u(k - 1) + \boldsymbol{\Gamma}_2 u(k).$$

In this case, we must eliminate $u(k - 1)$ from the right-hand side, which we do by defining a new state $x_{n+1}(k) = u(k - 1)$. We have thus an increased dimension of the state, and the equations are

$$\begin{bmatrix} \mathbf{x}(k + 1) \\ x_{n+1}(k + 1) \end{bmatrix} = \begin{bmatrix} \boldsymbol{\Phi} & \boldsymbol{\Gamma}_1 \\ 0 & 0 \end{bmatrix} \begin{bmatrix} \mathbf{x}(k) \\ x_{n+1}(k) \end{bmatrix} + \begin{bmatrix} \boldsymbol{\Gamma}_2 \\ 1 \end{bmatrix} u(k),$$

$$y(k) = \begin{bmatrix} \mathbf{H} & 0 \end{bmatrix} \begin{bmatrix} \mathbf{x} \\ x_{n+1} \end{bmatrix}. \tag{6.106}$$

For our final case, we consider $\ell > 1$. In this case, the equations are

$$\mathbf{x}(k + 1) = \boldsymbol{\Phi}\mathbf{x}(k) + \boldsymbol{\Gamma}_1 u(k - \ell) + \boldsymbol{\Gamma}_2 u(k - \ell + 1),$$

and we must eliminate the past controls up to $u(k)$. To do this we introduce ℓ new states such that

$$x_{n+1}(k) = u(k - \ell), \; x_{n+2}(k) = u(k - \ell + 1), \; \ldots, \; x_{n+\ell}(k) = u(k - 1).$$

The structure of the equations is

$$\begin{bmatrix} \mathbf{x}(k + 1) \\ x_{n+1}(k + 1) \\ x_{n+2}(k + 1) \\ \cdot \\ \cdot \\ \cdot \\ x_{n+\ell}(k + 1) \end{bmatrix} = \begin{bmatrix} \boldsymbol{\Phi} & \boldsymbol{\Gamma}_1 & \boldsymbol{\Gamma}_2 & 0 & \cdots & 0 \\ 0 & 0 & 1 & 0 & \cdots & 0 \\ 0 & 0 & 0 & 1 & \cdots & 0 \\ & & & & & \\ & & & & 1 & \\ & & & & & \\ 0 & 0 & 0 & & \cdots & 0 \end{bmatrix} \begin{bmatrix} \mathbf{x}(k) \\ x_{n+1}(k) \\ x_{n+2}(k) \\ \cdot \\ \cdot \\ \cdot \\ x_{n+\ell}(k) \end{bmatrix} + \begin{bmatrix} 0 \\ 0 \\ 0 \\ \cdot \\ \cdot \\ \cdot \\ 1 \end{bmatrix} u(k), \tag{6.107}$$

$$y(k) = [\mathbf{H} \quad 0] \begin{bmatrix} \mathbf{x} \\ x_{n+1} \\ \cdot \\ \cdot \\ \cdot \\ x_{n+\ell} \end{bmatrix}.$$

This final situation is easily visualized in terms of a block diagram, as shown in Fig. 6.16.

The numerical considerations of these computations are centered in the approximation to the infinite sum for Ψ given by (6.102) or, for $a = T$, by (6.16). The problem is that if $\mathbf{F}T$ is large, then it takes many terms before $(\mathbf{F}T)^N/N!$ becomes small, and acceptable accuracy is realized. Källström (1973) has analyzed a technique used by Kalman and Englar (1966), which has been found effective by Moler and Van Loan (1978). The basic idea comes from (6.6) with $t_2 - t_0 = 2T$ and $t_1 - t_0 = T$, namely

$$(e^{\mathbf{F}T})^2 = e^{\mathbf{F}T}e^{\mathbf{F}T} = e^{\mathbf{F}2T}. \tag{6.108}$$

Thus, if T is too large, we can compute the series for $T/2$ and square the result. If $T/2$ is too large, we compute the series for $T/4$, and so on, until we find a k such that $T/2^k$ is *not* too large. We need a test for deciding on the value of k. We propose to approximate the series for Ψ which may be written

$$\Psi\left(\frac{T}{2^k}\right) = \sum_{j=0}^{N-1} \frac{[\mathbf{F}(T/2^k)]^j}{(j+1)!} + \sum_{j=N}^{\infty} \frac{(\mathbf{F}T/2^k)^j}{(j+1)!} = \hat{\Psi} + \mathbf{R}.$$

We will select k, the factor which decides how much the sample period is divided down to yield a small remainder term $\mathbf{R}$. Källström suggests that we estimate the size of $\mathbf{R}$ by the size of the first term ignored in $\hat{\Psi}$, namely

$$\hat{\mathbf{R}} \cong (\mathbf{F}T)^N/(N+1)!2^{Nk}.$$

A simpler method is to select k such that the size of $\mathbf{F}T$ divided by 2^k is less

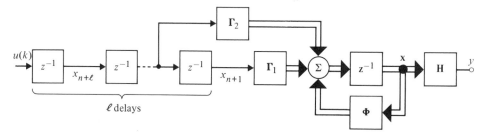

Fig. 6.16 Block diagram of system with delay of more than one period. Double line indicates vector valued variables.

than 1. In this case, the series for $FT/2^k$ will surely converge. The rule is to select k such that

$$2^k > \|\mathbf{F}T\| \triangleq \max_j \sum_{i=1}^{n} |F_{ij}|T.$$

Taking the log of both sides, we find

$$k > \log_2 \|\mathbf{F}T\|,$$

from which we select

$$k = \max(\lceil \log_2 \|\mathbf{F}T\|, 0), \tag{6.109}$$

where the symbol $\lceil x$ means the smallest integer greater than x. The maximum of this integer and zero is taken because it is possible that $\|\mathbf{F}T\|$ is already so small that its log is negative, in which case we want to select $k = 0$.

Having selected k, we now have the problem of computing $\hat{\boldsymbol{\Psi}}(T)$ from $\hat{\boldsymbol{\Psi}}(T/2^k)$. Our original concept was based on the series for $\boldsymbol{\Phi}$ which satisfied (6.108). To obtain the suitable formula for $\boldsymbol{\Psi}$, we use the relation between $\boldsymbol{\Phi}$ and $\boldsymbol{\Psi}$ given by (6.16) as follows to obtain the "doubling" formula for $\boldsymbol{\Psi}$:

$$\boldsymbol{\Phi}(2T) = \boldsymbol{\Phi}(T)\boldsymbol{\Phi}(T),$$
$$\mathbf{I} + 2T\,\mathbf{F}\boldsymbol{\Psi}(2T) = [\mathbf{I} + T\,\mathbf{F}\boldsymbol{\Psi}(T)][\mathbf{I} + T\,\mathbf{F}\boldsymbol{\Psi}(T)]$$
$$= \mathbf{I} + 2T\,\mathbf{F}\boldsymbol{\Psi}(T) + T^2\mathbf{F}^2\,\boldsymbol{\Psi}^2(T);$$

therefore

$$2T\,\mathbf{F}\boldsymbol{\Psi}(2T) = 2T\,\mathbf{F}\boldsymbol{\Psi}(T) + T^2\mathbf{F}^2\,\boldsymbol{\Psi}^2(T).$$

This is equivalent to

$$\boldsymbol{\Psi}(2T) = \left(\mathbf{I} + \frac{T\,\mathbf{F}}{2}\,\boldsymbol{\Psi}(T)\right)\boldsymbol{\Psi}(T),$$

which is the form to be used. The program logic for computing $\boldsymbol{\Psi}$ is shown in Fig. 6.17.

---

1. Select $\mathbf{F}$ and T.
2. Comment: Compute $\|\mathbf{F}T\|$.
3. $V \leftarrow \max_j \{\Sigma_i |F_{ij}|\} \times T$
4. $k \leftarrow$ smallest nonnegative integer greater than $\log_2 V$.
5. Comment: compute $\boldsymbol{\Psi}(T/2^k)$.
6. $T_1 \leftarrow T/2^k$
7. $\mathbf{I} \leftarrow$ Identity
8. $\boldsymbol{\Psi} \leftarrow \mathbf{I}$
9. $j \leftarrow 11$
10. If $j = 1$, go to step 14.

11. $\boldsymbol{\Psi} \leftarrow \mathbf{I} + \dfrac{\mathbf{F}T_1}{j}\,\boldsymbol{\Psi}$
12. $j \leftarrow j - 1$
13. Go to step 10.
14. Comment: now double $\boldsymbol{\Psi}$ k times.
15. If $k = 0$, stop.
16. $\boldsymbol{\Psi} \leftarrow \left(\mathbf{I} + \dfrac{\mathbf{F}T}{2^{k+1}}\,\boldsymbol{\Psi}\right)\boldsymbol{\Psi}$
17. $k \leftarrow k - 1$
18. Go to step 15.

Fig. 6.17 Logic for a program to compute $\boldsymbol{\Psi}$ using automatic time scaling.

Appendix B to Chapter 6

Ackermann's Formula for Pole-Placement Design

We are given a plant with equations of motion

$$\mathbf{x}_{k+1} = \mathbf{\Phi}\mathbf{x}_k + \mathbf{\Gamma}u_k$$

and propose to use a feedback control law

$$u = -\mathbf{K}\mathbf{x}$$

so that the closed-loop characteristic equation is

$$\det\left[z\mathbf{I} - \mathbf{\Phi} + \mathbf{\Gamma}\mathbf{K}\right] = \alpha_c(z). \tag{6.110}$$

First we wish to select $\alpha_c(z)$, which determines the placement of the closed-loop poles, and then solve for $\mathbf{K}$ so that (6.110) is satisfied. Our method will be based on the transformation of the plant equations to control canonical form.

First, we consider the effects of an arbitrary nonsingular transformation of state[26]

$$\mathbf{x} = \mathbf{T}\mathbf{w}. \tag{6.111}$$

From (6.111) it follows that the equations of motion in the new states are

$$\begin{aligned}
\mathbf{x}_{k+1} &= \mathbf{T}\mathbf{w}_{k+1} \\
&= \mathbf{\Phi}\mathbf{x}_k + \mathbf{\Gamma}u_k \\
&= \mathbf{\Phi}\mathbf{T}\mathbf{w}_k + \mathbf{\Gamma}u_k, \\
w_{k+1} &= \mathbf{T}^{-1}\mathbf{\Phi}\mathbf{T}\mathbf{w}_k + \mathbf{T}^{-1}\mathbf{\Gamma}u_k \\
&= \mathbf{A}\mathbf{w}_k + \mathbf{B}u_k.
\end{aligned} \tag{6.112}$$

We can see a useful expression for the transformation by considering the controllability matrix, $\mathscr{C}$, defined as the composite matrix with columns composed of powers of $\mathbf{\Phi}$ times $\mathbf{\Gamma}$ as follows (the order of the system is n):

$$\mathscr{C}_\mathbf{x} = [\mathbf{\Gamma} \vdots \mathbf{\Phi}\mathbf{\Gamma} \vdots \mathbf{\Phi}^2\mathbf{\Gamma} \vdots \sim \vdots \mathbf{\Phi}^{n-1}\mathbf{\Gamma}]. \tag{6.113}$$

We can, in the obvious way, also define $\mathscr{C}_\mathbf{w}$ in terms of $\mathbf{A}$ and $\mathbf{B}$ as

$$\mathscr{C}_\mathbf{w} = [\mathbf{B} \vdots \mathbf{A}\mathbf{B} \vdots \sim \vdots \mathbf{A}^{n-1}\mathbf{B}]. \tag{6.114}$$

Now consider the effects of the transformation (6.111) on this matrix by making the substitution from (6.112):

$$\mathbf{B} = \mathbf{T}^{-1}\mathbf{\Gamma}, \qquad \mathbf{A} = \mathbf{T}^{-1}\mathbf{\Phi}\mathbf{T}.$$

Thus we find that[27]

$$\begin{aligned}
\mathscr{C}_\mathbf{w} &= [\mathbf{T}^{-1}\mathbf{\Gamma} \vdots \mathbf{T}^{-1}\mathbf{\Phi}\mathbf{T}\mathbf{T}^{-1}\mathbf{\Gamma} \vdots \sim \ldots] \\
&= \mathbf{T}^{-1}\mathscr{C}_\mathbf{x}.
\end{aligned} \tag{6.115}$$

---

[26] The variable $\mathbf{w}$ in (6.111) is not to be confused with a plant disturbance signal.
[27] Note that $\mathbf{A}^2 = (\mathbf{T}^{-1}\mathbf{\Phi}\mathbf{T})(\mathbf{T}^{-1}\mathbf{\Phi}\mathbf{T}) = \mathbf{T}^{-1}\mathbf{\Phi}^2\mathbf{T}$, and in fact, $\mathbf{A}^k = \mathbf{T}^{-1}\mathbf{\Phi}^k\mathbf{T}$.

In terms of $\mathbf{T}$, (6.115) is

$$\mathbf{T} = \mathscr{C}_x \mathscr{C}_w^{-1}. \tag{6.116}$$

From (6.115) and (6.116) we can draw several important conclusions. First, from (6.115) we can see that if $\mathscr{C}_x$ is nonsingular, then for any nonsingular $\mathbf{T}$, $\mathscr{C}_w$ is also nonsingular. Thus controllability is neither gained nor lost by a nonsingular change of variables. From (6.116) we can go the other way. Suppose we wish to find a transformation which will put our equations of motion as given by $\mathbf{\Phi}$, $\mathbf{\Gamma}$ into the control canonical form as defined by Fig. 2.12 and Eq. (6.24). As we shall shortly see, $\mathscr{C}_w$ in that case is *always* nonsingular. From (6.116) we conclude that $\mathbf{T}$ will exist as a nonsingular transformation if and only if $\mathscr{C}_x$ is nonsingular. We conclude:

We can transform $[\mathbf{\Phi}, \mathbf{\Gamma}]$ to the control canonical form if and only if $\mathscr{C}(\mathbf{\Phi}, \mathbf{\Gamma})$ is nonsingular.

Suppose we now look more closely at the control canonical form, and treat specifically the third-order case, as in (6.24), although the results will be true for any n.

$$\mathbf{A} = \mathbf{\Phi}_c = \begin{bmatrix} a_1 & a_2 & a_3 \\ 1 & 0 & 0 \\ 0 & 1 & 0 \end{bmatrix}, \qquad \mathbf{B} = \mathbf{\Gamma}_c = \begin{bmatrix} 1 \\ 0 \\ 0 \end{bmatrix}. \tag{6.117}$$

The controllability matrix, by direct computation, is

$$\mathscr{C}_w = \begin{bmatrix} 1 & a_1 & a_1^2 + a_2 \\ 0 & 1 & a_1 \\ 0 & 0 & 1 \end{bmatrix}. \tag{6.118}$$

Since this matrix is triangular with all 1's on its main diagonal, it is nonsingular, as we observed above. Also note, since we will need this later, that the last row of $\mathscr{C}_w$ is the unit vector with all 0's except in the last place, where a 1 is found.

As described in Section 6.3, the design of a control law for state $\mathbf{w}$ when the equations of motion are in control canonical form is trivial. The characteristic equation of $\mathbf{\Phi}_c$ is

$$z^3 - a_1 z^2 - a_2 z - a_3 = 0, \tag{6.119}$$

and the characteristic equation for the closed loop ($\mathbf{w}$ feedback) comes from

$$\mathbf{\Phi}_c - \mathbf{\Gamma}_c \mathbf{K}$$

and has coefficients

$$z^3 - (a_1 - K_{w1})z^2 - (a_2 - K_{w2})z - (a_3 - K_{w3}) = 0.$$

To match the closed-loop characteristic equation

$$\alpha_c(z) = z^3 - \alpha_1 z^2 - \alpha_2 z - \alpha_3,$$

we need only make

$$a_1 - K_{w1} = \alpha_1, \qquad a_2 - K_{w2} = \alpha_2, \qquad a_3 - K_{w3} = \alpha_3,$$

or, in general,

$$\mathbf{a} - \mathbf{K_w} = \boldsymbol{\alpha}, \tag{6.120}$$

where $\mathbf{a}$ and $\boldsymbol{\alpha}$ are row vectors of coefficients of the characteristic polynomials of the open and closed loops, respectively.

We need to find a relation between these polynomial coefficients and the matrix $\boldsymbol{\Phi}$. The requirement is fulfilled by the Cayley-Hamilton theorem which states that if a matrix is substituted into its own characteristic polynomial, the result is zero. For $\boldsymbol{\Phi}_c$, we have

$$\boldsymbol{\Phi}_c^n - a_1\boldsymbol{\Phi}_c^{n-1} - a_2\boldsymbol{\Phi}_c^{n-2} - \cdots - a_n\mathbf{I} = 0. \tag{6.121}$$

Now suppose we form the polynomial $\alpha_c(\boldsymbol{\Phi})$, which is the *closed*-loop characteristic polynomial with the matrix $\boldsymbol{\Phi}$ substituted for the complex variable z:

$$\alpha_c(\boldsymbol{\Phi}) = \boldsymbol{\Phi}^n - \alpha_1\boldsymbol{\Phi}^{n-1} - \alpha_2\boldsymbol{\Phi}^{n-2} - \cdots - \alpha_n\mathbf{I}. \tag{6.122}$$

If we solve (6.121) for $\boldsymbol{\Phi}^n$ and substitute into (6.122) we get the interesting result that

$$\alpha_c(\boldsymbol{\Phi}_c) = (a_1 - \alpha_1)\boldsymbol{\Phi}_c^{n-1} + (a_2 - \alpha_2)\boldsymbol{\Phi}_c^{n-2} + \cdots + (a_n - \alpha_n)\mathbf{I}. \tag{6.123}$$

Furthermore, because $\boldsymbol{\Phi}_c$ is such a special form, we observe that something special happens if we multiply (6.123) by the transpose of the nth unit vector, $\mathbf{e}_n^T = [0 \quad 0 \ldots 0 \quad 1]$. We note that

$$\mathbf{e}_n^T\boldsymbol{\Phi}_c = [0 \ldots 1 \quad 0] = \mathbf{e}_{n-1}^T, \tag{6.124}$$

as we may see from (6.117). And, multiplying this vector by $\boldsymbol{\Phi}_c$ again will select row $n - 1$ which is

$$\begin{aligned}(\mathbf{e}_n^T\boldsymbol{\Phi}_c)\boldsymbol{\Phi}_c = [0 \ldots 1 \quad 0]\boldsymbol{\Phi}_c &= [0 \quad 0 \ldots 1 \quad 0 \quad 0] \\ &= \mathbf{e}_{n-2}^T. \end{aligned} \tag{6.125}$$

Continuing in this way, the successive unit vectors are generated until we compute $\mathbf{e}_1$ as

$$\begin{aligned}\mathbf{e}_n^T\boldsymbol{\Phi}_c^{n-1} &= [1 \quad 0 \ldots 0] \\ &= \mathbf{e}_1^T. \end{aligned} \tag{6.126}$$

Therefore, if we multiply (6.123) by $\mathbf{e}_n^T$, we find

$$\begin{aligned}\mathbf{e}_n^T\alpha_c(\boldsymbol{\Phi}_c) &= (a_1 - \alpha_1)\mathbf{e}_1^T + (a_2 - \alpha_2)\mathbf{e}_2^T + \cdots + (a_n - \alpha_n)\mathbf{e}_n^T \\ &= [K_{w1} \quad K_{w2} \cdots K_{wn}] \\ &= \mathbf{K_w}!!, \end{aligned} \tag{6.127}$$

where we use (6.120), which relates the gain $\mathbf{K_w}$ to the a's and α's.

We now have in (6.127) a compact expression for the gains for the system in control canonical form. We need to find the expression for $\mathbf{K_x}$, the gain to use on the *original* states. We return to the transformation $\mathbf{T}$ to find the necessary modifications. If $u = -\mathbf{K_w}\mathbf{w}$, then $u = -\mathbf{K_w}\mathbf{T}^{-1}\mathbf{x}$, so that

$$
\begin{aligned}
\mathbf{K_x} &= \mathbf{K_w}\mathbf{T}^{-1} \\
&= \mathbf{e}_n^T\alpha_c(\mathbf{\Phi}_c)\mathbf{T}^{-1} \\
&= \mathbf{e}_n^T\alpha_c(\mathbf{T}^{-1}\mathbf{\Phi}\mathbf{T})\mathbf{T}^{-1} \\
&= \mathbf{e}_n^T\mathbf{T}^{-1}\alpha(\mathbf{\Phi}),
\end{aligned} \tag{6.128}
$$

where in the last step we use the facts that $(\mathbf{T}^{-1}\mathbf{\Phi}\mathbf{T})^k$ is $\mathbf{T}^{-1}\mathbf{\Phi}^k\mathbf{T}$ and that α_c is a polynomial, i.e., a sum of powers of $\mathbf{\Phi}_c$. But what is $\mathbf{T}^{-1}$? From (6.115) we see that

$$
\mathbf{T}^{-1} = \mathscr{C}_w\mathscr{C}_x^{-1},
$$

so that

$$
\mathbf{K_x} = \mathbf{e}_n^T(\mathscr{C}_w\mathscr{C}_x^{-1})\alpha_c(\mathbf{\Phi}), \tag{6.129}
$$

and now we use the observation made earlier that the last row of $\mathscr{C}_w$, which is $\mathbf{e}_n^T\mathscr{C}_w$, is again $\mathbf{e}_n^T$! Our final result is Ackermann's formula

$$
\mathbf{K_x} = \mathbf{e}_n^T\mathscr{C}_x^{-1}\alpha_c(\mathbf{\Phi}). \tag{6.130}
$$

We note again that solving a specific set of equations is easier, by a factor of at least n, than computing the inverse. Thus we need the vector $\mathbf{b}^T$ so that

$$
\mathbf{e}_n^T\mathscr{C}_x^{-1} = \mathbf{b}^T.
$$

We solve

$$
\mathbf{b}^T\mathscr{C}_x = \mathbf{e}_n^T, \tag{6.131}
$$

and then

$$
\mathbf{K_x} = \mathbf{b}^T\alpha_c(\mathbf{\Phi}). \tag{6.132}
$$

The program logic is given in Fig. 6.5.

PROBLEMS AND EXERCISES

✓ **6.1** Compute $\mathbf{\Phi}$ by changing states so that the system matrix is diagonal.

a) Compute $e^{\mathbf{A}T}$ where

$$
\mathbf{A} = \begin{bmatrix} -1 & 0 \\ 0 & -2 \end{bmatrix},
$$

using the infinite series of (6.14).

b) Show that if $\mathbf{F} = \mathbf{T}^{-1}\mathbf{A}\mathbf{T}$ for some nonsingular transformation matrix $\mathbf{T}$, then $e^{\mathbf{F}T} = \mathbf{T}^{-1}e^{\mathbf{A}T}\mathbf{T}$.

c) Show that if

$$F = \begin{bmatrix} -3 & -2 \\ 1 & 0 \end{bmatrix},$$

there exists a T so that $T^{-1}AT = F$. [*Hint:* Write $T^{-1}A = FT^{-1}$, assume four unknowns for the elements of T^{-1}, and solve. For more general interest, write T^{-1} as a matrix of columns and show that the columns are eigenvectors of F.

d) Compute e^{FT}.

✓ **6.2** Compute Φ by Laplace transforms. We have shown in (6.4) that the solution to $\dot{x} = Fx$ is $x(t) = \Phi(t)x_0$. Thus, if $x_0 = [1 \ \ 0 \ \ 0]^T$, then the solution $x(t)$ will be the first column of $\Phi(t)$; if $x_0 = [0 \ \ 1 \ \ 0]^T$, we obtain the second column, and so on. Furthermore, if we envision an analog computer or block-diagram realization of $\dot{x} = Fx + x_0\delta(t)$, then we can compute $X(s)$, the Laplace transform of the solution to an impulse input which sets up x_0.

a) Draw a block-diagram realization of $\dot{x} = Fx + x_0\delta(t)$ for

$$F = \begin{bmatrix} -3 & -2 \\ 1 & 0 \end{bmatrix} \quad \text{and} \quad x_0 = (1 \ \ 0)^T.$$

b) Compute $X(s)$, the solution for the system of part (a).
c) Repeat (b) for $x_0 = (0 \ \ 1)^T$ and write $\Phi(s)$.
d) Solve by inverse transforms for $\Phi(t)$ and verify that this solution is the same as that given in Problem 6.1(d) if we let $t = T$.

✓ **6.3** Compute $G(z)$ from (6.18) for

$$\Phi = \begin{bmatrix} a_1 & a_2 & a_3 \\ 1 & 0 & 0 \\ 0 & 1 & 0 \end{bmatrix}, \quad \Gamma = \begin{bmatrix} 1 \\ 0 \\ 0 \end{bmatrix}, \quad H = [b_1 \ \ b_2 \ \ b_3].$$

Why is this form for Φ and Γ called control canonical form?

✓ **6.4** a) Write a computer program to compute Φ and Γ from F, G and sample period T, using the logic of Fig. 6.3, given that F may be as large as 10×10. Use the computer language of your control laboratory or as given by your instructor.

(b) Use your program to compute Φ, and Γ if

i) $F = \begin{bmatrix} -1 & 0 \\ 0 & -2 \end{bmatrix}, \quad G = \begin{bmatrix} 1 \\ 1 \end{bmatrix}, \quad T = 0.2$ sec,

ii) $F = \begin{bmatrix} -3 & -2 \\ 1 & 0 \end{bmatrix}, \quad G = \begin{bmatrix} 1 \\ 0 \end{bmatrix}, \quad T = 0.2$ sec.

✓ **6.5** a) For

$$\Phi = \begin{bmatrix} 1 & T \\ 0 & 1 \end{bmatrix}, \quad \Gamma = \begin{bmatrix} T^2/2 \\ T \end{bmatrix}$$

find a transform matrix T so that if $x = Tw$, then the equations in w will be in control canonical form.

b) Compute K_w, the gain, such that if $u = -K_w w$, the characteristic equation will be $\alpha_c(z) = z^2 - 1.6z + 0.7$.

c) Use T from part (a) to compute K_x, the gain in the x-states.

6.6 a) Show that the equations for the current estimator can be written in standard state form

$$\xi_{k+1} = A\xi_k + By_k, \qquad u = C\xi_k + Dy_k,$$

where $\xi_k = \hat{x}_k - Ly_k$ and $A = (I - LH)(\Phi - \Gamma K)$, $B = AL$, $C = -K$, $D = -KL$.

b) Use the results of (6.19) to show that the controller based on a current estimator always has a zero at $z = 0$ for any choice of control law K or estimator law L.

6.7 Design the antenna of Example 2 in Appendix A by state-variable pole and zero assignment.

a) Write the equations in state form with $x_1 = y$ and $x_2 = \dot{y}$. Give the matrices F, G, and H. Let $a = 0.1$.

b) Let $T = 1$ and design K for equivalent poles at $s = -1/2 \pm j(\sqrt{3}/2)$. Plot the step response of the resulting design.

c) Design a prediction estimator with L selected so that $\alpha_e(z) = z^2$; that is, poles are at the origin.

d) Use the estimated states for computing the control and introduce the reference input so as to leave the state estimate undisturbed. Plot the step response from this reference input and from a wind gust (step) disturbance.

e) Plot the root locus of the closed-loop system with respect to the plant gain and mark the locations of the closed-loop poles.

6.8 In Problem 5.7 we described an experiment in magnetic levitation described by the equations

$$\ddot{x} = 1000x + 20u.$$

Let the sampling time be 0.01 sec.

a) Use pole placement to design this system to meet the specifications that settling time is less than 0.25 sec and overshoot to an initial offset in x is less than 20%.

b) Design a reduced-order estimator for $\dot{x}$ for this system such that the error-settling time will be less than 0.08 sec.

c) Plot step responses of x, $\tilde{x}$, and u for an initial x displacement.

d) Plot the root locus for changes in the plant gain and mark the design pole locations.

e) Introduce a command reference with feedforward so that the estimate of $\dot{x}$ is not forced by r. Measure or compute the frequency response from r to system error $r - x$ and give the highest frequency for which the error amplitude is less than 20% of the command amplitude.

7 / Quantization Effects

7.1 INTRODUCTION

Thus far we have considered the impact of the digital computer as a discrete time device on control designs. Now we turn to consideration of the fact that numbers in the computer are taken in, stored, calculated, and put out with finite accuracy. The technique of representing a real number by a digital value of finite accuracy affects digital control in two principal ways. First, the variable values such as e, u, and the internal states used in the difference equations are not exact and thus errors are introduced into the output of the computer. As we shall see, these errors can often be analyzed as if there were noise sources in the computer. Second, the coefficients such as a_i and b_i of the difference equations are not exact, and this causes the machine to solve a slightly different problem than it was designed to do. As a result, the system dynamics are not exactly as designed. Our purpose in this chapter is to explore methods which can be used for the analysis of these two effects: finite accuracy of variables and finite accuracy of coefficients.

7.2 DETERMINISTIC ANALYSIS
OF ROUNDOFF VARIABLES

In our first analysis of finite accuracy we will assume that the computer represents numbers with fixed, rather than floating, points.[1] In such a scheme, addition and subtraction are done without error except that the sum may overflow the limits of the representation. Overflow must be avoided by proper amplitude scaling or analyzed as a major nonlinearity. In the case of multiplication, a double-length result is produced and the machine must reduce the number of bits in the product to fit it into the standard word size. This is often done by ignoring the least signifi-

---

[1] A brief discussion of number representations used in computers is given in the appendix to Chapter 7.

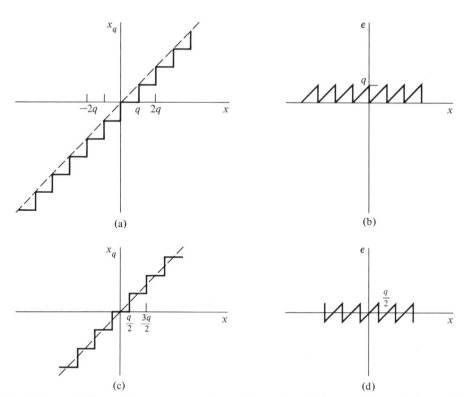

Fig. 7.1 Plot of effects of number truncation. (a) Plot of variable vs. truncated values. (b) Plot of error due to truncation. (c) Plot of variable vs. rounded values. (d) Roundoff error.

cant half of the product, a process called *truncation*. The same process would also follow division. If we assume that the numbers are represented as a fraction (less than 1 in magnitude) with ℓ bits, exclusive of sign, the least significant bit kept represents the magnitude $2^{-\ell}$. The result will be accurate to this value (the part thrown away could be almost as large as $2^{-\ell}$). Thus a plot of the "true" value of a variable x versus the stored value would look like Fig. 7.1, where the accuracy is decided by the quantization size, q (which is $2^{-\ell}$ under the conditions mentioned above).[2] A common alternative to truncation is roundoff; the result is the same as truncation if the first bit lost is a "0" but it is increased by $2^{-\ell}$ if the first bit lost is a "1." The process is the same as is common with ordinary base-ten numbers where 5.125 is rounded to 5.13 but 5.124 becomes 5.12 to two (decimal) places of accuracy. An input-output plot of rounding is shown in Fig. 7.1(c), and the corresponding error in 7.1(d).

---

[2] The plots in Fig. 7.1 assume that the integers are represented in two's complement; see the appendix to Chapter 7.

We can represent either process by the equation

$$x = x_q + \epsilon, \tag{7.1}$$

where ϵ is the error caused by the truncation or roundoff of x into the digital representation x_q. Since the error due to truncation equals a constant plus roundoff error, we will assume roundoff in our analysis unless otherwise stated. Clearly the process of analog-to-digital conversion also introduces a similar effect, although usually with a different q than that resulting from multiplication or division. The analysis of the effects of roundoff depends on the model we take for ϵ. We will analyze three such models which we may classify as (a) worst case, (b) steady-state worst case, (c) stochastic.

In this section we consider the two deterministic cases, (a) and (b). In the final section of the chapter we analyze quantization as a signal-dependent gain, and the possibility that the variables tend to a limit cycle rather than a steady-state constant.

The worst-case analysis is due to Bertram (1958). He takes the pessimistic view that the roundoff is done in such a way as to cause the maximum harm, and his analysis bounds the maximum error that can possibly result from truncation or roundoff. First we analyze the case shown in Fig. 7.2, in which a single case of quantization is assumed to occur somewhere in an otherwise linear constant system.

There is a transfer function from the point of roundoff to the output which we will call $H_1(z)$, and we can describe the situation by the equations

$$Y(z) = H(z)U(z),$$
$$\hat{Y}(z) = H(z)U(z) + H_1(z)E(z; x),$$
$$Y - \hat{Y} = \tilde{Y} = -H_1(z)E(z; x). \tag{7.2}$$

We write E as a function of some state variable x to emphasize that (7.2) is *not* linear because we do not know how to compute E until we have the values of x. However, we are not looking for the exact value of $\tilde{Y}$ but an upper bound on $\tilde{y}(k)$.

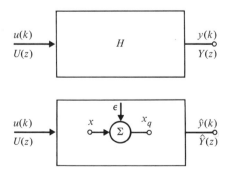

Fig. 7.2 A linear system and the introduction of one source of roundoff errors.

Since we wish to find a bound on the time signal $\bar{y}(k)$, we need the convolution equivalent of (7.2). Thus the relation between $\bar{y}(n)$ and $\epsilon(n)$ is the convolution sum given by

$$\bar{y}(n) = - \sum_{k=0}^{n} h_1(k)\epsilon(n - k; x).$$ (7.3)

If we examine Fig. 7.1(d), we can see that whatever the exact values of ϵ may be, its magnitude is bounded by $q/2$. We can use this information to determine in short steps the bound on $\bar{y}$ as we did in Chapter 2 in the discussion of BIBO[3] stability:

$$|\bar{y}| \le \left| \sum_{0}^{n} h\epsilon \right|.$$

The sum is bounded by the sum of the magnitudes of each term,

$$\le \sum_{0}^{n} |h\epsilon|,$$

which is the same as

$$\le \sum_{0}^{n} |h| \, |\epsilon|,$$

but by Fig. 7.1(d), the error is less than $q/2$,

$$\le \sum_{0}^{n} |h| \frac{q}{2},$$

and the sum can only get larger if we increase the number of terms:

$$|\bar{y}(n)| \le \sum_{0}^{\infty} |h| \frac{q}{2}.$$ (7.4)

As an example, we can apply (7.4) to a simple case. Consider the first-order system

$$y(n + 1) = \alpha y(n) + u(n)$$ (7.5)

and assume that the roundoff occurs in the computation of the product αy. Then we have

$$\hat{y}(n + 1) = \alpha \hat{y}(n) + \epsilon(n) + u(n),$$ (7.6)

and thus

$$\bar{y}(n + 1) = \alpha \bar{y}(n) - \epsilon(n).$$ (7.7)

---

[3] Bounded Input/Bounded Output.

For the error system, the unit pulse response is $H_1(k) = \alpha^k$, and (7.4) is easily computed as follows:

$$|\bar{y}| \leq \frac{q}{2} \sum_{k=0}^{\infty} \alpha^k \leq \frac{q}{2} \frac{1}{1 - \alpha}. \tag{7.8}$$

Regrettably, for second- and higher-order systems, the components of the unit pulse response usually have different signs so that the sum of the magnitudes of $h(k)$ is best done by computer. However, we can draw one very important conclusion from the general result represented by (7.4): roundoff of variables cannot cause the output of a stable linear system to become unbounded. We know this because, if the system is stable, then $\Sigma|h_i| < \infty$ for all possible transfer functions, and thus all stable systems have a finite (stable) error due to roundoff of variables. Roundoff can, however, cause variables to oscillate or otherwise fluctuate. Under no circumstances, however, can the output error exceed the value given by (7.4).

The steady-state worst case was analyzed by Slaughter (1964) in the context of digital control, and by Blackman (1965) in digital filters. In this perspective, we view the roundoff to cause some transient errors of no special concern, but we do assume that all variables eventually become constants, and we wish to know how large the steady-state error may be as a result of roundoff. We consider again the situation shown in Fig. 7.2 and thus Eq. (7.3). In this case, however, we assume that (7.3) reaches a steady state at which time ϵ is constant and in the range $-q/2 \leq \epsilon \leq q/2$. Then (7.3) reduces to

$$\bar{y}_{ss}(\infty) = \sum_{0}^{\infty} h_1(n)\epsilon_{\infty},$$

and in the worst (steady-state) case,

$$|\bar{y}_{ss}(\infty)| \leq \left| \sum h_1(n) \right| \frac{q}{2}. \tag{7.9}$$

There is one nice thing about (7.9): the sum is the value of the transfer function $H_1(z)$ at $z = 1$, and we can write

$$|\bar{y}_{ss}(\infty)| \leq |H_1(1)| \frac{q}{2}. \tag{7.10}$$

For multiple sources, these equations can be extended in an obvious way. For example, if we have K sources of roundoff, (7.4) becomes

$$|\bar{y}| \leq \left\{ \sum_{n=0}^{\infty} |h_1(n)| + \sum_{n=0}^{\infty} |h_2(n)| + \cdots \right\} \frac{q}{2}$$

$$\leq \sum_{j=1}^{K} \sum_{n=0}^{\infty} |h_j(n)| \frac{q}{2}. \tag{7.11}$$

And likewise (7.10) is extended to

$$|y_{ss}(\infty)| \leq [|H_1(1)| + |H_2(1)| + \cdots + |H_K(1)|]\frac{q}{2}. \qquad (7.12)$$

It is easily possible to express (7.12) in terms of a state-variable formulation of the equations of motion. Suppose the equations from the point of the quantization to the output are given by

$$\mathbf{x}(k + 1) = \mathbf{\Phi x}(k) + \mathbf{\Gamma}_1\epsilon(k),$$
$$\tilde{y}(k) = \mathbf{Hx}(k) + J\epsilon(k). \qquad (7.13)$$

The assumptions of the worst-case steady-state error are that $\mathbf{x}(k + 1) = \mathbf{x}(k) = \mathbf{x}_\infty$ and $\epsilon(k) = \epsilon_\infty$. Then

$$\mathbf{x}_\infty = \mathbf{\Phi x}_\infty + \mathbf{\Gamma}_1\epsilon_\infty, \qquad \tilde{y} = \mathbf{Hx}_\infty + J\epsilon_\infty. \qquad (7.14)$$

Solving for $\tilde{y}$, we find

$$\tilde{y} = [\mathbf{H}[\mathbf{I} - \mathbf{\Phi}]^{-1}\mathbf{\Gamma}_1 + J]\epsilon_\infty,$$

which is bounded by

$$|\tilde{y}| \leq |[\mathbf{H}[\mathbf{I} - \mathbf{\Phi}]^{-1}\mathbf{\Gamma}_1 + J]|\frac{q}{2}. \qquad (7.15)$$

The major advantage of the steady-state result is the vastly simpler forms of (7.12) and (7.15) as compared to (7.11). Unfortunately, (7.12) does not always hold because of the assumption of a constant quantization error of $q/2$ in the steady-state,[4] and yet the Bertram upper bound given by (7.11) is often excessively pessimistic. However, in some cases where special circuits would be necessary should a signal overflow, the absolute bound is used to select q so that no further decisions are required. See the paper on a digital filter for spectrum analysis by Schmidt (1978).

7.3 STOCHASTIC ANALYSIS OF ROUNDOFF VARIABLES

The third model for roundoff error is that of a stochastic variable. The analysis will follow Widrow (1956) and has two parts: development of a stochastic model for $\epsilon(k)$ and analysis of the response of a linear system to a stochastic process which has the given characteristics. Since the development of the model requires use of sophisticated concepts of stochastic processes and since the model can be given a very reasonable heuristic justification without this mathematical apparatus, we develop the model heuristically and proceed with an analysis of the

---

[4] Note that (7.12) is *not* a bound on the error. It is, however, a valuable estimate which is easy to compute for complex systems.

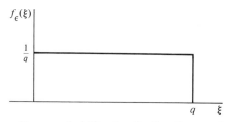

Fig. 7.3 Plot of the uniform probability density function used as a model for $\epsilon(n)$.

response. A review of the essential facts from the theory of probability and stochastic processes is to be found in Appendix D at the end of the book.

First, then, we give a heuristic argument for a stochastic model of truncation error. We begin with examination of Fig. 7.1, where we have a plot of the output versus the input of the truncation operation and a sketch of error versus amplitude of the input signal x. If we imagine that $x(n)$ is a random variable which takes values at successive sampling instants in a scattered way across the scale of values, then it seems reasonable to suppose that the sequence of errors $\epsilon(n)$ would be scattered over the range of possible values, which is to say, over the range from 0 to q. Furthermore, since the "teeth" in the sawlike plot of ϵ versus x are linear and contain no flat places which would signal a preference for one value of ϵ over another, it seems reasonable that the values of $\epsilon(n)$ are equally likely to be anywhere in the range $0 \le \epsilon \le q$.[5] The reflection of this argument in terms of probability is to assert that we can model ϵ as a random variable having a uniform probability density from 0 to q. A plot of the density is shown in Fig. 7.3. From this density we can immediately compute the mean and variance of ϵ as follows:

$$\mu_\epsilon = \mathcal{E}\epsilon = \int_{-\infty}^{\infty} \xi f_\epsilon(\xi)\, d\xi = \int_0^q \xi \frac{1}{q}\, d\xi = \frac{1}{q}\frac{\xi^2}{2}\Big|_0^q = \frac{q}{2}, \tag{7.16}$$

and

$$\sigma_\epsilon^2 = \mathcal{E}(\epsilon - \mu_\epsilon)^2 = \int_{-\infty}^{\infty}\left(\xi - \frac{q}{2}\right)^2 f_\epsilon(\xi)\, d\xi = \int_0^q \left(\xi - \frac{q}{2}\right)^2 \frac{1}{q}\, d\xi$$

$$= \frac{1}{q}\frac{(\xi - q/2)^3}{3}\Big|_0^q = \frac{1}{q}\left[\frac{1}{3}(q/2)^3 - \frac{1}{3}(-q/2)^3\right]$$

$$= \frac{1}{3q}\left[\frac{q^3}{8} + \frac{q^3}{8}\right] = \frac{q^2}{12}. \tag{7.17}$$

As to time dependence, we again appeal to the assumption that $x(n)$ is moving at random over a substantial number of "teeth" in the error plot given in Fig. 7.1(b). If $x(n + 1)$ has a good likelihood of being moved a substantial distance compared to q from $x(n)$, then the error deviation $\epsilon(n + 1) - q/2$ has little chance

---

[5] For roundoff, the density would have limits $-q/2 \le \epsilon \le q/2$.

of being dependent on the previous error deviation, $\epsilon(n) - q/2$. Thus we assume the following model for $\epsilon(n)$, for the case of truncation as sketched in Fig. 7.1:

$$\epsilon(n) = w(n) + \mu_\epsilon, \tag{7.18}$$

where $w(n)$ has the autocorrelation function

$$
\begin{aligned}
R_w(n) &= q^2/12 & (n = 0), \\
&= 0 & (n \neq 0),
\end{aligned}
\tag{7.19}
$$

and the mean value μ_ϵ is a constant $q/2$.

With the model given by (7.18) and (7.19) we can compute the mean and variance of the system error due to roundoff by either transform or state-space methods. In either case, we propose that analysis of the response to the mean μ_ϵ be computed separately from the response to the white noise component $w(n)$. By transform methods, the fundamental equation is the relation

$$\mathcal{S}_{\bar{y}}(z) = H_1(z)H_1(z^{-1})\mathcal{S}_w(z), \tag{7.20}$$

which is derived in Appendix D. For the present case, we have

$$\mathcal{S}_w(z) \overset{\Delta}{=} \mathfrak{z}\{R_w(n)\} = \sum_{n=-\infty}^{\infty} R_w(n)z^{-n} = R_w(0) = \frac{q^2}{12}. \tag{7.21}$$

Thus

$$\mathcal{S}_y(z) = H_1(z)H_1(z^{-1})\frac{q^2}{12}. \tag{7.22}$$

In Eq. (7.22), $H_1(z)$ is the transfer function from the point of introduction of the truncation noise to the output of interest. To compute the variance in y due to this noise, we use the fact that the variance in y due to a zero mean noise is equal to the autocorrelation of y at $n = 0$, namely

$$\sigma_{\bar{y}}^2 = R_{\bar{y}}(0),$$

and furthermore, from the spectral analysis,

$$R_{\bar{y}}(n) = \mathfrak{z}^{-1}\{\mathcal{S}_y(z)\}.$$

Therefore, using the integral form of the inverse transform, we have

$$
\begin{aligned}
\sigma_{\bar{y}}^2 = R_{\bar{y}}(0) &= \frac{1}{2\pi j} \oint \mathcal{S}_{\bar{y}}(z)\frac{dz}{z} \\
&= \frac{1}{2\pi j} \oint H_1(z)H_1(z^{-1})\frac{q^2}{12}\frac{dz}{z}.
\end{aligned}
\tag{7.23}
$$

Integrals of the form of (7.23) have been tabulated in terms of the coefficients of the polynomial factors, for example by Jury (1964) and Astrom (1970).

To apply this analysis to the example of (7.5), we have

$$H_1(z) = \frac{1}{z - \alpha},$$

and hence

$$S_{\tilde{y}}(z) = \frac{1}{z - \alpha} \frac{1}{z^{-1} - \alpha} \frac{q^2}{12}. \tag{7.24}$$

With the spectrum of (7.24), application of Cauchy's integral formula to evaluate (7.23) is elementary. The contour contains a single pole at $z = \alpha$, and the residue, found by partial-fraction expansion, gives

$$\sigma_{\tilde{y}}^2 = \frac{q^2}{12} \frac{1}{1 - \alpha^2}. \tag{7.25}$$

The mean-value response occurs when the signal in (7.5) is taken to be the constant $\mu_\epsilon = q/2$. Then we have

$$\mu_{\tilde{y}} = \alpha\mu_{\tilde{y}} + \frac{q}{2}$$

$$= \frac{q}{2} \frac{1}{1 - \alpha},$$

and the total mean square response is

$$\begin{aligned}
\overline{\tilde{y}^2} &= \sigma_{\tilde{y}}^2 + \mu_{\tilde{y}}^2 \\
&= \frac{q^2}{12} \frac{1}{1 - \alpha^2} + \frac{q^2}{4} \frac{1}{(1 - \alpha)^2} \\
&= \frac{q^2}{12} \left\{ \frac{4 + 2\alpha}{(1 - \alpha)^2(1 + \alpha)} \right\}.
\end{aligned} \tag{7.26}$$

Equation (7.26) gives the noise magnification factor of the simple first-order filter to truncation errors. By contrast, the worst-case bound is given by (7.8), which is the same as the steady-state dead band and may be written

$$|\tilde{y}|^2 \le q^2 \frac{1}{(1 - \alpha)^2}$$

$$\le \frac{q^2}{12} \left[\frac{12}{(1 - \alpha)^2} \right]. \tag{7.27}$$

A sketch of (7.26) and (7.27) versus α is given in Fig. 7.4.

We now turn to state-variable analysis and consider the equations (for the zero-mean component)

$$\begin{aligned}
\mathbf{x}(n + 1) &= \mathbf{\Phi}\mathbf{x}(n) + \mathbf{\Gamma}_1 w(n), \\
\tilde{y} &= \mathbf{H}_1\mathbf{x}(n) + J_1 w(n).
\end{aligned} \tag{7.28}$$

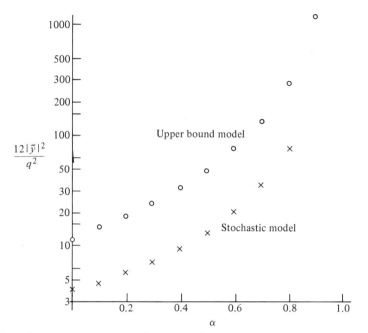

Fig. 7.4 Plot of mean square error estimates made by worst-case and stochastic models of roundoff in a simple system.

We define the covariance of the state as

$$\mathbf{R}_x(n) = \mathscr{E}\mathbf{x}(n)\mathbf{x}^T(n),$$

and, at $n + 1$,

$$\begin{aligned}
\mathbf{R}_x(n + 1) &= \mathscr{E}\mathbf{x}(n + 1)\mathbf{x}^T(n + 1) \\
&= \mathscr{E}\{(\mathbf{\Phi}\mathbf{x}(n) + \mathbf{\Gamma}_1 w(n))(\mathbf{\Phi}\mathbf{x}(n) + \mathbf{\Gamma}_1 w(n))^T\} \\
&= \mathbf{\Phi}\mathbf{R}_x(n)\mathbf{\Phi}^T + \mathbf{\Gamma}_1 \mathbf{R}_w(n)\mathbf{\Gamma}_1^T.
\end{aligned} \tag{7.29}$$

Equation (7.29) can be used with an initial value of $\mathbf{R}_x$ to compute the transient development of state covariance toward the steady state. The transform method gives only the steady state, the equivalent of letting $\mathbf{R}_x(n + 1) = \mathbf{R}_x(\infty)$. In this case, (7.29) reduces to the equation (called the Lyapunov equation for its occurrence in Lyapunov's stability studies)

$$\mathbf{R}_x(\infty) = \mathbf{\Phi}\mathbf{R}_x(\infty)\mathbf{\Phi}^T + \mathbf{\Gamma}_1 \mathbf{R}_w \mathbf{\Gamma}^T. \tag{7.30}$$

Several numerical methods for the solution of (7.30) have been developed, some based on solving (7.29) until $\mathbf{R}_x$ no longer changes and others based on converting (7.30) into a set of linear equations in the coefficients of $\mathbf{R}_x$ and solving these equations by numerical linear algebra. For the first-order example of (7.5), hand calculations are suitable. We have

$$x(n + 1) = \alpha x(n) + \epsilon(n)$$

and thus

$$\Phi = \alpha, \qquad \Gamma_1 = 1, \qquad R_w = \frac{q^2}{12}$$

so that (7.30) is merely

$$R_x = \alpha R_x \alpha + (1) \frac{q^2}{12} (1),$$

$$R_x = \frac{q^2}{12} \frac{1}{1 - \alpha^2}, \qquad (7.31)$$

which is recognized as the same as (7.25). If $\tilde{y}$ is not the state, we need to add to (7.30) the relation

$$\tilde{y} = \mathbf{H}_1 \mathbf{x},$$
$$\mathscr{E}(\tilde{y}\tilde{y}^T) = \mathscr{E}\mathbf{H}_1 \mathbf{x}\mathbf{x}^T \mathbf{H}_1^T,$$

which is to say

$$R_y = \mathbf{H}_1 R_x \mathbf{H}_1^T. \qquad (7.32)$$

Note that we can use (7.29) and (7.30) to compute the covariance of the state due to roundoff error at several locations simply by taking $\mathbf{w}$ to be a column matrix and $\mathbf{R_w}$ to be a square diagonal matrix of covariances of the components of $\mathbf{w}$. In the multiple-source case, Γ_1 is a matrix of dimension $n \times p$, where there are n states and p sources.

7.4 EFFECTS OF ROUNDOFF OF PARAMETERS

We have thus far analyzed the effects of roundoff on the variables such as y and u in (7.5). However, to do the calculation for a controller, the computer must also store the equation coefficients, and if the machine uses fixed point arithmetic, the parameter values must also be truncated or rounded off to the accuracy of the machine. This means that although we might design the program to solve

$$y(n + 1) = \alpha y(n) + u(n),$$

it will actually solve

$$\hat{y}(n + 1) = (\alpha + \delta\alpha)\hat{y}(n) + u + \epsilon. \qquad (7.33)$$

In this section, we give methods for the analysis of the effects of the parameter error $\delta\alpha$.

The principal concern with parameter variations is that the dynamic response and especially the stability of the system will be altered when the parameters are altered. To study this problem, we need look at the characteristic equation and ask what effect a parameter change has on the characteristic roots. For example, for the first-order system described by (7.33), the characteristic equation is

$$z - (\alpha + \delta\alpha) = 0,$$

and it is immediately obvious that if we want a pole at $z = 0.995$, it will be necessary to store α to three decimal places and that the limit on $\delta\alpha$ for stability is 0.005 since a variation of this magnitude results in a pole on the unit circle.[6]

To study these matters, we consider the characteristic equation and ask how a particular root changes when a particular parameter changes. The general study of root variation with parameter change is the root locus; however, we will obtain results of substantial value by a linearized sensitivity analysis. We can compare the direct and the cascade realizations of Chapter 2, for example, to see which is to be preferred from a sensitivity point of view. In the direct realization, the characteristic equation is

$$z^n - \alpha_1 z^{n-1} - \cdots - \alpha_n = 0. \tag{7.34}$$

This equation has roots at $\lambda_1, \lambda_2, \ldots, \lambda_n$, where $\lambda_i = r_i e^{j\theta_i}$. We assume that one of the α's, say α_k, is subject to error $\delta\alpha_k$, and we wish to compute the effect this has on λ_j and especially the effect on r_j so that stability can be checked. For this purpose, we can conveniently write (7.34) as a polynomial $P(z, \alpha)$ which depends on z and α_k. At $z = \lambda_j$, the polynomial is zero, so we have

$$P(\lambda_j, \alpha_k) = 0. \tag{7.35}$$

If α_k is changed to $\alpha_k + \delta\alpha_k$, then λ_j also changes, and the new polynomial is

$$P(\lambda_j + \delta\lambda_j, \alpha_k + \delta\alpha_k) = P(\lambda_j, \alpha_k) + \frac{\partial P}{\partial z}\bigg|_{z=\lambda_j} \delta\lambda_j + \frac{\partial P}{\partial \alpha_k} \delta\alpha_k + \cdots$$

$$= 0, \tag{7.36}$$

where the dots are terms of higher order in $\delta\lambda$ and $\delta\alpha$. By (7.35) we see that the first term on the right-hand side of (7.36) is zero. If $\delta\lambda_j$ and $\delta\alpha_k$ are both small, then the higher-order terms are also negligible. Thus the change in λ_j is given to first order by the derivative

$$\delta\lambda_j \cong - \frac{\partial P/\partial \alpha_k}{(\partial P/\partial z)}\bigg|_{z=\lambda_j} \delta\alpha_k. \tag{7.37}$$

---

[6] Note, however, that if we use feedback $1 - \beta$ and add the 1-term separately then $\beta = 1 - \alpha = 0.005$, and the product by 10^{-2} can also be obtained without error; so the accuracy requirements on β are much less than those on α.

We can evaluate the partial derivatives in (7.37) from (7.34) and the equivalent form

$$P(z, \alpha_k) = (z - \lambda_1)(z - \lambda_2) \cdots (z - \lambda_n). \tag{7.38}$$

First, using (7.34), we compute

$$\left. \frac{\partial P}{\partial \alpha_k} \right|_{z=\lambda_j} = -\lambda_j^{n-k}, \tag{7.39}$$

and next, using (7.38), we compute[7]

$$\left. \frac{\partial P}{\partial z} \right|_{z=\lambda_j} = \prod_{\ell \neq j} (\lambda_j - \lambda_\ell). \tag{7.40}$$

Thus (7.37) reduces to

$$\delta\lambda_j = + \frac{\lambda_j^{n-k}}{\displaystyle\prod_{\ell \neq j} (\lambda_j - \lambda_\ell)} \delta\alpha_k. \tag{7.41}$$

We can say a few things about root sensitivity by examining (7.41). Consider first the numerator term, which varies with k, the index number of the parameter whose variation we are considering. Since we are dealing with a stable system, the magnitude of λ_j is less than one, so the larger the power of λ_j^{n-k} the smaller the variation. We conclude that the most sensitive parameter is α_n, the constant term in (7.34). However, for values of λ_j near the unit circle, the relative sensitivity decreases slowly as k gets smaller. The denominator of (7.41) is the product of vectors from the characteristic roots to λ_j. This means that if all the roots are in a cluster, then the sensitivity is high, and if possible, the roots should be kept far apart. For example, if we wish to construct a digital low-pass filter with a narrow pass band and sharp cutoff, then the system will have many poles in a cluster near $z = 1$. If we implement such a filter in the control canonical form (Fig. 2.12), then the sensitivity given by (7.41) will have many factors in the denominator, all small. However, if we implement the *same* filter in the cascade or parallel forms, then the sensitivity factor will have only one term. Mantey (1968) quotes an example of a narrow-bandpass filter of six poles for which the parallel realization was less sensitive by a factor of 10^{-5}. In other words, it would take 17 bits of additional accuracy to implement this example in direct form over that required for the parallel or cascade form! The papers collected by Rabiner and Rader (1972) contain many other interesting results in this area.

---

[7] If $P(z)$ has only one root at λ_j.

7.5 LIMIT CYCLES AND DITHER

As a final study of the effects of finite word length in the realization of digital filters and compensators we present a view of the quantizer as a signal-dependent gain and analyze more closely the motions permitted by this nonlinearity. One type of motion is an output which persists in spite of there being no input and which eventually becomes periodic: such a motion is called a *limit cycle*.

To analyze a nonlinear system we must develop new tools since superposition and transforms do not apply. One such tool is to use graphic methods to solve the equations. For example, suppose we have the first-order equation

$$y(k + 1) = Q\{\alpha y(k)\}. \tag{7.42}$$

If we plot y versus αy as in Fig. 7.5, we can also plot the function $Q(\alpha y)$ and trace the trajectory of the solution beginning at the point a on the y-axis. Across from a at point b, we plot the value αy from the line with slope $1/\alpha$. Below b at c is the quantization of αy, and hence the next value of y, shown as d. We thus conclude that the trajectory may be found by starting with a point on the $(1/\alpha)$-line, dropping to the $Q\{\ \}$ staircase, projecting left to the $(1/\alpha)$-line again, dropping again, and so on, as shown by the dashed lines and arrowheads. Note that the path will always end on the segment of zero amplitude where $\alpha y \leq q$. Now, however, suppose the initial value of y is negative. The trajectory is plotted in Fig. 7.6, where projection is up to Q and to the right to the $(1/\alpha)$-line. Note that this time the trajectory gets stuck at the point s, and the response does *not* go to zero. The point s is an equilibrium or stationary point of the equation. If the initial value of y had a magnitude smaller than s, then the motion would have moved down to the next

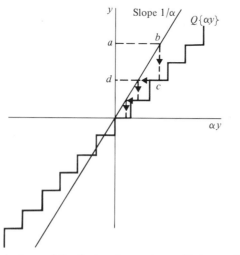

Fig. 7.5 Plot of trajectory of the first-order system with truncation quantization.

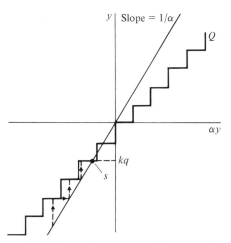

Fig. 7.6 Plot of trajectory of the first-order system with truncation quantization and negative initial condition.

equilibrium where the $(1/\alpha)$-line and $Q\{\ \}$ intersect. It is clear that s represents the largest value of y which is at equilibrium and that this value depends on α. We should be able to find a relation between kq, the largest value of y at equilibrium, and α, the time constant of the filter. In fact, from inspection of Fig. 7.6 we see that the largest value of y at a stationary point is a value $y = -kq$ such that

$$-kq\alpha < -kq + q, \qquad \text{or} \qquad k\alpha > k - 1, \qquad \text{or}$$

$$k < \frac{1}{1 - \alpha}, \qquad \text{or} \qquad |y| < \frac{q}{1 - \alpha}. \tag{7.43}$$

The last result is the same as (7.8), the worst-case bound. Thus we find in the first-order system that the worst case is going to be often realized.

In order to extrapolate these results to a higher-order system we must find a more subtle way to look at them. Such a viewpoint is provided by the observation that at the equilibrium points of Fig. 7.6 the gain of the quantizer is exactly $1/\alpha$. If we consider the quantizer as a variable gain, then the equilibrium occurs at those points where the combined gain of quantizer and parameter α are unity, which, for a linear system, would correspond to a pole at $z = 1$. The extrapolation of this idea to higher-order systems is to examine the range of possible "equivalent gains" of the quantizers and to conjecture that the limiting motion will be no larger than the largest signal for which the linear system with the resulting equivalent gain(s) has a pole on the unit circle.

We will illustrate the conjecture by means of the second-order control canonical form shown in Fig. 7.7(a) with the quantizer characteristic shown in Fig. 7.7(b) corresponding to roundoff rather than truncation. Note from Fig. 7.7(b) that the staircase of this quantizer is centered about the line of slope 1.0 passing

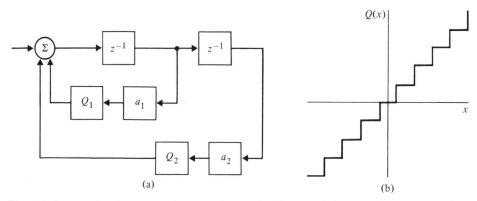

Fig. 7.7 A second-order system in control canonical form and the quantizer characteristic corresponding to roundoff.

through the origin.   The characteristic equation for this system, if quantization is ignored, is

$$z^2 - a_1z - a_2 = 0,$$
$$z^2 - 2r \cos \theta z + r^2 = 0.$$

Thus we see that the system will have complex roots on the unit circle only if the action of the quantizer is to make the effective value of $a_2 = -r^2 = -1.0$.   Following an analysis similar to (7.43) but taking account of the shift of the rounding quantizer, we find

$$-kqa_2 > kq - q/2,$$

or the amplitude is predicted to be less than

$$kq < \frac{1}{2} \frac{q}{1 - |a_2|}. \tag{7.44}$$

The effect of quantization on the a_1-term influences only the frequency of the oscillations in this model, but the equivalent gain of a_1 must be less than two.

Another benefit of the view of quantization as a variable gain is the idea that a second signal of high frequency and constant amplitude added to the input of the quantizer will probably destroy the limit cycle.   Such a signal is called a *dither* and its purpose is to make the effective gain of the quantizer 1.0 rather than something greater than 1.

Consider again the situation sketched in Fig. 7.6 with the signal stuck at s.   If an outside signal of amplitude $3q$ is added to y, then one may expect that while the output would contain a fluctuating component, the average value would drift toward zero rather than remain at s.   If the frequency of the dither is outside the desired pass band of the device, then the result is improved response; that is, a large constant bias or low-frequency self-sustained limit-cycle oscillation can sometimes be removed this way at the cost of a high-frequency noise which causes very low amplitude errors at the system output.

7.6 SUMMARY

In this chapter we studied some of the effects of the fact that numbers are stored in a digital computer with finite accuracy. The effects of finite accuracy on input, output, and state variables were analyzed for worst-case errors, and for worst-case errors with the added assumption that all the signals are in the constant steady state. The worst-case bound was found to be generally pessimistic. Also studied was a stochastic model of quantization errors, and methods were given for the computation of average values and variances of signals based on this model. These averages give a good approximation to observed performance in many cases. A general formula for the sensitivity of closed-loop system poles to parameter changes was derived in Section 7.4 and used to evaluate pole changes—the major consequence of storing controller parameters with finite accuracy. It was found that the direct controller (or observer) canonical forms are more sensitive than are cascade or parallel forms of implementation using second-order subblocks. In the final section, the phenomenon of oscillatory or limit-cycle behavior was discussed, and the technique of using a high-frequency external signal to dither the quantization and break up such motion was introduced.

Appendix to Chapter 7

Numbers in a Digital Computer

Usually, when we wish to represent a number, we use base (or radix) 10 and either fixed-point notation as $+10.739$ or floating (scientific) notation as $+0.10739 \times 10^{+2}$. The value of the number is understood to be a polynomial in powers of 10; in this case (fixed point),

$$n = 1 \times 10^1 + 0 \times 10^0 + 7 \times 10^{-1} + 3 \times 10^{-2} + 9 \times 10^{-3}.$$

Since data in a digital computer must be represented by variables which can assume only two values, the most common base for numbers is 2, although powers of 2 such as 4, 8, or 16 are also used. Using base 2 and a fixed-point format, an integer with four binary digits or bits is represented as $b_3 b_2 b_1 b_0$; for example, the integer 1 0 1 1 in base 2 has the value $1 \times 2^3 + 0 \times 2^2 + 1 \times 2 + 1 \times 2^0 = 11$ in base 10. To represent negative numbers using base 2, several schemes have been devised.

Two's Complement. In this scheme, the bit pattern

$$b_{m-1} b_{m-2} \ \cdot \ \cdot \ \cdot \ b_1 b_0$$

represents the value

$$n = -b_{m-1}2^{m-1} + \sum_{i=0}^{m-2} b_i 2^i.$$

Such a representation has the property that the leftmost bit is 0 for positive integers and 1 for negative values, so b_{m-1} acts as a sign bit. Numbers in two's complement code can be added as if they were positive integers and the result will be correct. Overflow occurs if numbers of the same sign are added and the result of integer addition gives a sign bit of the opposite sign. We represent the negative of a given number by replacing every bit by its complement (0 by 1, 1 by 0) and then add the value 1 to the result. For example:

$$
\begin{array}{ll}
0011 = 3 & 1100 = \text{Complement of 3} \\
+\underline{1101} = -3 & +\underline{1} \\
0000 = 0 & 1101 = \text{Two's complement of 3} \\
& = -3 \\
\\
0111 = 7 & 1000 \quad \text{Complement of 7} \\
& +\underline{1} \quad \text{Add 1} \\
& 1001 = -7 \text{ representation of } -7 \\
\\
0011 = 3 & 0011 = \text{Complement of } -4 \\
+\underline{1001} = -7 & +\underline{1} = \text{Add 1} \\
1100 = -4 & 0100 = +4
\end{array}
$$

One's Complement.   In this scheme the bit pattern

$$b_{m-1}b_{m-2} \ldots b_1 b_0$$

represents

$$n = -b_{m-1}(2^{m-1} - 1) + \sum_{i=0}^{m-2} b_i 2^i.$$

The advantage of the scheme is that the negative representation is found from the positive representation by complementing only. The addition of 1 as in the two's complement method is not needed. Again the b_{m-1} bit is a sign bit. There are two representations of zero, namely, all bits are 0 and all bits are 1. Arithmetic can be done with a binary-integer adder but because of the added value of -1 in the weight of the b_{m-1} bit, additional logic is required.

Sign and Magnitude.   In this representation, b_{m-1} is the sign (usually 0 for positive) and $|n| = \sum_{i=0}^{m-2} b_i 2^i$ is the magnitude. This format is recognized as the form used in base 10 for arithmetic done by hand.

Excess 2^m.   In this scheme, the pattern

$$b_{m-1} \ldots b_0$$

has the value

$$n = \sum_{i=0}^{m-1} b_i 2^i - 2^{m-1}.$$

For a floating point-representation, there are three numbers: f, the fraction or mantissa; e, the exponent; and β, the radix or base. The number is represented as

$$n = f \times \beta^e.$$

For storage in the computer, the values of f and e are given in fixed-point formats. A common choice for minicomputers is to use the two's complement approach for both f and e and to assume $\beta = 2$. The IBM 370 uses sign and magnitude for f, excess 64 for e, and $\beta = 16$.

Precision.   In fixed-point formats, the precision is determined by the total number of bits. The examples above have assumed that the radix point (decimal point, in base 10) is at the right-most position. By assuming the radix point to be between b_{m-1} and b_{m-2}, we have a system for representing fractions whose maximum magnitude is 1. For integers, the q-value is 1; for fractions, it would be $2^{-(m-1)}$.

For a floating-point representation, the precision is not absolute but is relative. If a real number n is represented by $\hat{n} = f \times \beta^e$ in a system with s bits in f (exclusive of sign) and m bits in e, then it can be shown that the relative error is bounded by

$$\frac{|n - \hat{n}|}{|n|} \leq \beta^{1-s}. \tag{7.45}$$

This result assumes that the mantissa is normalized so that, in two's complement, for example, b_{m-1} and b_{m-2} are opposite, i.e., if $b_{m-1} = 0$, then f is shifted and e is indexed until $b_{m-2} = 1$. The effect of a floating-point representation is that variable accuracy acts like a change in parameter values rather than independent noise.

PROBLEMS AND EXERCISES

7.1 a) For the second-order observer canonical form shown in Fig. 7.8, compute the transfer functions from the quantizers to the output, y. Note carefully how many need to be computed.

   b) If $b_1 = b_2 = 1$, $a_1 = 0.81$, $a_2 = -0.90$, what is the maximum steady-state error due to rounding offset of $\pm q/2$?

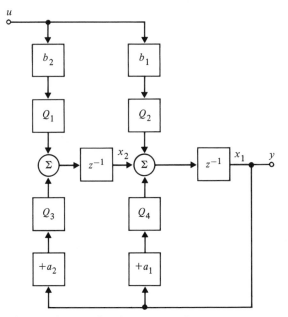

Fig. 7.8 A second-order system with quantization.

c) Show that the stochastic state-transition model to use Eq. (7.30) on this system has

$$\boldsymbol{\Phi} = \begin{bmatrix} +0.81 & 1 \\ -0.9 & 0 \end{bmatrix}, \quad \boldsymbol{\Gamma}_1 = \begin{bmatrix} 1 & 0 \\ 0 & 1 \end{bmatrix}, \quad R_w(\infty) = \frac{q^2}{6}, \quad \mathbf{H} = [1 \quad 0].$$

d) Assume that

$$\mathbf{R} = \begin{bmatrix} a & b \\ b & c \end{bmatrix},$$

solve (7.30) for this case, and compute $R_y(0)$ from (7.32).

7.2 a) Consider implementation of the discrete compensation

$$D(z) = \frac{z^3}{(z - 0.5)(z - 0.55)(z - 0.6)(z - 0.65)}$$

as a cascade of first-order filters with coefficients stored to finite accuracy. What is the smallest gain perturbation which could cause instability? Which parameter would be most sensitive?

b) Draw an implementation of $D(z)$ in control canonical form with parameters a_1, a_2, a_3, and a_4. Compute the first-order sensitivities of these parameters according to (7.41). Which parameter is most sensitive? Compare the sensitivities of root locations for the cascade and the control forms for these nominal root positions.

7.3 a) Show that the equivalent gain of the round-off quantizer is given by

$$G_q \le \frac{2k + 2}{2k + 1}, \qquad k = 0, 1, \ldots$$

for a signal of amplitude less than or equal to $kq + q/2$.

b) According to the equivalent-gain hypothesis of limit-cycle behavior, what is the largest-amplitude limit cycle you would expect for a second-order control canonical form system with complex roots at a radius of (0.9)?

7.4 Derive (7.45) by assuming that $n - \hat{n}$ is as large as possible for a given exponent and that n is as *small* as possible for the same exponent.

8 / System Identification

8.1 INTRODUCTION AND PROBLEM DEFINITION

In order to design controls for a dynamic system it is necessary to have a model which will adequately describe the system's motion. The information available to the designer for this purpose is typically of two kinds. First, there is the knowledge of physics, chemistry, biology, and the other basic sciences which have, over the years, developed equations of motion which explain the dynamic response of rigid bodies, electric circuits and motors, fluids, and many other constituents of systems to be controlled. However, it is often the case that, because of extremely complex physical phenomena such as the force on an airplane control surface mounted on the trailing edge of a wing or the heat of combustion of a fossil fuel of unknown detailed composition, the laws of science are either too complicated or inadequate to give a satisfactory description of the dynamic plant. In these circumstances the designer may turn to the second source of information about the dynamics, which is the data taken from experiments directly conducted to excite the plant and to measure the response. The process of constructing models and estimating the best values of unknown parameters from experimental data is called *system identification*. In obtaining models for control, our motivation is very different from that of modeling as practiced in the basic sciences. In pure science one aims to develop models of nature as it is, and may, for example, seek to describe the forces on an airfoil moving in an ideal compressible fluid; in control, one wishes to understand the forces on the wing and to design a strategy suitable for controlling those forces so the flight of aircraft will be simultaneously safe, comfortable, and efficient. In developing techniques for identification, we will be strongly influenced by the control design objectives for which our models will be used.

In this elementary treatment, we will make the very strong assumptions that our models need apply only to slowing changing variables which make small excursions: in short, our models will be stationary and linear. Formally, one proceeds as follows with the process of linearization and small-signal approximations. We begin with the assumption that our plant dynamics are adequately described by a set of ordinary differential equations in state variable form as

$$\dot{x}_1 = f_1(x_1, \ldots, x_n, u_1, \ldots, u_m, t),$$
$$\dot{x}_2 = f_2(x_1, \ldots, x_n, u_1, \ldots, u_m, t),$$
$$\cdot$$
$$\cdot$$
$$\cdot$$
$$\dot{x}_n = f_n(x_1, \ldots, x_n, u_1, \ldots, u_m, t), \qquad (8.1)$$
$$y_1 = h_1(x_1, \ldots, x_n, u_1, \ldots, u_m t),$$
$$\cdot$$
$$\cdot$$
$$\cdot$$
$$y_p = h_p(x_1, \ldots, x_n, u_1, \ldots, u_m, t),$$

or, more compactly in matrix notation, we assume that our plant dynamics are described by

$$\dot{\mathbf{x}} = \mathbf{f}(\mathbf{x}, \mathbf{u}, t),$$
$$\mathbf{x}(t_0) = \mathbf{x}_0, \qquad (8.2)$$
$$\mathbf{y} = \mathbf{h}(\mathbf{x}, \mathbf{u}, t).$$

The assumption of stationarity is here reflected by the approximation that $\mathbf{f}$ and $\mathbf{h}$ do not change significantly from their initial values at t_0. Thus we may set

$$\dot{\mathbf{x}} = \mathbf{f}(\mathbf{x}, \mathbf{u}, t_0)$$

or, simply,

$$\dot{\mathbf{x}} = \mathbf{f}(\mathbf{x}, \mathbf{u}), \qquad \mathbf{y} = \mathbf{h}(\mathbf{x}, \mathbf{u}). \qquad (8.3)$$

The assumption of small signals may be reflected by taking $\mathbf{x}$ and $\mathbf{u}$ to be always close to their reference values $\mathbf{x}_0$, $\mathbf{u}_0$, and these values, furthermore, to be an equilibrium point of (8.1), where

$$\mathbf{f}(\mathbf{x}_0, \mathbf{u}_0) = \mathbf{0}. \qquad (8.4)$$

Now, if $\mathbf{x}$ and $\mathbf{u}$ are "close" to $\mathbf{x}_0$ and $\mathbf{u}_0$, they may be written as $\mathbf{x} = \mathbf{x}_0 + \delta\mathbf{x}$; $\mathbf{u} = \mathbf{u}_0 + \delta\mathbf{u}$ and these substituted into (8.3). The fact that $\delta\mathbf{x}$ and $\delta\mathbf{u}$ are small is now used to motivate an expansion of (8.3) about $\mathbf{x}_0$ and $\mathbf{u}_0$ and to suggest that only terms in the first power of the small quantities $\delta\mathbf{x}$ and $\delta\mathbf{u}$ need to be retained. We thus have a vector equation and need the expansion of a vector-valued function of a vector variable,

$$\frac{d}{dt}(\mathbf{x}_0 + \delta\mathbf{x}) = \mathbf{f}(\mathbf{x}_0 + \delta\mathbf{x}, \mathbf{u}_0 + \delta\mathbf{u}). \qquad (8.5)$$

If we go back to (8.1) and do the expansion of the components f_i one at a time, it is tedious but simple to verify that (8.5) can be written as[1]

$$\delta\dot{\mathbf{x}} = \mathbf{f}(\mathbf{x}_0, \mathbf{u}_0) + \mathbf{f}_{,\mathbf{x}}(\mathbf{x}_0, \mathbf{u}_0)\delta\mathbf{x} + \mathbf{f}_{,\mathbf{u}}(\mathbf{x}_0, \mathbf{u}_0)\delta\mathbf{u} + \cdots, \tag{8.6}$$

where we define the partial derivative of a scalar f_1 with respect to the vector $\mathbf{x}$ by a subscript notation:

$$f_{1,\mathbf{x}} \triangleq \left(\frac{\partial f_1}{\partial x_1} \frac{\partial f_1}{\partial x_2} \cdots \frac{\partial f_1}{\partial x_n}\right). \tag{8.7}$$

The row vector in (8.7) is called the *gradient* of the scalar f_1 with respect to the vector $\mathbf{x}$. If $\mathbf{f}$ is a vector, we define its partial derivates with respect to the vector $\mathbf{x}$ as the matrix (called the Jacobian) composed of *rows* of gradients. In the subscript notation, if we mean to take the partial of *all* components, we omit the specific subscript such as 1 or 2 but hold its "place" by use of the comma:

$$\mathbf{f}_{,\mathbf{x}} = \begin{bmatrix} \dfrac{\partial f_1}{\partial \mathbf{x}} \\[2mm] \dfrac{\partial f_2}{\partial \mathbf{x}} \\[2mm] \cdot \\ \cdot \\ \cdot \\ \dfrac{\partial f_n}{\partial \mathbf{x}} \end{bmatrix}$$

$$= \begin{bmatrix} \dfrac{\partial f_i}{\partial x_j} \end{bmatrix} \quad \text{in row } i \text{ and column } j. \tag{8.8}$$

Now, to return to (8.6), we note that by (8.4) we chose $\mathbf{x}_0$, $\mathbf{u}_0$ to be an equilibrium point, so the first term on the right of (8.6) is zero, and since the terms beyond those shown depend on higher powers of the small signals $\delta\mathbf{x}$ and $\delta\mathbf{u}$, we are led to the approximation

$$\delta\dot{\mathbf{x}} \cong \mathbf{f}_{,\mathbf{x}}(\mathbf{x}_0, \mathbf{u}_0)\delta\mathbf{x} + \mathbf{f}_{,\mathbf{u}}(\mathbf{x}_0, \mathbf{u}_0)\delta\mathbf{u},$$
$$\mathbf{y} = \mathbf{h}_{,\mathbf{x}}\,\delta\mathbf{x} + \mathbf{h}_{,\mathbf{u}}\,\delta\mathbf{u}. \tag{8.9}$$

But now the notation is overly clumsy and we drop the δ-part of $\delta\mathbf{x}$ and $\delta\mathbf{u}$ and define the constant matrices

$$\mathbf{F} = \mathbf{f}_{,\mathbf{x}}(\mathbf{x}_0, \mathbf{u}_0), \qquad \mathbf{G} = \mathbf{f}_{,\mathbf{u}}(\mathbf{x}_0, \mathbf{u}_0),$$
$$\mathbf{H} = \mathbf{h}_{,\mathbf{x}}(\mathbf{x}_0, \mathbf{u}_0), \qquad \mathbf{J} = \mathbf{h}_{,\mathbf{u}}(\mathbf{x}_0, \mathbf{u}_0),$$

to obtain the form we have used in earlier chapters:

$$\dot{\mathbf{x}} = \mathbf{Fx} + \mathbf{Gu}, \qquad \mathbf{y} = \mathbf{Hx} + \mathbf{Ju}. \tag{8.10}$$

---

[1] Note that $d/dt\ \mathbf{x}_0 = \mathbf{0}$ because our "reference trajectory" $\mathbf{x}_0$ is a constant here.

We will go even further and restrict ourselves to the case of single input and single output and for discrete time. We write the model as

$$\mathbf{x}_{k+1} = \mathbf{\Phi}\mathbf{x}_k + \mathbf{\Gamma}u_k,$$
$$y_k = \mathbf{H}\mathbf{x}_k + Ju_k, \tag{8.11}$$

from which the transfer function is

$$\mathscr{H}(z) = Y(z)/U(z) = \mathbf{H}(z\mathbf{I} - \mathbf{\Phi})^{-1}\mathbf{\Gamma} + J. \tag{8.12}$$

Although we have agreed to restrict ourselves to the case where linear and stationary models are satisfactory, we still do not have a well-defined problem of identification. We mentioned earlier that our purpose in making the identification was to construct a model suitable for control system design. If we examine the design methods described in the earlier chapters, we find that these data are fundamentally grouped in two categories. For design via root locus or pole assignment we require a transfer function or state-variable description from which we can obtain the poles and zeros of the plant. As we have seen, if we have a transfer function in pole-zero form, we can find an equivalent state-variable description, and conversely, if we have the state-variable description as $\mathbf{F}$, $\mathbf{G}$, $\mathbf{H}$, $\mathbf{J}$ matrices, we can (with some computational effort) obtain the poles and zeros. These equivalent models are described by numbers which specify the elements of $\mathbf{F}$, $\mathbf{G}$, $\mathbf{H}$, and $\mathbf{J}$ or, equivalently, the numbers which specify poles and zeros. In either case we call these numbers the *parameters* of the model and the category of such models a *parametric description* of the plant.

In contrast to parametric models, the frequency response methods of Nyquist, Bode, and Nichols require the *curve* of the transfer function $\mathscr{H}(e^{j\omega T}) = Y(j\omega)/U(j\omega)$ as a function of ω. We often will take a state description $\mathbf{F}$, $\mathbf{G}$, $\mathbf{H}$, $\mathbf{J}$ and from these compute $\mathscr{H}(j\omega)$, or, again, we will be given the poles and zeros of $\mathscr{H}$ from which the curve readily follows. The fact remains, however, that the requirement of the *design* methods, which depend on such things as phase margin and gain margin, is the functional curve of $\mathscr{H}(j\omega)$ versus ω, no more. We call such a model a nonparametric model because in principle there may be *no* finite number of parameters which describe this curve.

Because two of our design methods (root-locus and pole-assignment) require a parametric model and the third method (frequency response) can *use* a parametric model, we will restrict our discussion here to identification of discrete parametric models.

To be more specific, suppose we use a transfer function description and select the parameters as the coefficients of the numerator and the denominator polynomials. We have, for the third-order case with one unit delay,

$$\mathscr{H}(z, \boldsymbol{\theta}) = \frac{b_1 z^{-1} + b_2 z^{-2} + b_3 z^{-3}}{1 - a_1 z^{-1} - a_2 z^{-2} - a_3 z^{-3}}, \tag{8.13}$$

and the parameter vector $\boldsymbol{\theta}$ is

$$\boldsymbol{\theta} = (a_1\, a_2\, a_3\, b_1\, b_2\, b_3)^T. \tag{8.14}$$

We imagine that we observe a set of input sample values $\{u(k)\}$ and a set of corresponding output sample values $\{y(k)\}$, and that these come from a plant which may be described by the transfer function (8.13) for some "true" value of the parameters, $\boldsymbol{\theta}^0$. Our task—the identification problem—is to compute from these $\{u(k)\}$ and $\{y(k)\}$ an estimate $\hat{\boldsymbol{\theta}}$ which is a "good" approximation to $\boldsymbol{\theta}^0$.[2]

To repeat the formulations from a different point of view, suppose we postulate a (discrete) state-variable description of the plant and take the parameters to be the elements of the matrices $\boldsymbol{\Phi}$, $\boldsymbol{\Gamma}$, $\mathbf{H}$. We assume $J = 0$. Then:

$$\begin{aligned} \mathbf{x}_{k+1} &= \boldsymbol{\Phi}(\boldsymbol{\theta}_1)\mathbf{x}_k + \boldsymbol{\Gamma}(\boldsymbol{\theta}_1)u_k, \\ y_k &= \mathbf{H}(\boldsymbol{\theta}_1)\mathbf{x}_k. \end{aligned} \tag{8.15}$$

For the third-order model, there are nine elements in $\boldsymbol{\Phi}$, three elements each in $\boldsymbol{\Gamma}$ and $\mathbf{H}$ for a total of fifteen parameters in (8.15), where six were enough in (8.14) for an equivalent description. If the six element $\boldsymbol{\theta}$ in (8.14) and the fifteen element $\boldsymbol{\theta}_1$ in (8.15) have the same transfer function, then we say they are *equivalent* parameters. Any set of u_k and y_k generated by one can be generated by the other, so, as far as control of y based on input u is concerned, there is no difference between $\boldsymbol{\theta}$ and $\boldsymbol{\theta}_1$ even though they have very different elements. This means, of course, that the state-variable description has nine parameters that are in some sense redundant and may be chosen rather arbitrarily. In fact, we have already seen in Chapter 6 that the definition of the state is not unique and that if we were to change the state in (8.15) by the substitution $\boldsymbol{\xi} = \mathbf{T}\mathbf{x}$, we would have the same transfer function $\mathcal{H}(z)$. It is just the nine elements of $\mathbf{T}$ which represent the excess parameters in (8.15); we should select these in a way which makes our task as easy as possible. The standard, even obvious, way to do this is to *define* our state so that $\boldsymbol{\Phi}$, $\boldsymbol{\Gamma}$, $\mathbf{H}$ are in accordance with one of the *canonical* forms of Chapter 2 for transfer functions. For example, in observer canonical form we would have

$$\boldsymbol{\Phi} = \begin{bmatrix} a_1 & 1 & 0 \\ a_2 & 0 & 1 \\ a_3 & 0 & 0 \end{bmatrix}, \qquad \boldsymbol{\Gamma} = \begin{bmatrix} b_1 \\ b_2 \\ b_3 \end{bmatrix}, \qquad \mathbf{H} = [1 \ \ 0 \ \ 0], \tag{8.16}$$

and we see immediately that these matrices are just functions of the $\boldsymbol{\theta}$ of (8.14), and the equivalence of $\boldsymbol{\theta}_1$ to $\boldsymbol{\theta}$ is obvious.

But now let us repeat, in this context, the comments made earlier about the difference between identification for control and modeling for science. In taking the canonical form represented by (8.16), we have almost surely scrambled into

---

[2] For the moment we assume that we know the number of states required to describe the plant. Estimation of n will be considered later.

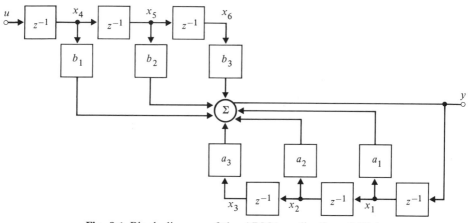

Fig. 8.1 Block diagram of the ARMA realization of $\mathcal{H}(z)$.

the a_i and b_i a jumble of the *physical* parameters of masses, force coefficients, gravity force, coefficients of friction, and so on. While the *physical* nature of the problem may be best described and understood in terms of these numbers, it is the a_i and b_i that best serve our purpose of control, and it is with these that we will be concerned here.

Before leaving the matter of canonical forms, it is appropriate at this point to introduce another form called in the past the "direct" form and more recently the ARMA[3] model, and which has one feature especially appropriate for identification. A block diagram of the ARMA realization of $\mathcal{H}(z)$ is shown in Fig. 8.1. Elementary calculations left to the reader will show that the transfer function of this block diagram is given by (8.13). The state equations are more interesting. Here we find six states and the matrices

$$\boldsymbol{\Phi} = \begin{bmatrix} a_1 & a_2 & a_3 & b_1 & b_2 & b_3 \\ 1 & 0 & 0 & 0 & 0 & 0 \\ 0 & 1 & 0 & 0 & 0 & 0 \\ 0 & 0 & 0 & 0 & 0 & 0 \\ 0 & 0 & 0 & 1 & 0 & 0 \\ 0 & 0 & 0 & 0 & 1 & 0 \end{bmatrix}, \quad \boldsymbol{\Gamma} = \begin{bmatrix} 0 \\ 0 \\ 0 \\ 1 \\ 0 \\ 0 \end{bmatrix},$$

$$\mathbf{H} = \begin{bmatrix} a_1 & a_2 & a_3 & b_1 & b_2 & b_3 \end{bmatrix}. \tag{8.17}$$

From the point of view of state-space analysis, the system described by (8.17) is seen to have six states to describe a third-order transfer function and thus to be "nonminimal." In fact, the $\boldsymbol{\Phi}$-matrix can be shown to have three poles at $z = 0$

---

[3] ARMA is an acronym for AutoRegressive Moving Average and comes from study of random processes generated by white noise filtered through the transfer function $\mathcal{H}(z)$ of (8.13). We will have more to say about this after the element of randomness is introduced.

that are *not observable* for any values of a_i or b_i. However, the system does have one remarkable property: the state is given by

$$\mathbf{x}_k = [y_{k-1} \quad y_{k-2} \quad y_{k-3} \quad u_{k-1} \quad u_{k-2} \quad u_{k-3}]^T. \tag{8.18}$$

In other words, the state is exactly given by the past inputs and outputs so that, if we have the set of $\{u_k\}$ and $\{y_k\}$, we have the state also, since it is merely a listing of six members of the set.

All the action, as it were, takes place in the output equation which is

$$
\begin{aligned}
y(k) &= \mathbf{H}\mathbf{x}(k) \\
&= a_1 y(k-1) + a_2 y(k-2) + a_3 y(k-3) + b_1 u(k-1) \\
&\quad + b_2 u(k-2) + b_3 u(k-3).
\end{aligned} \tag{8.19}
$$

There is no need to carry any other equation along since the state equations are trivially related to this output equation. We will use the ARMA model in some of our later formulations for identification.

Thus we conclude that within the class of discrete parametric models we wish to select a model which has the fewest number of parameters and yet will be equivalent to the assumed plant description. A model whose parameters are uniquely determined by the observed data is highly desirable; a model which will make subsequent control design simple is also often selected.

Having selected the class of models described by our assumed plant description, we now turn to the techniques for selecting the particular estimate, $\hat{\theta}$, which best represents the given data. For this we require some idea of goodness of fit of a proposed value of θ to the true θ^0. Because, by the very nature of the problem, θ^0 is unknown, it is unrealistic to define a direct parameter error between θ and θ^0. We must define the error in a way which can be computed from $\{u_k\}$ and $\{y_k\}$. Three criteria which have been proposed and studied extensively are *equation error, output error,* and *output-prediction error.*

For equation error, we need complete equations of motion as given, for example, by a state-variable description. To be just a bit general for the moment, suppose we have a nonlinear continuous time description with parameter vector θ which we can write as

$$\dot{\mathbf{x}} = \mathbf{f}(\mathbf{x}, \mathbf{u}; \theta).$$

We assume, first, that we know the form of the vector functions $\mathbf{f}$ but not the particular parameters θ^0, which describe the plant. We assume, second, that we are able to measure not only the controls, $\mathbf{u}$, but also the states, $\mathbf{x}$, and the state derivatives $\dot{\mathbf{x}}$. We thus know *everything* about the equations but the particular parameter values. We can, therefore, form an error comprised of the extent to which these equations of motion fail to be true for a specific value of θ when used with the specific actual data $\mathbf{x}_a$, $\dot{\mathbf{x}}_a$, and $\mathbf{u}_a$. We write

$$\dot{\mathbf{x}}_a - \mathbf{f}(\mathbf{x}_a, \mathbf{u}_a; \theta) = \mathbf{e}(t; \theta)$$

and $e(t, \theta^0) = 0$ where θ^0 are the true plant parameters. The vector $e(t, \theta)$ are the *equation errors*, and we should form some nonnegative function such as

$$J(\theta) = \int_0^T e^T(t, \theta)e(t; \theta) \, dt \tag{8.20}$$

and search over θ until we find $\hat{\theta}$ such that $J(\hat{\theta}) = 0$, at which time we will have a parameter set $\hat{\theta}$ which is equivalent to θ^0. If we have selected a unique parametrization, then only one parameter set will make $e(t; \theta^0) = 0$ and so we will have $\hat{\theta} = \theta^0$.[4]

The assumption that we have enough instruments to measure the total state and all state derivatives is a strong assumption and often not realistic in continuous model identification. However, in discrete linear models there is one case where it is immediate, and that is the case of an ARMA model. The reason for this is not hard to find: in an ARMA model the state is no more than recent values of input and output! To be explicit about it, let us write the linear discrete-model equation error, which is

$$x_a(k + 1) - \Phi x_a(k) - \Gamma u_a(k) = e(k; \theta), \tag{8.21}$$

where Φ and Γ are functions of the parameters θ. Now let us substitute the values from (8.17), the ARMA model:

$$\begin{bmatrix} x_1(k+1) \\ x_2(k+1) \\ x_3(k+1) \\ x_4(k+1) \\ x_5(k+1) \\ x_6(k+1) \end{bmatrix} - \begin{bmatrix} a_1 & a_2 & a_3 & b_1 & b_2 & b_3 \\ 1 & 0 & 0 & 0 & 0 & 0 \\ 0 & 1 & 0 & 0 & 0 & 0 \\ 0 & 0 & 0 & 0 & 0 & 0 \\ 0 & 0 & 0 & 1 & 0 & 0 \\ 0 & 0 & 0 & 0 & 1 & 0 \end{bmatrix} \begin{bmatrix} x_1 \\ x_2 \\ x_3 \\ x_4 \\ x_5 \\ x_6 \end{bmatrix} - \begin{bmatrix} 0 \\ 0 \\ 0 \\ 1 \\ 0 \\ 0 \end{bmatrix} u = \begin{bmatrix} e_1 \\ e_2 \\ e_3 \\ e_4 \\ e_5 \\ e_6 \end{bmatrix}. \tag{8.22}$$

When we make the substitution from (8.18) we find that, for any θ (which is to say, for any values of a_i and b_i) the elements of equation error are *all zero* except e_1. And this element of error is given by

$$x_1(k + 1) - a_1 x_1(k) - a_2 x_2(k) - a_3 x_3(k) - b_1 x_4(k) - b_2 x_5(k) - b_3 x_6(k) = e_1(k; \theta) \tag{8.23a}$$

or

$$y_a(k) - a_1 y_a(k - 1) - a_2 y_a(k - 2) - a_3 y_a(k - 3) - b_1 u_a(k - 1)$$
$$- b_2 u_a(k - 2) - b_3 u_a(k - 3) = e_1(k; \theta). \tag{8.23b}$$

The performance measure (8.20) becomes

$$J(\theta) = \sum_{k=0}^N e_1^2(k; \theta). \tag{8.24}$$

---

[4] Methods to search (8.20) for the minimizing $\hat{\theta}$ are the subject of nonlinear programming, which is discussed in Luenberger (1973). We give a brief discussion later in this chapter connected with maximum likelihood estimates.

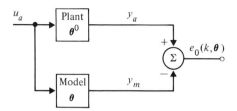

Fig. 8.2 Block diagram showing the formulation of output error.

Again, we place the subscript a on the observed data to emphasize the fact that these are *actual* data which were produced via the plant with (we assume) parameter values $\theta^0 = [a_1^0\, a_2^0\, a_3^0\, b_1^0\, b_2^0\, b_3^0]$, and in (8.23) and (8.24) an error results because θ differs from θ^0.

As we have seen, the general case of equation error requires measurement of all elements of state and state derivatives. In order to avoid these assumptions, a criterion based on *output error* may be used. This error is formulated in Fig. 8.2. Here we see that no attempt is made to measure the entire state of the plant, but rather the estimated parameter θ is used in a model to produce the model output $y_m(k)$, which is a function, therefore, of θ in the same way that the actual output is a function of the true parameter value θ^0. Now, in the scalar output case, e will be a scalar, and we can form, for example, the criterion function

$$J(\boldsymbol{\theta}) = \sum_{k=1}^{N} e_0^2(k;\, \boldsymbol{\theta}) \qquad (8.25)$$

and search for that $\hat{\boldsymbol{\theta}}$ which makes $J(\hat{\boldsymbol{\theta}})$ as small as possible. To illustrate the difference between output error and equation error in one case, consider again the ARMA model for which we have the model

$$y_m(k) = a_1 y_m(k-1) + a_2 y_m(k-2) + a_2 y_m(k-3) + b_1 u_a(k-1) \\ + b_2 u_a(k-2) + b_3 u_a(k-3)$$

and then the output error

$$e(k;\, \boldsymbol{\theta}) = y_a(k) - y_m(k) = y_a(k) - a_1 y_m(k-1) - a_2 y_m(k-2) - a_3 y_m(k-3) \\ - b_1 u_a(k-1) - b_2 u_a(k-2) - b_3 u_a(k-3). \qquad (8.26)$$

If we compare (8.26) with (8.23) we see that the equation-error formulation has past values of the *actual* output while the output-error formulation uses past values of the *model* output. Presumably the equation error is in some way better since it takes more knowledge into account, but we are not well situated to study the matter at this point.

The third approach to developing an error signal by which a parameter search can be structured is the output prediction error. Roughly speaking, the situation is this. When we discussed observers in Chapter 6 we made the point that a simple model is not a good basis for an observer because the errors cannot be controlled. When we add random effects into our data collection scheme, as we are

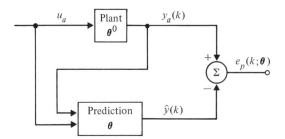

Fig. 8.3 Block diagram showing the generation of output prediction error.

about to do, much the same complaint can be raised about the generation of output errors. Instead of working with a simple model, with output $y_m(k)$, we are led to consider an output *predictor* which, within the confines of known structures, will do the best possible job of *predicting* $y(k)$ based on previous observations. We will see that in one special circumstance the ARMA model error as given by (8.23) has the least prediction error, but this is not true in general. Following our development of the observer, we may suspect that a *corrected* model would be appropriate, but we must delay any results in that area until further developments. For the moment, and for the purpose of displaying the output prediction error formulation, we have the situation pictured in Fig. 8.3. Again, having defined an error which depends on $\boldsymbol{\theta}$, we can formulate a performance criterion and search for that $\hat{\boldsymbol{\theta}}$ which is the best estimate of $\boldsymbol{\theta}^0$ within the context of the defined performance. Already, to illustrate this use of error we have defined a $J(\boldsymbol{\theta})$ in (8.20), (8.24), and (8.25), in each case a sum of squares. The choice of a performance criterion is guided both by the computational difficulties it imposes in the effort to obtain $\hat{\boldsymbol{\theta}}$ and by the properties of the estimate which results—by how hard $\hat{\boldsymbol{\theta}}$ is to find and how "good" it is when found. Among the criteria most widely used are:

1. Least-Squares Estimate (LS),
2. Best Linear Unbiased Estimate (BLUE),
3. Maximum Likelihood Estimate (MLE).

The latter two require a stochastic element to be introduced into the model, and our understanding of least squares is enhanced by a stochastic perspective. However, before we introduce random noise into the picture, it is informative to consider deterministic least squares. Thus we will discuss least squares first, then introduce random noise and discuss the formulation of random errors, and finally define and discuss the BLUE and the MLE, their calculation and their properties.

8.2 LEAST SQUARES

To begin our discussion of least squares, we will take the ARMA model and equation error which leads us to (8.23), repeated below for the nth-order case, but with the subscript a understood but omitted:

$$y(k) - a_1y(k-1) - a_2y(k-2) - \cdots - a_ny(k-n)$$
$$- b_1u(k-1) - \cdots - b_nu(k-n) = e(k; \boldsymbol{\theta}). \quad (8.27)$$

We assume that we observe the set of inputs and outputs

$$\{u(0), u(1), \ldots, u(N), y(0), y(1), \ldots, y(N)\}$$

and wish to compute values for

$$\boldsymbol{\theta} = [a_1 \cdots a_n b_1 \cdots b_n]^T$$

which will best fit the observed data. Since $y(k)$ depends on past data back to n periods earlier, the first error we can form is $e(n; \boldsymbol{\theta})$. Suppose we define the *vector* of errors by writing (8.27) over and over for $k = n, n+1, \ldots, N$. The results would be

$$y(n) = \boldsymbol{\psi}^T(n)\boldsymbol{\theta} + e(n; \boldsymbol{\theta}),$$
$$y(n+1) = \boldsymbol{\psi}^T(n+1)\boldsymbol{\theta} + e(n+1; \boldsymbol{\theta}),$$
$$\cdot$$
$$\cdot \qquad\qquad\qquad\qquad\qquad (8.28)$$
$$\cdot$$
$$y(N) = \boldsymbol{\psi}^T(N)\boldsymbol{\theta} + e(N; \boldsymbol{\theta}),$$

where we have used the fact that the state of the ARMA model is

$$\boldsymbol{\psi}(k) = [y(k-1)y(k-2) \ldots u(k-1) \ldots u(k-n)]^{T.5}$$

To make the error even more compact, we introduce another level of matrix notation and define

$$\mathbf{Y}(N) = [y(n) \ldots y(N)]^T,$$
$$\boldsymbol{\Psi}(N) = [\boldsymbol{\psi}(n)\boldsymbol{\psi}(n+1) \ldots \boldsymbol{\psi}(N)]^T,$$
$$\boldsymbol{\epsilon}(N; \boldsymbol{\theta}) = [e(n) \ldots e(N)]^T, \qquad\qquad (8.29)$$
$$\boldsymbol{\theta} = [a_1 \ldots a_n b_1 \ldots b_n]^T.$$

Note that $\boldsymbol{\Psi}(N)$ is a *matrix* with $2n$ columns and $N - n + 1$ rows. In terms of these definitions, we can write the equation errors as

$$\mathbf{Y} = \boldsymbol{\Psi}\boldsymbol{\theta} + \boldsymbol{\epsilon}(N; \boldsymbol{\theta}). \qquad\qquad (8.30)$$

Least squares is a prescription that one should take that value of $\boldsymbol{\theta}$ which makes the sum of the squares of the $e(k)$ as small as possible. In terms of (8.28) we define

$$J(\boldsymbol{\theta}) = \sum_{k=n}^{N} e^2(k; \boldsymbol{\theta}), \qquad\qquad (8.31a)$$

---

[5] We use $\boldsymbol{\psi}$ for the state here rather than $\mathbf{x}$ as in (8.23) because we will soon be developing equations for the evolution of $\boldsymbol{\theta}(k)$, estimates of the parameters, which will be states of our identification dynamic system. In Chapter 9 we will wish to compare these $\boldsymbol{\theta}(k)$-equations to equations in another $\mathbf{x}$ and do not wish to confuse these with $\boldsymbol{\psi}$ as given above.

and in terms of (8.30), this is

$$J(\boldsymbol{\theta}) = \boldsymbol{\epsilon}^T(N; \boldsymbol{\theta})\boldsymbol{\epsilon}(N; \boldsymbol{\theta}). \tag{8.31b}$$

We want to find $\hat{\boldsymbol{\theta}}_{LS}$, the least squares estimate of $\boldsymbol{\theta}^0$, which is that $\boldsymbol{\theta}$ having the property

$$J(\hat{\boldsymbol{\theta}}_{LS}) \leq J(\boldsymbol{\theta}). \tag{8.32}$$

But $J(\boldsymbol{\theta})$ is a quadratic function of the $2n$ parameters in $\boldsymbol{\theta}$, and, from calculus, we take the result that a necessary condition on $\hat{\boldsymbol{\theta}}_{LS}$ is that the partial derivatives of J with respect to $\boldsymbol{\theta}$ at $\boldsymbol{\theta} = \hat{\boldsymbol{\theta}}_{LS}$ should be zero. This we do as follows:

$$\begin{aligned} J_{\boldsymbol{\theta}} &= \boldsymbol{\epsilon}^T\boldsymbol{\epsilon} \\ &= (\mathbf{Y} - \boldsymbol{\Psi}\boldsymbol{\theta})^T(\mathbf{Y} - \boldsymbol{\Psi}\boldsymbol{\theta}) \\ &= \mathbf{Y}^T\mathbf{Y} - \boldsymbol{\theta}^T\boldsymbol{\Psi}^T\mathbf{Y} - \mathbf{Y}^T\boldsymbol{\Psi}\boldsymbol{\theta} + \boldsymbol{\theta}^T\boldsymbol{\Psi}^T\boldsymbol{\Psi}\boldsymbol{\theta}, \end{aligned}$$

and applying the rules developed above for derivatives of scalars with respect to vectors,[6] we obtain

$$J_{\boldsymbol{\theta}} = \left[\frac{\partial J}{\partial \boldsymbol{\theta}_i}\right] = -2\mathbf{Y}^T\boldsymbol{\Psi} + 2\boldsymbol{\theta}^T\boldsymbol{\Psi}^T\boldsymbol{\Psi}. \tag{8.33}$$

If we take the transpose of (8.33) and let $\boldsymbol{\theta} = \hat{\boldsymbol{\theta}}_{LS}$, we must get zero; thus

$$\boldsymbol{\Psi}^T\boldsymbol{\Psi}\hat{\boldsymbol{\theta}}_{LS} = \boldsymbol{\Psi}^T\mathbf{Y}. \tag{8.34}$$

These equations are called the *normal equations* of the problem, and their solution will provide us with the least squares $\hat{\boldsymbol{\theta}}_{LS}$.

Do the equations have a unique solution? The answer depends mainly on how $\boldsymbol{\theta}$ was selected and what input signals $\{u(k)\}$ were used. Recall that earlier we saw that a general third-order state model had fifteen parameters, but that only six of these were needed to completely describe the input-output dependency. If we stayed with the fifteen element $\boldsymbol{\theta}$, the resulting normal equations could not have a unique solution. To obtain a unique parameter set, we must select a canonical form having a minimal number of parameters, such as the observer or ARMA forms. By way of definition, a parameter $\boldsymbol{\theta}$ having the property that one and only one value of $\boldsymbol{\theta}$ makes $J(\boldsymbol{\theta})$ a minimum is said to be "identifiable." Two parameters having the property that $J(\boldsymbol{\theta}_1) = J(\boldsymbol{\theta}_2)$ are said to be "equivalent."

As to the selection of the inputs $u(k)$, let us consider an absurd case. Suppose $u(k) \equiv c$ for all k—a step function input. Now suppose we look at (8.28) again,

---

[6] The reader who has not done so before should write out $\mathbf{a}^T\mathbf{Q}\mathbf{a}$ for a 3×3 case and verify that $\partial \mathbf{a}^T\mathbf{Q}\mathbf{a}/\partial \mathbf{a} = 2\mathbf{a}^T\mathbf{Q}$.

for the third-order case to be specific. The errors are

$$y(3) = a_1 y(2) + a_2 y(1) + a_3 y(0) + b_1 c + b_2 c + b_3 c + e(3),$$

$$\cdot$$
$$\cdot$$
$$\cdot$$

$$\tag{8.34}$$

$$y(N) = a_1 y(N-1) + a_2(N-2) + a_3(N-3) + b_1 c + b_2 c + b_3 c + e(N).$$

It is obvious that in (8.34) the parameters b_1, b_2, and b_3 always appear as the sum $b_1 + b_2 + b_3$ and that separation of them is not possible when a constant input is used. Somehow the constant u fails to "excite" all the dynamics of the plant. This problem has been studied especially by Professor Astrom and his colleagues and students at Lund University, and the property of "persistently exciting" has been defined by them to describe a sequence $\{u(k)\}$ which fluctuates enough to avoid the possibility that only linear combinations of elements of θ will show up in the error and normal equations [Astrom (1966)]. Without being more specific at this point, we may say that an input is *persistently exciting* if the lower right $(n \times n)$-matrix component of $\Psi^T \Psi$ (which depends only on $\{u(k)\}$) is nonsingular.

For the moment, then, we will *assume* that the $u(k)$ are persistently exciting and that the θ are identifiable and consequently that $\Psi^T \Psi$ is nonsingular. We can then write the explicit solution

$$\hat{\theta}_{LS} = (\Psi^T \Psi)^{-1} \Psi^T Y. \tag{8.35}$$

It should be especially noted that although we are here mainly interested in identification of parameters to describe dynamic systems, the solution (8.35) derives entirely from the error equations (8.30) and the sum of squares criterion (8.31). Least squares is used for all manner of curve fitting, including nonlinear least squares when the error is a nonlinear function of the parameters. Numerical methods for solving for the least-squares solution to (8.30) without ever explicitly forming the product $\Psi^T \Psi$ have been extensively studied [see Golub (1965) and Strang (1976)].

The performance measure (8.31) is essentially based on the view that all the errors are equally important. This is not necessarily so, and a very simple addition can take account of known differences in the errors. We may know, for example, that data taken later in the experiment were much more in error than data taken early on and it would seem reasonable to weight the errors accordingly. Such a scheme is referred to as weighted least squares and is based on the performance criterion

$$J(\theta) = \sum_{k=n}^{N} w(k) e^2(k; \theta) = \epsilon^T W \epsilon. \tag{8.36}$$

In (8.36) we take the weighting function $w(k)$ to be positive and, presumably, to be small where the errors are expected to be large, and vice versa. In any event, calculation of the normal equations from (8.36) follows at once and gives

$$\boldsymbol{\Psi}^T \mathbf{W} \boldsymbol{\Psi} \hat{\boldsymbol{\theta}}_{\text{WLS}} = \boldsymbol{\Psi}^T \mathbf{W} \mathbf{Y},$$

and, subject to the coefficient matrix being nonsingular, we have

$$\hat{\boldsymbol{\theta}}_{\text{WLS}} = (\boldsymbol{\Psi}^T \mathbf{W} \boldsymbol{\Psi})^{-1} \boldsymbol{\Psi}^T \mathbf{W} \mathbf{Y}. \tag{8.37}$$

We note that (8.36) reduces to ordinary least squares when $\mathbf{W} = \mathbf{I}$, the identity matrix. Another common choice for $w(k)$ in (8.36) is $w(k) = (1 - \gamma)\gamma^{N-k}$. This choice weights the recent (k near N) observations more than the past (k near n) ones and corresponds to a first-order filter operating on the squared error. The factor $1 - \gamma$ causes the gain of the equivalent filter to be 1 for constant errors. As γ nears 1, the filter memory becomes long, and noise effects are reduced, while for smaller γ, the memory is short, and the estimate can track the changes which may occur in $\boldsymbol{\theta}$ if the computation is done over and over as N increases, a process we call *recursive least squares*.

8.3 RECURSIVE LEAST SQUARES

The weighted least-squares calculation for $\hat{\boldsymbol{\theta}}_{\text{WLS}}$ given in (8.37) is referred to as a "batch" calculation, since by the definition of the several entries, the formula presumes that one has a batch of data of length N from which the matrices $\mathbf{Y}$ and $\boldsymbol{\Psi}$ are composed according to the definitions in (8.29), and from which, with the addition of the weighting matrix $\mathbf{W}$, the normal equations are solved. There are times when the data are acquired sequentially rather than in a batch, and other times when one wishes to examine the nature of the solution as more data are included to see whether perhaps some improvement in the parameter estimates continues to be made or whether any surprises occur such as a sudden change in $\boldsymbol{\theta}$ or a persistent drift in one or more of the parameters. In short, one wishes sometimes to do a visual and experimental examination of the new estimate if one or several more data points are included in the computed values of $\hat{\boldsymbol{\theta}}_{\text{WLS}}$.

The equations of (8.37) can be put into a better form for sequential processing of the type described. We begin with (8.37) as solved for N data points and consider the consequences of taking one more observation. We need to consider the structure of $\boldsymbol{\Psi}^T \mathbf{W} \boldsymbol{\Psi}$ and $\boldsymbol{\Psi}^T \mathbf{W} \mathbf{Y}$ as one more datum is added. Consider first $\boldsymbol{\Psi}^T \mathbf{W} \boldsymbol{\Psi}$. To be specific about the weights, we will assume $w = a\gamma^{N-k}$. Then, if $a = 1$ and $\gamma = 1$, we have ordinary least squares, and if $a = 1 - \gamma$, we have exponentially weighted least squares. From (8.29) we have, for data up to time $N + 1$,

$$\boldsymbol{\Psi}^T = [\boldsymbol{\psi}(n) \ . \ . \ . \ \boldsymbol{\psi}(N) \, \boldsymbol{\psi}(N + 1)]$$

and

$$\boldsymbol{\Psi}^T \mathbf{W} \boldsymbol{\Psi} = \sum_{k=n}^{N+1} \boldsymbol{\psi}(k) w(k) \boldsymbol{\psi}^T(k) = \sum_{k=n}^{N+1} \boldsymbol{\psi}(k) a\gamma^{N+1-k} \boldsymbol{\psi}^T(k),$$

which can be written in two terms as

$$\mathbf{\Psi}^T\mathbf{W}\mathbf{\Psi} = \sum_{k=n}^{N} \psi(k)a\gamma\gamma^{N-k}\psi^T(k) + \psi(N + 1)a\psi^T(N + 1)$$

$$= \gamma\mathbf{\Psi}^T(N)\mathbf{W}(N)\mathbf{\Psi}(N) + \psi(N + 1)a\psi^T(N + 1). \qquad (8.38)$$

From the solution (8.37) we see that the inverse of the matrix in (8.38) will be required,[7] and for convenience and by convention we *define* the $2n \times 2n$ matrix $\mathbf{P}$ as

$$\mathbf{P}(N + 1) = [\mathbf{\Psi}^T(N + 1)\mathbf{W}\mathbf{\Psi}(N + 1)]^{-1}. \qquad (8.39)$$

Then we see that (8.38) may be written as

$$\mathbf{P}(N + 1) = [\gamma\mathbf{P}^{-1}(N) + \psi(N + 1)a\psi^T(N + 1)]^{-1}, \qquad (8.40)$$

and we need the inverse of a sum of two matrices. This is a well-known sort of problem, and a formula may be found attributed to Householder (1964) and known in the present context as the *matrix inversion lemma:*

$$(\mathbf{A} + \mathbf{BCD})^{-1} = \mathbf{A}^{-1} - \mathbf{A}^{-1}\mathbf{B}(\mathbf{C}^{-1} + \mathbf{DA}^{-1}\mathbf{B})^{-1}\mathbf{DA}^{-1}. \qquad (8.41)$$

To apply (8.41) to (8.40) we make the associations

$$\mathbf{A} = \gamma\mathbf{P}^{-1}(N),$$
$$\mathbf{B} = \psi(N + 1) \equiv \psi,$$
$$\mathbf{C} = w(N + 1) \equiv a,$$
$$\mathbf{D} = \psi^T(N + 1) \equiv \psi^T,$$

and find at once that

$$\mathbf{P}(N + 1) = \frac{\mathbf{P}(N)}{\gamma} - \frac{\mathbf{P}(N)}{\gamma}\psi\left(\frac{1}{a} + \psi^T\frac{\mathbf{P}(N)}{\gamma}\psi\right)^{-1}\psi^T\frac{\mathbf{P}(N)}{\gamma}. \qquad (8.42)$$

In the solution we also need $\mathbf{\Psi}^T\mathbf{W}\mathbf{Y}$ which we write as

$$\mathbf{\Psi}^T\mathbf{W}\mathbf{Y} = [\psi(n) \cdots \psi(N) \quad \psi(N + 1)]\begin{bmatrix} a\gamma^{N+1-n} & & & \bigcirc \\ & \cdot & & \\ & & \cdot & \\ & & & \cdot a\gamma & \\ \bigcirc & & & & a \end{bmatrix}\begin{bmatrix} y(n) \\ \cdot \\ \cdot \\ \cdot \\ y(N) \\ y(N + 1) \end{bmatrix}, \qquad (8.43)$$

---

[7] We assume for this discussion the existence of the inverse.

which can be expressed in two terms as

$$\mathbf{\Psi}^T\mathbf{WY}(N + 1) = \gamma\mathbf{\Psi}^T\mathbf{WY}(N) + \mathbf{\psi}(N + 1)ay(N + 1). \tag{8.44}$$

If we now substitute the expression for $\mathbf{P}(N + 1)$ from (8.42) and for $\mathbf{\Psi WY}(N + 1)$ from (8.44) into (8.37), we find [letting $\mathbf{P}(N) = P$, $\mathbf{\psi}(N + 1) = \mathbf{\psi}$, and $y(N + 1) = y$ for notational convenience]:

$$\hat{\boldsymbol{\theta}}_{\mathrm{WLS}}(N + 1) = \left[\frac{\mathbf{P}}{\gamma} - \frac{\mathbf{P}}{\gamma}\mathbf{\psi}\left(\frac{1}{a} + \mathbf{\psi}^T\frac{\mathbf{P}}{\gamma}\mathbf{\psi}\right)^{-1}\mathbf{\psi}^T\frac{\mathbf{P}}{\gamma}\right][\gamma\mathbf{\Psi WY}(N) + \mathbf{\psi}ay]. \tag{8.45}$$

When we multiply the factors in (8.45), we see that the term $\mathbf{P\Psi WY}(N) = \hat{\boldsymbol{\theta}}_{\mathrm{WLS}}(N)$ so that (8.45) reduces to

$$\hat{\boldsymbol{\theta}}_{\mathrm{WLS}}(N + 1) = \hat{\boldsymbol{\theta}}_{\mathrm{WLS}}(N) + \frac{\mathbf{P}}{\gamma}\mathbf{\psi}ay - \frac{\mathbf{P}}{\gamma}\mathbf{\psi}\left(\frac{1}{a} + \mathbf{\psi}^T\frac{\mathbf{P}}{\gamma}\mathbf{\psi}\right)^{-1}\mathbf{\psi}^T\hat{\boldsymbol{\theta}}_{\mathrm{WLS}}$$

$$- \frac{\mathbf{P}}{\gamma}\mathbf{\psi}\left(\frac{1}{a} + \mathbf{\psi}^T\frac{\mathbf{P}}{\gamma}\mathbf{\psi}\right)^{-1}\mathbf{\psi}^T\frac{\mathbf{P}}{\gamma}\mathbf{\psi}ay. \tag{8.46}$$

If we now insert the identity

$$\left(\frac{1}{a} + \mathbf{\psi}^T\frac{\mathbf{P}}{\gamma}\mathbf{\psi}\right)^{-1}\left(\frac{1}{a} + \mathbf{\psi}^T\frac{\mathbf{P}}{\gamma}\mathbf{\psi}\right)$$

between the $\mathbf{\psi}$ and the a in the second term on the right of (8.46), we can combine the two terms which multiply y to reduce (8.46) to

$$\hat{\boldsymbol{\theta}}_{\mathrm{WLS}}(N + 1) = \hat{\boldsymbol{\theta}}_{\mathrm{WLS}}(N) + \mathbf{L}(N + 1)(y(N + 1) - \mathbf{\psi}^T\hat{\boldsymbol{\theta}}_{\mathrm{WLS}}(N)), \tag{8.47}$$

where we have defined

$$\mathbf{L}(N + 1) = \frac{\mathbf{P}}{\gamma}\mathbf{\psi}\left(\frac{1}{a} + \frac{\mathbf{\psi}^T\mathbf{P}\mathbf{\psi}}{\gamma}\right)^{-1}. \tag{8.48}$$

Equations (8.42), (8.47), and (8.48) combined constitute an algorithm for computing $\hat{\boldsymbol{\theta}}$ recursively. To collect these, we proceed as follows:

1. Select a and γ and N.
2. Comment: $a = \gamma = 1$ is ordinary least squares; $a = 1 - \gamma$ and $0 < \gamma < 1$ is exponentially weighted least squares.
3. Select initial values for $\mathbf{P}(N)$ and $\hat{\boldsymbol{\theta}}(N)$.
4. Collect $y(0), \ldots, y(N)$ and $u(0), \ldots, u(N)$ and form $\mathbf{\psi}^T(N + 1)$.
5. Let $k \leftarrow N$.
6. $\mathbf{L}(k + 1) \leftarrow \dfrac{\mathbf{P}(k)}{\gamma}\mathbf{\psi}(k + 1)\left(\dfrac{1}{a} + \mathbf{\psi}^T(k + 1)\dfrac{\mathbf{P}(k)}{\gamma}\mathbf{\psi}(k + 1)\right)^{-1}$
7. Collect $y(k + 1)$ and $u(k + 1)$.
8. $\hat{\boldsymbol{\theta}}(k + 1) \leftarrow \hat{\boldsymbol{\theta}}(k) - \mathbf{L}(k + 1)(y(k + 1) - \mathbf{\psi}^T(k + 1)\hat{\boldsymbol{\theta}}(k))$

9. $\mathbf{P}(k + 1) \leftarrow \dfrac{1}{\gamma}[\mathbf{I} - \mathbf{L}(k + 1)\boldsymbol{\psi}^T(k + 1)]\mathbf{P}(k)$

10. Form $\boldsymbol{\psi}(k + 2)$.

11. Let $k \leftarrow k + 1$.

12. Go to step 6. $\hspace{6cm}$ (8.49)

Especially pleasing is the form of step 8, the "update" formula for the next value of the estimate. We see that the term $\boldsymbol{\psi}^T\hat{\boldsymbol{\theta}}(N)$ is the output to be expected at the time $N + 1$ based on the previous data, $\boldsymbol{\psi}(N + 1)$, and the previous estimate, $\hat{\boldsymbol{\theta}}(N)$. Thus the next estimate of $\boldsymbol{\theta}$ is given by the old estimate corrected by a term linear in the error between the observed output, $y(N + 1)$, and the predicted output, $\boldsymbol{\psi}^T\hat{\boldsymbol{\theta}}(N)$. The gain of the correction, $\mathbf{L}(N + 1)$, is given by (8.47) and (8.48). Note especially that in (8.49) no matrix inversion is required, but only division by

$$\frac{1}{a} + \boldsymbol{\psi}^T \frac{\mathbf{P}}{\gamma} \boldsymbol{\psi},$$

a scalar. However, one should not take the implication that (8.49) is without numerical difficulties, but their study is beyond the scope of this text.

We still have the question of initial conditions to resolve. Two possibilities are commonly recommended.

1. Collect a batch of $N > 2n$ data values and solve the batch formula (8.37) once for $\mathbf{P}(N)$, $\mathbf{L}(N + 1)$, and $\hat{\boldsymbol{\theta}}(N)$ and enter these values in (8.49) at step 3.

2. Set $\hat{\boldsymbol{\theta}}(N) = \mathbf{0}$, $\mathbf{P}(N) = \alpha\mathbf{I}$, where α is a large scalar. The suggestion has been made that an estimate of a suitable α is

$$\alpha = (10) \frac{1}{N + 1} \sum_{i=0}^{N} y^2(i).[8]$$

The steps in (8.49) update the least-squares estimate of the parameters $\boldsymbol{\theta}$ when one more pair of data points are taken. With only modest effort we can extend these formulas to include the case of vector or multivariable observations wherein the data $\mathbf{y}(k)$ are a vector of p simultaneous observations. We assume that parameters $\boldsymbol{\theta}$ have been defined such that the system may be described by

$$\mathbf{y}(k) = \boldsymbol{\psi}^T(k)\boldsymbol{\theta} + \mathbf{e}(k; \boldsymbol{\theta}). \hspace{3cm} (8.50)$$

One such set of parameters is defined by the multivariable ARMA model

$$\mathbf{y}(k) = -\sum_{i=1}^{n} a_i\mathbf{y}(k - i) + \sum_{i=1}^{n} \mathbf{B}_i\mathbf{u}(k - i), \hspace{2cm} (8.51)$$

---

[8] Soderstrom, Ljung, and Gustavsson (1974).

where the a_i are scalars, the $\mathbf{B}_i$ are $p \times m$ matrices, $\boldsymbol{\theta}$ is $(n + nmp) \times 1$, and the $\boldsymbol{\psi}(k)$ are now $(n + nmp) \times p$ matrices. If we define $\boldsymbol{\Psi}$, $\mathbf{Y}$, and $\boldsymbol{\epsilon}$ as in (8.29), the remainder of the batch formula development proceeds exactly as before, leading to (8.35) for the least-squares estimates and (8.37) for the weighted least-squares estimates with little more than a change in the definition of the elements in the equations. We need to modify (8.36) to $J = \Sigma \mathbf{e}^T \mathbf{we}$ reflecting the fact that $\mathbf{e}(k)$ is now also a $p \times 1$ vector and the $\mathbf{w}(k)$ are $p \times p$ nonsingular matrices.

To compute the recursive estimate equations, we need to repeat (8.38) and the development following (8.38) with the new definitions. Only minor changes are required. For example, in (8.42) we must replace $1/a$ by $\mathbf{a}^{-1}$. The resulting equations are, in the format of (8.49),

$$\mathbf{L}(N + 1) = \frac{\mathbf{P}}{\gamma}\,\boldsymbol{\psi}\,\left(\mathbf{a}^{-1} + \boldsymbol{\psi}^T\,\frac{\mathbf{P}}{\gamma}\,\boldsymbol{\psi}\right)^{-1}, \tag{8.52a}$$

$$\mathbf{P}(N + 1) = \frac{1}{\gamma}\,(\mathbf{I} - \mathbf{L}(N + 1)\boldsymbol{\psi}^T)\mathbf{P}, \tag{8.52b}$$

$$\hat{\boldsymbol{\theta}}_{\text{WLS}}(N + 1) = \hat{\boldsymbol{\theta}}_{\text{WLS}} + \mathbf{L}(N + 1)[\mathbf{y}(N + 1) - \boldsymbol{\psi}^T\hat{\boldsymbol{\theta}}_{\text{WLS}}(N)]. \tag{8.52c}$$

8.4 STOCHASTIC LEAST SQUARES

Thus far we have presented the least-squares method with no comment about the possibility that the data may in fact be subject to random effects. Because such effects are very common, and often are the best available vehicle for describing the differences between an ideal model and real-plant signal observations, it is essential that some account of such effects be included in our calculations. We will begin with an analysis of the most elementary of cases and add realism (and difficulties) as we go along. Appendix D provides a brief catalog of results we will need from probability, statistics, and stochastic processes.

As a start, we consider the case of a deterministic model with random errors in the data. We consider the equations which generate the data to be[9]

$$y(k) = \mathbf{a}^T\boldsymbol{\theta}^0 + v(k), \tag{8.53}$$

which, in matrix notation, becomes

$$\mathbf{Y} = \mathbf{A}\boldsymbol{\theta}^0 + \mathbf{V}. \tag{8.54}$$

The $v(k)$ in (8.53) are assumed to be random variables with zero mean (one can always subtract a known mean) and known covariance. In particular, we assume

---

[9] There is no connection between the a in (8.53) and the a used as part of the weights in Section 8.3.

that the actual data are generated from (8.54) with $\boldsymbol{\theta} = \boldsymbol{\theta}^0$ and

$$
\begin{aligned}
\mathcal{E}v(k) &= 0, \\
\mathcal{E}v(k)v(j) = \sigma^2\delta_{kj} &= 0 \qquad (k \neq j) \\
&= \sigma^2 \qquad (k = j),
\end{aligned} \tag{8.55}
$$

or

$$
\mathcal{E}\mathbf{V}\mathbf{V}^T = \sigma^2\mathbf{I}.
$$

As an example of equations of the type considered here, suppose we have a physics experiment in which observations are made of the positions of a mass which moves without forces but with unknown initial position and velocity. We assume then that the motion will be a line, which may be written as

$$
y(t) = a^0 + b^0 t + n(t). \tag{8.56}
$$

If we take observations at times $t_0, t_1, t_2, \ldots, t_N$, then

$$
\mathbf{Y} = [y(t_0) \ldots y(t_N)]^T, \qquad \mathbf{A} = \begin{pmatrix} t_0 & t_1 & \cdots & t_N \\ 1 & 1 & & 1 \end{pmatrix}^T,
$$

the unknown parameters are

$$
\boldsymbol{\theta} = [a \quad b]^T,
$$

and the noise is

$$
v(k) = n(t_k).
$$

We assume that the observations are without systematic bias, i.e., the noise has zero average value, and that we can estimate the noise "intensity," or mean-square value, σ^2. We assume zero error in the clock times, t_k, so $\mathbf{A}$ is known.

The (stochastic) least-squares problem is to find $\hat{\boldsymbol{\theta}}$ which will minimize

$$
\begin{aligned}
J(\boldsymbol{\theta}) &= (\mathbf{Y} - \mathbf{A}\boldsymbol{\theta})^T(\mathbf{Y} - \mathbf{A}\boldsymbol{\theta}) \\
&= \sum_{k=0}^{N} e^2(k; \boldsymbol{\theta}).
\end{aligned} \tag{8.57}
$$

Note that the errors, $e(k; \boldsymbol{\theta})$, depend both on the random noise and on the choice of $\boldsymbol{\theta}$, and $e(k; \boldsymbol{\theta}^0) = v(k)$. Now the solution will be a random variable because the data, $\mathbf{Y}$, on which it is based, is random. However, for specific actual data, (8.57) represents the same quadratic function of $\boldsymbol{\theta}$ we have seen before and the same form of the solution results, save only the substitution of $\mathbf{A}$ for $\boldsymbol{\Psi}$, namely,

$$
\hat{\boldsymbol{\theta}}_{\text{LS}} = (\mathbf{A}^T\mathbf{A})^{-1}\mathbf{A}^T\mathbf{Y}. \tag{8.58}
$$

Now, however, we would not expect to find zero for the sum of the errors given by (8.57), even if we should determine $\hat{\boldsymbol{\theta}} = \boldsymbol{\theta}^0$ exactly (which of course we won't).

Because of the random effects, then, we must generalize our concepts of what constitutes a "good" estimate. There are three features of a stochastic estimate which we will use. The first of these is consistency.

An estimate $\hat{\boldsymbol{\theta}}$ of a parameter $\boldsymbol{\theta}^0$ is said to be *consistent* if, in the long run, the difference between $\hat{\boldsymbol{\theta}}$ and $\boldsymbol{\theta}^0$ becomes negligible. In statistics and probability, there are several formal ways by which one can define a negligible difference. For our purposes we will use the mean-square criterion by which we measure the difference between $\hat{\boldsymbol{\theta}}$ and $\boldsymbol{\theta}^0$ by the sum of the squares of the parameter error, $\hat{\boldsymbol{\theta}} - \boldsymbol{\theta}^0$. We make explicit the dependence of $\hat{\boldsymbol{\theta}}$ on the length of the data by writing $\hat{\boldsymbol{\theta}}(N)$ and say that an estimate is consistent if

$$\lim_{N\to\infty} \mathscr{E}(\hat{\boldsymbol{\theta}}(N) - \boldsymbol{\theta}^0)^T(\hat{\boldsymbol{\theta}}(N) - \boldsymbol{\theta}^0) = 0,$$

$$\lim_{N\to\infty} \text{tr } \mathscr{E}(\hat{\boldsymbol{\theta}}(N) - \boldsymbol{\theta}^0)(\hat{\boldsymbol{\theta}}(N) - \boldsymbol{\theta}^0)^T = 0. \qquad (8.59)^{[10]}$$

If (8.59) is true, we say that $\hat{\boldsymbol{\theta}}(N)$ converges to $\boldsymbol{\theta}^0$ in the mean-square sense as N approaches infinity. The expression in (8.59) can be made more explicit in the case of the least-squares estimate of (8.58). We have

$$\begin{aligned}
\hat{\boldsymbol{\theta}}_{\text{LS}} - \boldsymbol{\theta}^0 &= (\mathbf{A}^T\mathbf{A})^{-1}\mathbf{A}^T(\mathbf{A}\boldsymbol{\theta}^0 + \mathbf{V}) - \boldsymbol{\theta}^0 \\
&= \boldsymbol{\theta}^0 + (\mathbf{A}^T\mathbf{A})^{-1}\mathbf{A}^T\mathbf{V} - \boldsymbol{\theta}^0 \\
&= (\mathbf{A}^T\mathbf{A})^{-1}\mathbf{A}^T\mathbf{V}
\end{aligned}$$

and

$$\begin{aligned}
\mathscr{E}(\hat{\boldsymbol{\theta}}_{\text{LS}} - \boldsymbol{\theta}^0)(\hat{\boldsymbol{\theta}}_{\text{LS}} - \boldsymbol{\theta}^0)^T &= \mathscr{E}\{(\mathbf{A}^T\mathbf{A})^{-1}\mathbf{A}^T\mathbf{V}\mathbf{V}^T\mathbf{A}(\mathbf{A}^T\mathbf{A})^{-1}\} \\
&= (\mathbf{A}^T\mathbf{A})^{-1}\mathbf{A}^T\mathscr{E}\mathbf{V}\mathbf{V}^T\mathbf{A}(\mathbf{A}^T\mathbf{A})^{-1} \\
&= \sigma^2(\mathbf{A}^T\mathbf{A})^{-1} \qquad (8.60)
\end{aligned}$$

In making the reductions shown in the development of (8.60) we have used the facts that $\mathbf{A}$ is a known matrix (not random) and that $\mathscr{E}\mathbf{V}\mathbf{V}^T = \sigma^2\mathbf{I}$. Continuing then with consideration of the consistency of the least-squares estimate, we see that $\hat{\boldsymbol{\theta}}_{\text{LS}}$ is consistent if

$$\lim_{N\to\infty} \text{tr } \sigma^2(\mathbf{A}^T\mathbf{A})^{-1} = 0. \qquad (8.61)$$

As an example of consistency, we consider the most simple of problems, the observation of a constant. Perhaps we have the mass of the earlier problem, but with zero velocity. Thus

$$y(k) = a^0 + v(k), \qquad k = 0, 1, \ldots, N, \qquad (8.62)$$

---

[10] The trace (tr) of a matrix $\mathbf{A}$ with elements a_{ij} is the sum of the diagonal elements, or trace $\mathbf{A} = \Sigma_{i=1}^n a_{ii}$.

and

$$\mathbf{A} = [1 \quad 1 \ . \ . \ . \ 1]^T, \qquad \theta^0 = a^0,$$

$$\mathbf{A}^T\mathbf{A} = \sum_{k=0}^{N} 1 = N + 1, \qquad \mathbf{A}^T\mathbf{Y} = \sum_{k=0}^{N} y(k);$$

then

$$\hat{\boldsymbol{\theta}}_{\text{LS}}(N) = (N + 1)^{-1} \sum_{k=0}^{N} y(k), \tag{8.63}$$

which is the sample average.

Now, if we apply (8.61), we find

$$\lim_{N \to \infty} \text{tr } \sigma^2(N + 1)^{-1} = \lim_{N \to \infty} \frac{\sigma^2}{N + 1} = 0, \tag{8.64}$$

and we conclude that this estimate (8.63) is a *consistent* estimate of a^0. If we keep taking observations according to (8.62) and calculating the sum according to (8.63), we will eventually have a value which differs from a^0 by a negligible amount in the mean-square sense.

The estimate given by (8.63) is a batch calculation. It may be informative to apply the recursive algorithm of (8.49) to this trivial case just to see how the equations will look. Suppose we agree to start the equations with one stage of "batch" on, say, two observations, $y(0)$ and $y(1)$. We thus have $a = \gamma = 1$, for least squares, and

$$\begin{array}{ll}
P(1) = (A^TA)^{-1} = \frac{1}{2}, & \\
\psi(N) = 1, & \text{initial conditions} \\
\hat{\boldsymbol{\theta}}(1) = \frac{1}{2}(y(0) + y(1)), & \\
W = 1. &
\end{array} \tag{8.65}$$

The iteration in P is given by (since $\psi \equiv 1$)

$$P(N + 1) = P(N) - \frac{P(N)(1)(1)P(N)}{1 + (1)P(N)(1)}$$

$$= \frac{P(N)}{1 + P(N)}.$$

Thus the entire iteration is given by the initial conditions of (8.65) plus

1. Let $N = 1$.

2. $P(N + 1) = \dfrac{P(N)}{1 + P(N)}$.

3. $L(N + 1) = \dfrac{P(N)}{1 + P(N)}$.

4. $\hat{\boldsymbol{\theta}}(N + 1) = \hat{\boldsymbol{\theta}}(N) + L(N + 1)(y(N + 1) - \hat{\boldsymbol{\theta}}(N))$.

5. Let N be replaced by $N + 1$.

6. Go to step 2. $\tag{8.66}$

The reader can verify that in this simple case, $\mathbf{P}(N) = 1/(N + 1)$ and step 4 of the recursive equations gives the same (consistent) estimate as the batch formula (4.60). We also note that the matrix $\mathbf{P}$ is proportional to the variance in the error of the parameter estimate as expressed in (8.60).

While consistency is the first property one should expect of an estimate, it is after all an asymptotic property which describes a feature of $\hat{\boldsymbol{\theta}}$ as N grows without bound. A second property which may be evaluated is that of "bias." If $\hat{\boldsymbol{\theta}}$ is an estimate of $\boldsymbol{\theta}^0$, the bias in the estimate is the difference between the mean value of $\hat{\boldsymbol{\theta}}$ and the true value, $\boldsymbol{\theta}^0$. We have

$$\mathscr{E}\hat{\boldsymbol{\theta}}(N) - \boldsymbol{\theta}^0 = \mathbf{b}. \tag{8.67}$$

If $\mathbf{b} = \mathbf{0}$ for all N, the estimate is said to be *unbiased*. The least-squares estimate given by (8.35) is unbiased, which we can prove by direct calculation as follows. If we return to the development of the mean-square error in $\hat{\boldsymbol{\theta}}_{LS}$ given by (8.60) we find that

$$\hat{\boldsymbol{\theta}}_{LS} - \boldsymbol{\theta}^0 = (\mathbf{A}^T\mathbf{A})^{-1}\mathbf{A}^T\mathbf{V}, \tag{8.68}$$

and the bias is

$$\begin{aligned}
\mathscr{E}\hat{\boldsymbol{\theta}}_{LS} - \boldsymbol{\theta}^0 &= \mathscr{E}(\mathbf{A}^T\mathbf{A})^{-1}\mathbf{A}^T\mathbf{V} \\
&= (\mathbf{A}^T\mathbf{A})^{-1}\mathbf{A}^T\mathscr{E}\mathbf{V} \\
&= \mathbf{0}.
\end{aligned} \tag{8.69}$$

Actually, we would really like to have $\hat{\boldsymbol{\theta}}(N)$ be "close" to $\boldsymbol{\theta}^0$ for finite values of N and would, as a third property, like to treat the mean-square parameter error for finite numbers of samples. Unfortunately, it is difficult to obtain an estimate which has a minimum for the square of $\hat{\boldsymbol{\theta}} - \boldsymbol{\theta}^0$ without involving the value of $\boldsymbol{\theta}^0$ directly, which we do not know, else there is no point in estimating it. We can, however, find an estimate which is the best (in the mean-square parameter error sense) estimate which is also linear in $\mathbf{Y}$ and unbiased. The result is called a BLUE.

The development proceeds as follows. Since the estimate is to be a linear function of the data, we write[11]

$$\hat{\boldsymbol{\theta}} = \mathbf{L} \cdot \mathbf{Y}. \tag{8.70}$$

Since the estimate is to be unbiased, we require

$$\mathscr{E}\hat{\boldsymbol{\theta}} = \boldsymbol{\theta}^0$$

or

$$\mathscr{E}\mathbf{L}\mathbf{Y} = \mathscr{E}\mathbf{L}(\mathbf{A}\boldsymbol{\theta}^0 + \mathbf{V}) = \boldsymbol{\theta}^0.$$

---

[11] The matrix $\mathbf{L}$ used here for a linear dependence has no connection to the gain matrix used in the recursive estimate equations (8.52).

Thus

$$\mathbf{LA}\boldsymbol{\theta}^0 = \boldsymbol{\theta}^0,$$

or

$$\mathbf{LA} = \mathbf{I}. \tag{8.71}$$

We wish to find $\mathbf{L}$ so that an estimate of the form (8.70), subject to the constraint (8.71), makes the mean-square error

$$J(\mathbf{L}) = \text{tr } \mathscr{E}(\hat{\boldsymbol{\theta}} - \boldsymbol{\theta}^0)(\hat{\boldsymbol{\theta}} - \boldsymbol{\theta}^0)^T \tag{8.72}$$

as small as possible. Using (8.70) in (8.72), we have

$$J(\mathbf{L}) = \text{tr } \mathscr{E}\{(\mathbf{LY} - \boldsymbol{\theta}^0)(\mathbf{LY} - \boldsymbol{\theta}^0)^T\},$$

and using (8.54) for $\mathbf{Y}$

$$J(\mathbf{L}) = \text{tr } \mathscr{E}(\mathbf{LA}\boldsymbol{\theta}^0 + \mathbf{LV} - \boldsymbol{\theta}^0)(\mathbf{LA}\boldsymbol{\theta}^0 + \mathbf{LV} - \boldsymbol{\theta}^0)^T.$$

But from (8.71), this reduces to

$$\begin{aligned} J(\mathbf{L}) &= \text{tr } \mathscr{E}(\mathbf{LV})(\mathbf{LV})^T \\ &= \text{tr } \mathbf{LRL}^T, \end{aligned} \tag{8.73}$$

where we take the covariance for the noise to be

$$\mathscr{E}\mathbf{VV}^T = \mathbf{R}.$$

We now have the entirely deterministic problem of finding that $\mathbf{L}$ subject to (8.71) which makes (8.73) as small as possible. We solve the problem in an indirect way: we first find $\mathbf{L}$ when $\boldsymbol{\theta}^0$ is a scalar and there is no trace operation ($\mathbf{L}$ is a single row). From this scalar solution we *conjecture* what the multiparameter solution might be and demonstrate that it is in fact correct. First, the case when $\mathbf{L}$ is a row. We must introduce the constraint (8.71), and we do this by a Lagrange multiplier as in calculus and are led to find $\mathbf{L}$ such that

$$J(\mathbf{L}) = \mathbf{LRL}^T + \lambda(\mathbf{A}^T\mathbf{L}^T - \mathbf{I}) \tag{8.74}$$

is a minimum. The λ are the Lagrange multipliers. The necessary conditions on the elements of $\mathbf{L}$ are that $\partial J/\partial \mathbf{L}^T$ be zero. (We use $\mathbf{L}^T$, which is a column, to retain notation that is consistent with our earlier discussion of vector-matrix derivatives.)

$$\left.\frac{\partial J}{\partial \mathbf{L}^T}\right|_{\mathbf{L}=\hat{\mathbf{L}}} = 2\hat{\mathbf{L}}\mathbf{R} + \lambda\mathbf{A}^T = 0.$$

Thus the best value for $\mathbf{L}$ is given by

$$\begin{aligned} 2\hat{\mathbf{L}}\mathbf{R} &= -\lambda\mathbf{A}^T, \\ \hat{\mathbf{L}} &= -\tfrac{1}{2}\lambda\mathbf{A}^T\mathbf{R}^{-1}. \end{aligned} \tag{8.75}$$

Since the constraint (8.71) must be satisfied, $\hat{\mathbf{L}}$ must be such that

$$\hat{\mathbf{L}}\mathbf{A} = \mathbf{I}, \quad \text{or} \quad -\tfrac{1}{2}\lambda\mathbf{A}^T\mathbf{R}^{-1}\mathbf{A} = \mathbf{I}.$$

From this we conclude that

$$\lambda = -2(\mathbf{A}^T\mathbf{R}^{-1}\mathbf{A})^{-1}, \tag{8.76}$$

and substituting (8.76) back again in (8.75), we have, finally, that

$$\hat{\mathbf{L}} = (\mathbf{A}^T\mathbf{R}^{-1}\mathbf{A})^{-1}\mathbf{A}^T\mathbf{R}^{-1}, \tag{8.77}$$

and the BLUE is

$$\hat{\boldsymbol{\theta}}_B = \hat{\mathbf{L}}\mathbf{Y} = (\mathbf{A}^T\mathbf{R}^{-1}\mathbf{A})^{-1}\mathbf{A}^T\mathbf{R}^{-1}\mathbf{Y}. \tag{8.78}$$

If we look at (8.37), we immediately see that this is exactly weighted least squares with $\mathbf{W} = \mathbf{R}^{-1}$. What, in effect, we have done is give a reasonable—best linear unbiased—criterion for selection of the weights. If, as we assumed earlier, $\mathbf{R} = \sigma^2\mathbf{I}$, then the ordinary least-squares estimate is also the BLUE.

But we have jumped ahead of ourselves; we have yet to show that (8.77) or (8.78) is true for a matrix $\mathbf{L}$. Suppose we have another linear unbiased estimate, $\bar{\boldsymbol{\theta}}$. We can write, without loss of generality,

$$\bar{\boldsymbol{\theta}} = \mathbf{L}\mathbf{Y} = (\hat{\mathbf{L}} + \overline{\mathbf{L}})\mathbf{Y},$$

where $\hat{\mathbf{L}}$ is given by (8.77). Since $\bar{\boldsymbol{\theta}}$ is required to be unbiased, (8.71) requires that

$$\mathbf{L}\mathbf{A} = \mathbf{I}, \quad (\hat{\mathbf{L}} + \overline{\mathbf{L}})\mathbf{A} = \mathbf{I}, \quad \hat{\mathbf{L}}\mathbf{A} + \overline{\mathbf{L}}\mathbf{A} = \mathbf{I},$$
$$\mathbf{I} + \overline{\mathbf{L}}\mathbf{A} = \mathbf{I}, \quad \overline{\mathbf{L}}\mathbf{A} = \overline{\mathbf{0}}. \tag{8.79}$$

Now the mean-square error using $\bar{\boldsymbol{\theta}}$ is

$$J(\bar{\boldsymbol{\theta}}) = \text{tr } \mathscr{E}(\bar{\boldsymbol{\theta}} - \boldsymbol{\theta}^0)(\bar{\boldsymbol{\theta}} - \boldsymbol{\theta}^0)^T,$$

which can be written as

$$J(\bar{\boldsymbol{\theta}}) = \text{tr } \mathscr{E}(\bar{\boldsymbol{\theta}} - \hat{\boldsymbol{\theta}} + \hat{\boldsymbol{\theta}} - \boldsymbol{\theta}^0)(\bar{\boldsymbol{\theta}} - \hat{\boldsymbol{\theta}} + \hat{\boldsymbol{\theta}} - \boldsymbol{\theta}^0)^T$$
$$= \text{tr}\{\mathscr{E}(\bar{\boldsymbol{\theta}} - \hat{\boldsymbol{\theta}})(\bar{\boldsymbol{\theta}} - \hat{\boldsymbol{\theta}})^T + 2\mathscr{E}(\bar{\boldsymbol{\theta}} - \hat{\boldsymbol{\theta}})(\hat{\boldsymbol{\theta}} - \boldsymbol{\theta}^0)^T + \mathscr{E}(\hat{\boldsymbol{\theta}} - \boldsymbol{\theta}^0)(\hat{\boldsymbol{\theta}} - \boldsymbol{\theta}^0)^T\}. \tag{8.80}$$

We note that the trace of a sum is the sum of the traces of the components, and the trace of the last term is, by definition, $J(\hat{\mathbf{L}})$. Let us consider the second term of (8.80), namely

$$\text{term 2} \overset{\Delta}{=} \text{tr } \mathscr{E}(\bar{\boldsymbol{\theta}} - \hat{\boldsymbol{\theta}})(\hat{\boldsymbol{\theta}} - \boldsymbol{\theta}^0)^T = \text{tr } \mathscr{E}(\mathbf{L}\mathbf{Y} - \hat{\mathbf{L}}\mathbf{Y})(\hat{\mathbf{L}}\mathbf{Y} - \boldsymbol{\theta}^0)^T$$
$$= \text{tr } \mathscr{E}(\hat{\mathbf{L}}\mathbf{Y} + \overline{\mathbf{L}}\mathbf{Y} - \hat{\mathbf{L}}\mathbf{Y})(\hat{\mathbf{L}}\mathbf{Y} - \boldsymbol{\theta}^0)$$
$$= \text{tr } \mathscr{E}(\overline{\mathbf{L}}\mathbf{Y})(\hat{\mathbf{L}}\mathbf{Y} - \boldsymbol{\theta}^0)$$
$$= \text{tr } \mathscr{E}(\overline{\mathbf{L}}(\mathbf{A}\boldsymbol{\theta}^0 + \mathbf{V}))(\hat{\mathbf{L}}(\mathbf{A}\boldsymbol{\theta}^0 + \mathbf{V}) - \boldsymbol{\theta}^0)^T.$$

Now we use (8.79) to eliminate one term and (8.71) to eliminate another to the effect that

$$\text{term 2} = \text{tr } \mathscr{E}(\overline{\mathbf{L}}\mathbf{V})(\hat{\mathbf{L}}\mathbf{V})^T$$
$$= \text{tr } \overline{\mathbf{L}}\mathbf{R}\hat{\mathbf{L}}^T,$$

and using (8.77) for $\hat{\mathbf{L}}$, we have

$$\text{term 2} = \text{tr } \overline{\mathbf{L}}\mathbf{R}\{\mathbf{R}^{-1}\mathbf{A}(\mathbf{A}^T\mathbf{R}^{-1}\mathbf{A})^{-1}\}$$
$$= \text{tr } \overline{\mathbf{L}}\mathbf{A}(\mathbf{A}^T\mathbf{R}^{-1}\mathbf{A})^{-1},$$

but now from (8.79) we see that the term is zero. Thus we reduce (8.80) to

$$J(\overline{\boldsymbol{\theta}}) = \text{tr } \mathscr{E}(\overline{\boldsymbol{\theta}} - \hat{\boldsymbol{\theta}})(\overline{\boldsymbol{\theta}} - \hat{\boldsymbol{\theta}})^T + J(\hat{\boldsymbol{\theta}}).$$

Since the first term on the right is the sum of expected values of squares, it is zero or positive. Thus $J(\overline{\boldsymbol{\theta}}) \geq J(\hat{\boldsymbol{\theta}})$, and we have proved that (8.78) is really and truly BLUE.

To conclude this section on stochastic least squares, we may summarize our findings as follows.

If the data are described by

$$\mathbf{Y} = \mathbf{A}\boldsymbol{\theta}^0 + \mathbf{V},$$

and

$$\mathscr{E}\mathbf{V} = 0, \qquad \mathscr{E}\mathbf{V}\mathbf{V}^T = \mathbf{R},$$

then the least-squares estimate of $\boldsymbol{\theta}^0$ is given by

$$\hat{\boldsymbol{\theta}}_{\text{LS}} = (\mathbf{A}^T\mathbf{A})^{-1}\mathbf{A}^T\mathbf{Y}, \tag{8.81}$$

which is unbiased. If $\mathbf{R} = \sigma^2\mathbf{I}$, then the variance of $\hat{\boldsymbol{\theta}}_{\text{LS}}$, which is defined as $\mathscr{E}(\hat{\boldsymbol{\theta}}_{\text{LS}} - \boldsymbol{\theta}^0)(\hat{\boldsymbol{\theta}}_{\text{LS}} - \boldsymbol{\theta}^0)^T$, is

$$\text{var}(\hat{\boldsymbol{\theta}}_{\text{LS}}) = \sigma^2(\mathbf{A}^T\mathbf{A})^{-1}. \tag{8.82}$$

From this we showed that the least-squares estimation of a constant in noise is not only unbiased but consistent.

Furthermore, if we insist that the estimate be both a linear function of the data and unbiased, we showed that the BLUE is a weighted least squares given by

$$\hat{\boldsymbol{\theta}}_B = (\mathbf{A}^T\mathbf{R}^{-1}\mathbf{A})^{-1}\mathbf{A}^T\mathbf{R}^{-1}\mathbf{Y}. \tag{8.83}$$

The variance of $\hat{\boldsymbol{\theta}}_B$ is

$$\text{var}(\hat{\boldsymbol{\theta}}_B) = \mathscr{E}(\hat{\boldsymbol{\theta}}_B - \boldsymbol{\theta}^0)(\hat{\boldsymbol{\theta}}_B - \boldsymbol{\theta}^0)^T$$
$$= (\mathbf{A}^T\mathbf{R}^{-1}\mathbf{A})^{-1}. \tag{8.84}$$

Thus, if $\mathbf{R} = \sigma^2\mathbf{I}$, then the least-squares estimate is also the BLUE.

As another comment on the BLUE, we note that in a recursive formulation we take, according to (8.39),

$$\mathbf{P} = (\mathbf{A}^T\mathbf{R}^{-1}\mathbf{A})^{-1},$$

and so the matrix $\mathbf{P}$ is the variance of the estimate $\hat{\boldsymbol{\theta}}_B$. In the recursive equations given for vector measurements in (8.52), the weight matrix a becomes $\mathbf{R}_v^{-1}$, the inverse of the covariance of the single time measurement noise vector and, of course, $\gamma = 1$. Thus the BLUE version of (8.52a) is

$$\mathbf{L}(N + 1) = \mathbf{P}\boldsymbol{\psi}(\mathbf{R}_v + \boldsymbol{\psi}^T\mathbf{P}\boldsymbol{\psi})^{-1}.$$

Thus far we have considered only the least-squares estimation where $\mathbf{A}$, the coefficient of the unknown parameter, is known. However, as we have seen earlier, in the true identification problem, the coefficient of $\boldsymbol{\theta}$ is $\boldsymbol{\Psi}$, the state of the plant model, and if random noise is present, the elements of $\boldsymbol{\Psi}$ will be random variables also. Analysis in this case is complex, and the best we will be able to do is to quote results on consistency and bias before turning to the method of maximum likelihood. We can, however, illustrate the major features and the major difficulty of least-squares identification by analysis of a very simple case. Suppose we consider a first-order model with no control. The equations are taken to be

$$\begin{align}
y(k) &= a^0 y(k - 1) + v(k) + cv(k - 1), \\
\mathscr{E}v(k) &= 0, \\
\mathscr{E}v(k)v(j) &= \sigma^2 \quad k = j \\
&= 0 \quad\quad k \neq j.
\end{align} \tag{8.85}$$

In general, we would not know either the constant c or the noise intensity σ^2. We will estimate them later. For the moment, we wish to consider the effects of the noise on the least-squares estimation of the constant a^0. Thus we take

$$\begin{align}
\mathbf{Y} &= [y(1) \ \ldots \ y(N)]^T, \\
\boldsymbol{\Psi} &= [y(0) \ \ldots \ y(N - 1)]^T, \\
\theta &= a.
\end{align}$$

Then we have the simple sums

$$\boldsymbol{\Psi}^T\boldsymbol{\Psi} = \sum_{k=1}^{N} y(k - 1)y(k - 1), \qquad \boldsymbol{\Psi}^T\mathbf{Y} = \sum_{k=1}^{N} y(k - 1)y(k),$$

and the normal equations tell us that $\hat{\theta}$ must satisfy

$$\left[\sum_{k=1}^{N} y^2(k - 1)\right]\hat{\theta} = \sum_{k=1}^{N} y(k - 1)y(k). \tag{8.86}$$

Now we must quote a result from statistics.[12]   For signals such as $y(k)$ which are generated by white noise of zero mean having finite intensity passed through a stationary filter, we can define the autocorrelation function

$$R_y(j) = \mathscr{E}y(k)y(k + j) \tag{8.87}$$

and (in a suitable sense of convergence)

$$\lim_{N\to\infty} \frac{1}{N} \sum_{k=1}^{N} y(k)y(k + j) = R_y(j). \tag{8.88}$$

Thus, while the properties of $\hat{\theta}$ for finite N are difficult, we can say that the asymptotic least-squares estimate is given by the solution to

$$\lim_{N\to\infty} \left(\frac{1}{N} \sum_{k=1}^{N} y^2(k-1) \right) \hat{\theta} = \lim_{N\to\infty} \frac{1}{N} \sum_{k=1}^{N} y(k-1)y(k),$$

which is

$$R_y(0)\hat{\theta} = R_y(1),$$
$$\hat{\theta} = R_y(1)/R_y(0). \tag{8.89}$$

If we now return to the model from which we assume $y(k)$ to be generated, values for $R_y(0)$ and $R_y(1)$ can be obtained.   For example, if we multiply (8.85) by $y(k-1)$ and take the expected value, we find

$$\mathscr{E}y(k-1)y(k) = \mathscr{E}a^0 y(k-1)y(k-1) + \mathscr{E}v(k)y(k-1) + c\mathscr{E}v(k-1)y(k-1),$$
$$R_y(1) = a^0 R_y(0) + c\mathscr{E}v(k-1)y(k-1). \tag{8.90}$$

We set $\mathscr{E}v(k)y(k-1) = 0$ because $y(k-1)$ is generated by $v(j)$ occurring at or before time $k-1$ and y is independent of (uncorrelated with) the noise $v(j)$, which is in the future.   However, we still need to compute $\mathscr{E}v(k-1)y(k-1)$.   For this, we write (8.85) for $k-1$ as

$$y(k-1) = a^0 y(k-2) + v(k-1) + cv(k-2)$$

and multiply by $v(k-1)$ and take expected values,

$$\mathscr{E}y(k-1)v(k-1) = \mathscr{E}a^0 y(k-2)v(k-1) + \mathscr{E}v^2(k-1) + c\mathscr{E}v(k-2)v(k-1)$$
$$= \sigma^2.$$

We have, then,

$$R_y(1) = a^0 R_y(0) + c\sigma^2. \tag{8.91}$$

---

[12] See Appendix D.

Now we can substitute these values into (8.89) to obtain

$$\hat{\theta}(\infty) = \frac{R_y(1)}{R_y(0)}$$

$$= a^0 + c \frac{\sigma^2}{R_y(0)}. \tag{8.92}$$

If $c = 0$, then $\hat{\theta}(\infty) = a^0$, and we can say that the least-squares estimate is asymptotically unbiased. However, if $c \neq 0$, (8.92) shows that even in this simple case the least-squares estimate is asymptotically biased and cannot be consistent.[13] Primarily because of the bias introduced when the ARMA model has noise terms which are correlated from one equation to the next, as in (8.85) when $c \neq 0$, least squares is not a good scheme for constructing parameter estimates of dynamic models which include noise. Many alternatives have been studied, among the most successful being those based on maximum likelihood.

8.5 MAXIMUM LIKELIHOOD

The method of maximum likelihood requires that we introduce a probability density function for the random variables involved. Here we consider only the normal or gaussian distribution; the method is not restricted to the form of the density function in any way, however. A scalar random variable x is said to have a normal distribution if it has a density function given by

$$f_x(\xi) = \frac{1}{\sqrt{2\pi}\sigma} \exp \left[-\frac{1}{2} \frac{(\xi - \mu)^2}{\sigma^2} \right]. \tag{8.93}$$

It may be readily verified, perhaps by use of a table of definite integrals, that

$$\int_{-\infty}^{\infty} f_x(\xi) \, d\xi = 1,$$

$$\mathscr{E}x \triangleq \int_{-\infty}^{\infty} \xi f_x(\xi) \, d\xi = \mu,$$

$$\mathscr{E}(x - \mu)^2 \triangleq \int_{-\infty}^{\infty} (\xi - \mu)^2 f_x(\xi) \, d\xi = \sigma^2. \tag{8.94}$$

The number σ^2 is called the variance of x, written var(x), and σ is the standard deviation. For the case of a vector-valued set of n random variables with a joint distribution which is normal, we find that if we define the mean vector $\boldsymbol{\mu}$ and the

---

[13] Clearly, if the constant c is known we could subtract out the bias term, but such corrections are not very pleasing since the essential nature of noise is its unknown dependencies. In some instances it is possible to construct a term which asymptotically cancels the bias term, however, as discussed in Mendel (1973).

(nonsingular) covariance matrix $\mathbf{R}$ as

$$\mathscr{E}\mathbf{x} = \boldsymbol{\mu},$$
$$\mathscr{E}(\mathbf{x} - \boldsymbol{\mu})(\mathbf{x} - \boldsymbol{\mu})^T = \mathbf{R},$$

then

$$f_\mathbf{x}(\boldsymbol{\xi}) = \frac{1}{[(2\pi)^n \det \mathbf{R}]^{1/2}} \exp\left[-\frac{1}{2}(\boldsymbol{\xi} - \boldsymbol{\mu})^T \mathbf{R}^{-1}(\boldsymbol{\xi} - \boldsymbol{\mu})\right]. \tag{8.95}$$

If the elements of the vector $\mathbf{x}$ are mutually uncorrelated and have identical means, μ, and variances, σ^2, then $\mathbf{R} = \sigma^2 \mathbf{I}$, $\det \mathbf{R} = (\sigma^2)^n$, and (8.95) can be written as

$$f_\mathbf{x}(\boldsymbol{\xi}) = \frac{1}{(2\pi\sigma^2)^{n/2}} \exp\left[-\frac{1}{2\sigma^2}\sum_{i=1}^{n}(\xi_i - \mu)^2\right]. \tag{8.96}$$

Because the normal distribution is completely determined by the vector of means, $\boldsymbol{\mu}$, and the covariance matrix, $\mathbf{R}$, we often use the notation $N(\boldsymbol{\mu}, \mathbf{R})$ to designate the density (8.95).

The maximum-likelihood estimate is calculated on the basis of an assumed structure for the probability density function of the available observations. Suppose, for example, that the data consist of a set of observations having a density given by (8.96), but having an *unknown* mean value. The parameter is therefore $\theta^0 = \mu$.

The function[14]

$$f_\mathbf{x}(\boldsymbol{\xi}\,|\,\theta) = (2\pi\sigma^2)^{-n/2} \exp\left[-\frac{1}{2}\sum_{i=1}^{n}\frac{(\xi_i - \theta)^2}{\sigma^2}\right] \tag{8.97}$$

can be presented *as a function of* θ, as giving the density of a set of x_i for *any* value of the population mean θ. Since the probability that a particular x_i is in the range $a \le x_i \le b$ is given by

$$\Pr\{a \le x_i \le b\} = \int_a^b f_{x_i}(\xi_i\,|\,\theta)\,d\xi_i,$$

the density function is seen to be a measure of the "likelihood" for a particular value; when f is large in a neighborhood $\mathbf{x}^0$, we would expect to find many samples from the population with values near $\mathbf{x}^0$. As a function of the parameters, $\boldsymbol{\theta}$, the density function $f_\mathbf{x}(\boldsymbol{\xi}\,|\,\boldsymbol{\theta})$ is called the *likelihood function*. If the actual data come from a population with the density $f_\mathbf{x}(\boldsymbol{\xi}\,|\,\boldsymbol{\theta}^0)$, then one might expect the samples to reflect this fact and that a good estimate for $\boldsymbol{\theta}^0$ given the observations $\mathbf{x} = \{x_1, \ldots, x_n\}^T$ would be $\boldsymbol{\theta} = \hat{\boldsymbol{\theta}}$, where $\hat{\boldsymbol{\theta}}$ is such that the likelihood function $f_\mathbf{x}(\mathbf{x}\,|\,\boldsymbol{\theta})$ is as large as possible. Such an estimate is called the maximum-likelihood

---

[14] We read $f_x(\xi\,|\,\theta)$ as "the probability density of ξ given θ."

estimate, $\hat{\theta}_{\text{ML}}$. Formally, $\hat{\theta}_{\text{ML}}$ is such that

$$f_{\mathbf{x}}(\mathbf{x} \mid \hat{\boldsymbol{\theta}}_{\text{ML}}) \geq f_{\mathbf{x}}(\mathbf{x} \mid \boldsymbol{\theta}). \tag{8.98}$$

From (8.97) we can immediately compute $\hat{\theta}_{\text{ML}}$ for the mean μ, by setting the derivative of $f_{\mathbf{x}}(\mathbf{x} \mid \theta)$ with respect to θ equal to zero. First, we note that

$$\frac{d}{d\theta} \log f = \frac{1}{f} \frac{df}{d\theta}, \tag{8.99}$$

so that the derivative of the log of f is zero when $df/d\theta$ is zero.[15] Since the (natural) log of the normal density is a simpler function than the density itself, we will often deal with the log of f. In fact, the negative of the log of f is used so much, we will give it the functional designation

$$-\log f_{\mathbf{x}}(\mathbf{x} \mid \boldsymbol{\theta}) \overset{\Delta}{=} \ell(\mathbf{x} \mid \boldsymbol{\theta}).$$

Thus, from (8.97), for the scalar parameter μ,

$$\ell(\mathbf{x} \mid \theta) = +\frac{n}{2} \log(2\pi\sigma^2) + \frac{1}{2} \sum_{i=1}^{n} \frac{(x_i - \theta)^2}{\sigma^2} \tag{8.100}$$

and

$$\frac{\partial \ell}{\partial \theta} = -\frac{1}{2\sigma^2} \sum_{i=1}^{n} 2(x_i - \theta)$$

$$= -\frac{1}{\sigma^2} \left\{ \sum_{i=1}^{n} x_i - n\theta \right\}. \tag{8.101}$$

If we now set $\partial \ell / \partial \theta = 0$, we have

$$\sum_{i=1}^{n} x_i - n\hat{\theta}_{\text{ML}} = 0$$

or

$$\hat{\theta}_{\text{ML}} = \frac{1}{n} \sum_{i=1}^{n} x_i. \tag{8.102}$$

We thus find that for an unknown mean of a normal distribution, the maximum-likelihood estimate is the sample mean, which is also least squares, unbiased, consistent, and BLUE. We have studied this estimate before. However, we can go on to apply the principle of maximum likelihood to the problem of dynamic system identification.

Consider next the ARMA model with simple white-noise disturbances for which we write

$$y(k) = a_1 y(k-1) + \cdots + a_n y(k-n)$$
$$+ b_1 u(k-1) + \cdots + b_n u(k-n) + v(k), \tag{8.103}$$

---

[15] It is not possible for f to be zero in the neighborhood of its maximum.

and we assume that the distribution of $\mathbf{V} = [v(n) \ldots v(N)]^T$ is $N(0, \sigma^2\mathbf{I})$. Thus we are assuming that the $v(k)$ are each normally distributed with zero mean and variance σ^2 and furthermore that the covariance between $v(k)$ and $v(j)$ is zero for $k \neq j$. Suppose, now, that a sequence of the $y(k)$ are observed and that we wish to estimate from them the a_i, b_i and σ^2 by the method of maximum likelihood. We require the probability-density function of the observed $y(k)$ for known values of the parameters. From a look at Eq. (8.103), it is apparent that if we assume that y, u, a_i and b_i are all known, then we can compute $v(k)$ from these observed y and u and assumed (true) a_i and b_i, and the distribution of y is immediately determined by the distribution of v. Using the earlier notation, we define

$$\begin{aligned}
\mathbf{V}(N) &= [v(n) \ldots v(N)]^T, \\
\mathbf{Y}(N) &= [y(n) \ldots y(N)]^T, \\
\boldsymbol{\psi}(k) &= [y(k-1) \ldots y(k-n)\, u(k-1) \ldots u(k-n)]^T, \quad (8.104) \\
\boldsymbol{\Psi}(N) &= [\boldsymbol{\psi}(n) \ldots \boldsymbol{\psi}(N)]^T, \\
\boldsymbol{\theta}^0 &= [a_1 \ldots a_n\, b_1 \ldots b_n]^T.
\end{aligned}$$

Then (8.103) implies, again for the true parameters, that

$$\mathbf{Y}(N) = \boldsymbol{\Psi}\boldsymbol{\theta}^0 + \mathbf{V}(N). \qquad (8.105)$$

Equation (8.105) is an expression of the input-output relation of our plant equations of motion. To obtain the probability-density function $f(\mathbf{Y} \mid \boldsymbol{\theta}^0)$ as required for the method of maximum likelihood, we need only be able to compute the $v(k)$ or, in batch form, the $\mathbf{V}(N)$ from the $y(k)$ or $\mathbf{Y}(N)$ because we are given the probability density of $\mathbf{V}$. To compute $\mathbf{V}$ from $\mathbf{Y}$ requires the *inverse* model of our plant, which in this case is so trivial as to be almost missed[16]; namely, we solve (8.105) for $\mathbf{V}$ to obtain

$$\mathbf{V}(N) = \mathbf{Y}(N) - \boldsymbol{\Psi}(N)\boldsymbol{\theta}^0. \qquad (8.106)$$

Since we have assumed that the density function of $\mathbf{V}$ is $N(0, \sigma^2\mathbf{I})$, we can write instantly

$$f(\mathbf{Y} \mid \boldsymbol{\theta}^0) = (2\pi\sigma^2)^{-m/2} \exp\left[-\frac{1}{2} \frac{(\mathbf{Y} - \boldsymbol{\Psi}\boldsymbol{\theta}^0)^T(\mathbf{Y} - \boldsymbol{\Psi}\boldsymbol{\theta}^0)}{\sigma^2} \right], \qquad (8.107)$$

where $m = N - n + 1$, the number of samples in Y.

The likelihood function is by definition $f(\mathbf{Y} \mid \boldsymbol{\theta})$ which is to say, (8.107) with the $\boldsymbol{\theta}^0$ dropped and replaced by a general $\boldsymbol{\theta}$.

---

[16] We deliberately formulated our problem so that this inverse would be trivial. If we add plant and independent sensor noise to a state-variable description, the inverse requires a Kalman filter, to be discussed in Chapter 9. Since $v(k)$ is independent of all past u and y, for any θ, $e(k)$ is the prediction error.

As in the elementary example discussed above, we will consider the log of f as follows:

$$\ell(\mathbf{Y} \mid \boldsymbol{\theta}) = \log(2\pi\sigma^2)^{-m/2} + \log \left\{ \exp \left[-\frac{1}{2} \frac{(\mathbf{Y} - \boldsymbol{\Psi}\boldsymbol{\theta})^T(\mathbf{Y} - \boldsymbol{\Psi}\boldsymbol{\theta})}{\sigma^2} \right] \right\}$$

$$= -\frac{m}{2} \log 2\pi - \frac{m}{2} \log \sigma^2 - \frac{1}{2} \frac{(\mathbf{Y} - \boldsymbol{\Psi}\boldsymbol{\theta})^T(\mathbf{Y} - \boldsymbol{\Psi}\boldsymbol{\theta})}{\sigma^2}. \qquad (8.108)$$

Our estimates, $\hat{\boldsymbol{\theta}}_{\text{ML}}$ and $\hat{\sigma}^2_{\text{ML}}$, are those values of $\boldsymbol{\theta}$ and σ^2 which make $\ell(\mathbf{Y} \mid \boldsymbol{\theta})$ as small as possible. We find these estimates by setting to zero and partial derivatives of ℓ with respect to $\boldsymbol{\theta}$ and σ^2. These derivatives are (following our earlier treatment of taking partial derivatives)

$$\frac{1}{\hat{\sigma}^2} [\boldsymbol{\Psi}^T\boldsymbol{\Psi}\hat{\boldsymbol{\theta}}_{\text{ML}} - \boldsymbol{\Psi}^T\mathbf{Y}] = \mathbf{0}, \qquad (8.109a)$$

$$-\frac{m}{2\hat{\sigma}^2_{\text{ML}}} + \frac{\mathbf{Q}}{2\hat{\sigma}^4_{\text{ML}}} = 0, \qquad (8.109b)$$

where the quadratic term $\mathbf{Q}$ is defined as

$$\mathbf{Q} = (\mathbf{Y} - \boldsymbol{\Psi}\hat{\boldsymbol{\theta}}_{\text{ML}})^T(\mathbf{Y} - \boldsymbol{\Psi}\hat{\boldsymbol{\theta}}_{\text{ML}}). \qquad (8.109c)$$

We see immediately that (8.109a) is the identical "normal" equation of the least-squares method so that $\hat{\boldsymbol{\theta}}_{\text{ML}} = \hat{\boldsymbol{\theta}}_{\text{LS}}$ in this case. We thus know that $\hat{\boldsymbol{\theta}}_{\text{ML}}$ is asymptotically unbiased and consistent. The equations for $\hat{\sigma}^2_{\text{ML}}$ decouple from those for $\hat{\boldsymbol{\theta}}_{\text{ML}}$, and solving (8.109b), we obtain (again using earlier notation)

$$\hat{\sigma}^2_{\text{ML}} = \frac{\mathbf{Q}}{m} = \frac{1}{m} (\mathbf{Y} - \boldsymbol{\Psi}\hat{\boldsymbol{\theta}})^T(\mathbf{Y} - \boldsymbol{\Psi}\hat{\boldsymbol{\theta}})$$

$$= \frac{1}{N - n + 1} \sum_{k=n}^{N} e^2(k; \hat{\boldsymbol{\theta}}_{\text{ML}}). \qquad (8.110)$$

Thus far we have no new solutions except the estimate for σ^2 given in (8.110) but we have shown that the method of maximum likelihood gives the same solution for $\hat{\boldsymbol{\theta}}$ as the least-squares method for the model of (8.103).

Now let us consider the general model given by

$$y(k) = \sum_{i=1}^{n} a_i y(k - i) + \sum_{i=1}^{n} b_i u(k - i) + \sum_{i=1}^{n} c_i v(k - i) + v(k); \qquad (8.111)$$

the distribution of $\mathbf{V}(N) = [v(n) \ \ldots \ v(N)]^T$ is again taken to be normal and $\mathbf{V}$ is distributed according to the $N(0, \sigma^2\mathbf{I})$ density. The difference between (8.111) and (8.103), of course, is that in (8.111) we find past values of the noise $v(k)$ weighted by the c_i and, as we saw in (8.92), the least-squares estimate is biased if the c_i are nonzero.

Consider first the special case where only c_1 is nonzero of the c's and write the terms in (8.111) which depend on noise $v(k)$ on one side and define $z(k)$ as follows:

$$v(k) + c_1 v(k-1) = y(k) - \sum_{i=1}^{n} a_i y(k-i) - \sum_{i=1}^{n} b_i u(k-i)$$

$$= z(k). \tag{8.112}$$

Now let the reduced parameter vector be $\bar{\boldsymbol{\theta}} = [a_1 \ldots a_n \, b_1 \ldots b_n]^T$, composed of the a and b parameters but not including the c_1. In a natural way, we can write

$$\mathbf{Z}(N) = \mathbf{Y}(N) - \boldsymbol{\Psi}\bar{\boldsymbol{\theta}}^0. \tag{8.113}$$

However, since $z(k)$ is a sum of two random variables $v(k)$ and $v(k-1)$, each of which has a normal distribution, we know that $\mathbf{Z}$ is also normally distributed. Furthermore we can easily compute the mean and covariance of z:

$$\begin{aligned}
\mathscr{E}z(k) &= \mathscr{E}(v(k) + c_1 v(k-1)) = 0 \qquad \text{for all} \quad k, \\
\mathscr{E}z(k)z(j) &= \mathscr{E}(v(k) + c_1 v(k-1))(v(j) + c_1 v(j-1)) \\
&= \sigma^2(1 + c_1^2) \quad (k = j) \\
&= \sigma^2 c_1 \quad (k = j - 1) \\
&= \sigma^2 c_1 \quad (k = j + 1) \\
&= 0 \quad \text{elsewhere.}
\end{aligned}$$

Thus the structure of the covariance of $\mathbf{Z}(N)$ is

$$\mathbf{R} = \mathscr{E}\mathbf{Z}(N)\mathbf{Z}^T(N)$$

$$= \begin{bmatrix} 1 + c_1^2 & c_1 & 0 & & \cdots & & 0 \\ & & & & & & \\ c_1 & 1 + c_1^2 & c_1 & 0 & \cdots & & 0 \\ & & & & & & \\ 0 & c_1 & 1 + c_1^2 & c_1 & & 0 \ldots 0 \end{bmatrix} \sigma^2, \tag{8.114}$$

and, with the mean and covariance in hand, we can write the probability density of $\mathbf{Z}(N)$ as

$$g(\mathbf{Z}(N) \mid \boldsymbol{\theta}^0) = ((2\pi)^m \det \mathbf{R})^{-1/2} \exp(-\tfrac{1}{2}\mathbf{Z}^T\mathbf{R}^{-1}\mathbf{Z}), \tag{8.115}$$

where $m = N - n + 1$. But from (8.113) this is the likelihood function if we substitute for $\mathbf{Z}$ and $\boldsymbol{\theta}$ as follows:

$$f(\mathbf{Y} \mid \boldsymbol{\theta}) = [(2\pi)^m \det \mathbf{R}]^{-1/2} \exp[-\tfrac{1}{2}(\mathbf{Y} - \boldsymbol{\Psi}\bar{\boldsymbol{\theta}})^T\mathbf{R}^{-1}(\mathbf{Y} - \boldsymbol{\Psi}\bar{\boldsymbol{\theta}})]. \tag{8.116}$$

The log of f is again similar in form to previous results:

$$\ell(\mathbf{Y} \mid \boldsymbol{\theta}) = \frac{m}{2}\log 2\pi + \tfrac{1}{2}\log(\det \mathbf{R}) + \tfrac{1}{2}(\mathbf{Y} - \boldsymbol{\Psi}\bar{\boldsymbol{\theta}})^T\mathbf{R}^{-1}(\mathbf{Y} - \boldsymbol{\Psi}\bar{\boldsymbol{\theta}}). \tag{8.117}$$

The major point to be made about (8.117) is that the log likelihood function depends on the a's and b's through $\bar{\theta}$ and is thus *quadratic* in these parameters, but it depends on the c's (c_1 only in this special case) through both $\mathbf{R}^{-1}$ and det $\mathbf{R}$, a dependence which is definitely not quadratic. An explicit formula for the maximum likelihood estimate is thus not possible, and we must retreat to a numerical algorithm to compute $\hat{\theta}_{ML}$ in this case.

8.6 NUMERICAL SEARCH FOR THE MAXIMUM LIKELIHOOD ESTIMATE

Being unable to give a closed-form expression for the maximum likelihood estimate, we turn to an algorithm which can be used for the numerical search for $\hat{\theta}_{ML}$. We first formulate the problem from the assumed ARMA model of (8.111), once again forming the inverse system as

$$v(k) = y(k) - \sum_{i=1}^{n} a_i y(k - i) - \sum_{i=1}^{n} b_1 u(k - i) - \sum_{i=1}^{n} c_i v(k - i). \quad (8.118)$$

By assumption, $v(k)$ has a normal distribution with zero mean and unknown (scalar) variance R_v. As before, we define the successive outputs of this inverse as

$$\mathbf{V}(N) = [v(n)v(n + 1) \cdots v(N)]^T. \quad (8.119)$$

This multivariable random vector also has a normal distribution with zero mean and covariance;

$$\begin{aligned} \mathbf{R} &= R_v \mathbf{I}_m \\ &= \mathscr{E} \mathbf{V} \mathbf{V}^T, \end{aligned} \quad (8.120)$$

where I_m is the $m \times m$ identity matrix, and $m = N - n + 1$ is the number of elements in $\mathbf{V}$. Thus we can immediately write the density function as a function of the true parameters $\theta^0 = [a_1 \ldots a_n b_1 \ldots b_n c_1 \ldots c_n]^T$ as

$$f(\mathbf{V} \mid \boldsymbol{\theta}) = ((2\pi)^m R_v^m)^{-1/2} \exp\left[-\frac{1}{2} \sum_{k=n}^{N} \frac{v^2(k)}{R_v}\right]. \quad (8.121)$$

The (log) likelihood function is found by substituting arbitrary parameters, θ, in (8.121), and we use $e(k)$ as the output of (8.118) when $\theta \neq \theta^0$:

$$\ell(\mathbf{E} \mid \boldsymbol{\theta}) = \frac{m}{2} \log 2\pi + \frac{m}{2} \log \hat{R}_v + \frac{1}{2\hat{R}_v} \sum_{k=n}^{N} e^2(k). \quad (8.122)$$

As with (8.110) we can compute the estimate of $\hat{R}_v$ by direct calculation of

$$\partial \ell / \partial \hat{R}_v = 0,$$

which gives

$$\hat{R}_v = \frac{1}{N - n + 1} \sum_{k=n}^{N} e^2(k). \tag{8.123}$$

To compute the values for the a_i, b_i, and c_i, we need to construct a numerical algorithm which will be suitable for minimizing $\ell(\mathbf{E}|\boldsymbol{\theta})$. The study of such algorithms is extensive[17]; we will be content to present a frequently used one, based on a method due to Newton. The essential concept is that given the Kth estimate of $\hat{\boldsymbol{\theta}}$, we wish to find a $(K + 1)^{st}$ estimate which will make $\ell(\mathbf{E} \mid \hat{\boldsymbol{\theta}}(K + 1))$ smaller than $\ell(\mathbf{E}|\hat{\boldsymbol{\theta}}(K))$. The method is to expand ℓ about $\hat{\boldsymbol{\theta}}(K)$ and choose $\hat{\boldsymbol{\theta}}(K + 1)$ so that the quadratic terms—the first three terms in the expansion of ℓ—are minimized. Formally, we proceed as follows. Let $\hat{\boldsymbol{\theta}}(K + 1) = \hat{\boldsymbol{\theta}}(K) + \delta\hat{\boldsymbol{\theta}}$. Then

$$\begin{aligned}\ell(\mathbf{E} \mid \hat{\boldsymbol{\theta}}(K + 1)) &= \ell(\mathbf{E} \mid \hat{\boldsymbol{\theta}}(K) + \delta\hat{\boldsymbol{\theta}}) \\ &= c + \mathbf{g}^T\delta\hat{\boldsymbol{\theta}} + \tfrac{1}{2}\delta\hat{\boldsymbol{\theta}}^T Q \delta\hat{\boldsymbol{\theta}} + \cdots, \end{aligned} \tag{8.124}$$

where

$$c = \ell(\mathbf{E} \mid \hat{\boldsymbol{\theta}}(K)), \qquad \mathbf{g}^T = \frac{\partial\ell}{\partial\boldsymbol{\theta}}\bigg|_{\theta=\hat{\theta}(K)}, \qquad \mathbf{Q} = \frac{\partial^2\ell}{\partial\boldsymbol{\theta}\,\partial\boldsymbol{\theta}}\bigg|_{\theta=\theta(K)}. \tag{8.125}$$

We must return later to the computation of $\mathbf{g}^T$ and $\mathbf{Q}$ but let us first construct the algorithm. We would like to select $\delta\boldsymbol{\theta}$ so that the quadratic approximation to ℓ is as small as possible. The analytic condition for ℓ to be a minimum is that $\partial\ell/\partial\,\delta\boldsymbol{\theta} = 0$. We thus differentiate (8.124) and set the derivative to zero with the result (ignoring the higher-order terms in $\delta\boldsymbol{\theta}$)

$$\frac{\partial\ell}{\partial\,\delta\boldsymbol{\theta}} = \mathbf{g}^T + \delta\boldsymbol{\theta}^T\mathbf{Q}$$

$$= 0. \tag{8.126}$$

Solving (8.126) for $\delta\boldsymbol{\theta}$ we find

$$\delta\boldsymbol{\theta} = -\mathbf{Q}^{-1}\mathbf{g}. \tag{8.127}$$

We now use the $\delta\boldsymbol{\theta}$ found in (8.127) to compute $\hat{\boldsymbol{\theta}}(K + 1)$ as

$$\hat{\boldsymbol{\theta}}(K + 1) = \hat{\boldsymbol{\theta}}(K) - \mathbf{Q}^{-1}\mathbf{g}.$$

In terms of ℓ, as given in (8.125), the algorithm may be written as

$$\hat{\boldsymbol{\theta}}(K + 1) = \hat{\boldsymbol{\theta}}(K) - \left(\frac{\partial^2\ell}{\partial\boldsymbol{\theta}\,\partial\boldsymbol{\theta}}\right)^{-1}\left(\frac{\partial\ell}{\partial\boldsymbol{\theta}}\right)^T. \tag{8.128}$$

Our final task is to express the partial derivatives in (8.128) in terms of the observed signals y and u. To do this, we return to (8.122) and proceed formally, as

---

[17] See Luenberger (1973) for a lucid account of minimizing algorithms.

follows, taking $\hat{R}_v$ to be a constant since $\hat{R}_v$ is given by (8.123):

$$\frac{\partial \ell(\mathbf{E} \mid \boldsymbol{\theta})}{\partial \boldsymbol{\theta}} = \frac{\partial}{\partial \boldsymbol{\theta}} \left\{ \frac{m}{2} \log 2\pi + \frac{m}{2} \log \hat{R}_v + \frac{1}{2\hat{R}_v} \sum_{k=n}^{N} e^2(k) \right\}$$

$$= \frac{1}{\hat{R}_v} \sum_{k=n}^{N} e(k) \frac{\partial e(k)}{\partial \boldsymbol{\theta}}, \tag{8.129}$$

$$\frac{\partial^2 \ell}{\partial \boldsymbol{\theta} \, \partial \boldsymbol{\theta}} = \frac{\partial}{\partial \boldsymbol{\theta}} \left(+ \frac{1}{\hat{R}_v} \sum_{k=n}^{N} e(k) \frac{\partial e}{\partial \boldsymbol{\theta}} \right)$$

$$= \frac{1}{\hat{R}_v} \sum_{k=n}^{N} \left(\frac{\partial e}{\partial \boldsymbol{\theta}} \right)^T \frac{\partial e}{\partial \boldsymbol{\theta}} + \frac{1}{\hat{R}_v} \sum_{k=n}^{N} e(k) \frac{\partial^2 e}{\partial \boldsymbol{\theta} \, \partial \boldsymbol{\theta}}. \tag{8.130}$$

We note that (8.129) and the first term in (8.130) depend only on the derivative of e with respect to $\boldsymbol{\theta}$. Since our algorithm is only expected to produce an improvement in $\hat{\boldsymbol{\theta}}$, and since, near the minimum, we would expect the first derivative terms in (8.130) to dominate, we will simplify the algorithm to include only the first term in (8.130). Thus we need only compute $\partial e / \partial \boldsymbol{\theta}$. It is standard terminology to refer to these partial derivatives as the sensitivities of e with respect to $\boldsymbol{\theta}$.

We turn to the difference equations for $v(k)$, (8.118), substitute $e(k)$ for $v(k)$, and compute the sensitivities as follows:

$$\frac{\partial e(k)}{\partial a_i} = -y(k - i) - \sum_{j=1}^{n} c_j \frac{\partial e(k - j)}{\partial a_i}, \tag{8.131a}$$

$$\frac{\partial e(k)}{\partial b_i} = -u(k - i) - \sum_{j=1}^{n} c_j \frac{\partial e(k - j)}{\partial b_i}, \tag{8.131b}$$

$$\frac{\partial e(k)}{\partial c_i} = -e(k - i) - \sum_{j=1}^{n} c_j \frac{\partial e(k - j)}{\partial c_i}. \tag{8.131c}$$

If we consider (8.131a) for the moment for fixed i, we see that this is a constant-coefficient difference equation in the variable $\partial e(k) / \partial a_i$ with $y(k - i)$ as a forcing function. If we take the z-transform of this system and call $\partial e / \partial a_i \triangleq e_{a_i}$, then we find

$$E_{a_i}(z) = -z^{-i} Y(z) - \sum_{j=1}^{n} c_j z^{-j} E_a(z)$$

or

$$E_{a_i}(z) = \frac{-z^{-i}}{1 + \sum_{j=1}^{n} c_{j_i} z^{-j}} Y(z).$$

Thus the derivative of e with respect to a_2 is simply z^{-1} times the partial derivative of e with respect to a_1, and so on. In fact, we can realize all of these partial derivatives via the structure shown in Fig. 8.4.

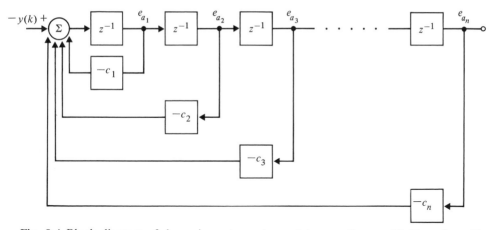

Fig. 8.4 Block diagram of dynamic system whose states are the sensitivities of e with respect to a_i.

In exactly analogous fashion, dynamic systems whose states are the sensitivities of e with respect to b_i and c_i can be constructed. With these, then, we have all the elements of an explicit algorithm which can be used to compute improvements in $\boldsymbol{\theta}$. The steps may be summarized as follows:

1. Select an initial parameter estimate, $\hat{\boldsymbol{\theta}}(0)$, based on analysis of the physical situation, the system step and frequency responses, cross correlation of input and output, and/or least squares.

2. Construct (compute) $e(k)$ from (8.118) with e substituted for v, compute the sensitivities from (8.131) using three structures as shown in Fig. 8.4, and simultaneously compute $\hat{R}_v \, \partial \ell / \partial \boldsymbol{\theta}$ from (8.129) and the first term of $\hat{R}_v \, \partial^2 \ell / \partial \boldsymbol{\theta} \, \partial \boldsymbol{\theta}$ from (8.130).

3. Compute $\hat{R}_v$ from (8.123) and solve for $\mathbf{g}$ and $\mathbf{Q}$.

4. Compute $\hat{\boldsymbol{\theta}}(K + 1) = \hat{\boldsymbol{\theta}}(K) - \mathbf{Q}^{-1}\mathbf{g}$.

5. If $[\hat{R}_v(K + 1) - \hat{R}_v(K)]/\hat{R}_v(K) < 10^{-4}$, stop. Else go back to step 2.

In step 5 of this algorithm it is suggested that when the sum of squares of the prediction errors, as given by $\hat{R}_v$, fails to be reduced by more than a relative amount of 10^{-4}, we should stop. This number, 10^{-4}, is suggested by a statistical test of the significance of the reduction. A discussion of such tests is beyond our scope here, but may be found in Astrom and Eykhoff (1971), and Kendal and Stuart (1967). If the order of the system, n, is not known, this entire process can be done for $n = 1, 2, 3, \ldots$, and a test similar to that of step 5 may be used to decide when further increases in n are not significant.

8.7 SUMMARY

In this chapter we have introduced some of the concepts of identification of dynamic systems. The central technique presented was the maximum likelihood method for parametric identification which was developed as a generalization of least squares, weighted least squares, and best linear unbiased estimates. The selection of parameters and the formulation of the equation error, output error, and prediction error were also described.

PROBLEMS AND EXERCISES

8.1 Data from a mass moving in a plane (x, y) without force are given below.

t	x	y
0	−0.426	1.501
1	0.884	1.777
2	2.414	2.530
3	2.964	3.441
4	3.550	3.762
5	5.270	4.328
6	5.625	4.016
7	7.188	5.368
8	8.129	5.736
9	9.225	6.499
10	10.388	6.233

Plot the points and use least squares to estimate the initial value of x, the initial value of y, and the velocity of the mass in x and in y. Draw the curve (line) that corresponds to your estimates.

8.2[18] A simple mechanical system without friction is sketched in Fig. 8.5.

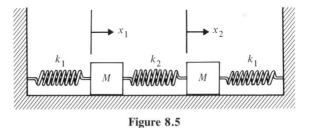

Figure 8.5

a) Show that this system is described by the equations

$$M\ddot{x}_1 + k_1x_1 + k_2(x_1 - x_2) = 0, \qquad x_1(0) = x_{10},$$
$$M\ddot{x}_2 + k_1x_2 + k_2(x_2 - x_1) = 0, \qquad x_2(0) = x_{20}.$$

---

[18] Parts of this problem require a computer for solution.

b) We wish to estimate the parameters k_1, k_2, M for measurements of $x_1(t)$. Note from (a) that if we divide by M in each equation, the only available parameters are $k_1/M = K_1$ and $k_2/M = K_2$. Show that if $x_{10} = x_{20}$ or if $x_{10} = -x_{20}$, it is impossible to estimate both K_1 and K_2 from $x_1(t)$. Show that if $x_{10} = 1$ and $x_{20} = 0$, it should be possible to estimate both K_1 and K_2 from $x_1(t)$. What do you think makes a "good" initial condition?

c) Compute a discrete equivalent system for the z-transform of $x_1(kT)$ in the form

$$X_1(z) = \frac{b_0 z^4 + b_1 z^3 + b_2 z^2 + b_3 z + b_4}{z^4 - a_1 z^3 - a_2 z^2 - a_3 z - a_4},$$

where the parameters a_i are functions of K_1, K_2 and sampling period T, and the b_i also involve $x_1(0)$ and $x_2(0)$. Give expressions for each a_i.

d) Formulate the least-squares problem to estimate the a_i from $x(kT)$. Show that it is possible to ignore the b_i.

e) Set up the equations of part (a) for $K_1 = 0.4$ and $K_2 = 0.6$ on the analog computer, sample x_1 at a rate of 0.5 sec for 30 samples, and compute $\hat{\theta} = [\hat{a}_1 \hat{a}_2 \hat{a}_3 \hat{a}_4]$. Assume all the noise comes from the A/D converter operating at 10 bits including sign and compute the predicted variance in this estimate.

f) Deliberately add noise of known variance to your data, recompute $\hat{\theta}$ 50 times, and compare the sample variance of $\hat{\theta}$ to the theoretical value.

g) Keep the number of samples fixed at 30 but vary the sample rate from 0.1 to 1 sec in 0.1 steps and compute

$$\frac{1}{4} \sum_{k=1}^{4} \left| \frac{\hat{\theta}_k - \theta_k^0}{\theta_k^0} \right|$$

for each sample period as a measure of total estimate accuracy. What do you conclude respecting selection of sample period for identification?

h) Keep the sample period fixed at 0.5 sec, but compute the estimate for varying numbers of samples. Consider at least 5, 10, 30, 100, and 300 samples. Compare the same criterion used in part (g) and give an explanation of the results.

i) Program the recursive least-squares algorithm of (8.49) for $a = \gamma = 1$ and compare your results to those of part (h). Use $\hat{\theta}(5) = 0$ and $P(5) = I$ for your initial conditions.

8.3 Give the transfer function $H(z)$ of the filter which corresponds to exponentially weighted least squares as a function of a and γ. Prove that $H(1) = 1$ if $a = 1 - \gamma$.

8.4 In Eq. (8.60) we showed that the variance in the least-squares estimate is $(A^T A)^{-1} A^T R A (A^T A)^{-1}$ and, in (8.84), we showed that the variance of the BLUE is $(A^T R^{-1} A)^{-1}$. Use the development of the proof of the BLUE following (8.79) to devise an expression for the excess variance of least squares over BLUE.

8.5 (Contributed by N. Gupta.) Show that the least-squares estimate $\hat{\theta} = (A^T A)^{-1} A^T Y$ results in the error squared $\mathcal{E}(Y - A\hat{\theta})^T (Y - A\hat{\theta}) = (m - n)\sigma^2$, where $\mathcal{E} e^2 = \sigma^2$, Y has m components and $\hat{\theta}$ has n components. What estimate of σ^2 does this result suggest? *Hint:* If $M = I - A(A^T A)^{-1} A^T$, then $M^2 = M = M^T$, and $\mathrm{tr}(AB) = \mathrm{tr}(BA)$ where $\mathrm{tr}(\cdot)$ is the trace of the matrix.

8.6 In (8.110) we showed that the maximum likelihood estimate of σ^2 is $(Y - A\hat{\theta})^T \times (Y - A\hat{\theta})/m$. Show that this estimate is biased.

8.7 a) Write a computer program to implement the search for the maximum likelihood estimate following the method of Section 8.6.

b) Simulate the system [described in Astrom and Eykhoff (1971)]

$$y_{k+1} = a_1 y_k + b_1 u_k + v_{k+1} + c_1 v_k,$$

where u_k and v_k are independent sequences of unit variances and $a_1 = 0.5$, $b_1 = 1.0$, $c_1 = 0.1$.

c) Compute the least-squares estimate of a_1 and b_1 from observation of 5, 50, and 500 samples. Compare variance and bias to the theory in each case.

d) Compute the maximum likelihood estimates of a_1, b_1, c_1, and σ_v^2 from 5, 50, and 500 samples and compare the estimates to the known true values.

8.8 Suppose we wish to identify a plant which is operating in a closed loop as

$$y_{k+1} = a y_k + b u_k + e_k,$$
$$u_k = -K y_k,$$

where e_k is white noise. Show that we cannot identify a and b from observation of y and u, even if K is known.

9 / Multivariable and Optimal Control

9.1 INTRODUCTION

The control design procedures described in Chapters 5 and 6 were applied to systems with one input and one output. The transfer-function approach in Chapter 5 is fundamentally limited to single input/output systems while the state-space methods of Chapter 6 were limited to single input/output to simplify the procedures. In fact, if one tries to apply the "pole-placement" approach of Chapter 6 to a multivariable (more than one input or output) system, all the procedures work but the gains, **K** or **L**, are not uniquely determined by the resulting equations. Until a design approach is available which intelligently uses this extra freedom, the pole-placement concept for estimator designs for systems with more than one output or for controller designs for systems with more than one input has limited value.

The subject of this chapter is the use of optimal control techniques as a tool in the design of multivariable systems. We have no illusions that some true "optimal" design is being achieved; rather, we will be transferring the designer's iteration on pole locations as used in Chapter 6 or compensation loop gain as used in Chapter 5, to iterations on elements in a cost function, "$\mathcal{J}$." The method will determine the (multivariable) control law which minimizes $\mathcal{J}$, but since $\mathcal{J}$ will have been largely arbitrarily selected in the first place, the design is at best only partially optimal. However, these designs will make coordinated use of all the control and output variables and can be organized to guarantee a stable system which can be logically changed to meet reasonable performance objectives.

9.2 DECOUPLING

The first step in any multivariable design should be an attempt to find an approximate model consisting of two or more single input/output models or else decouple the control law matrix **K** and the estimator law matrix **L**. This step will give better physical insight into the important feedback variables and may lead to a

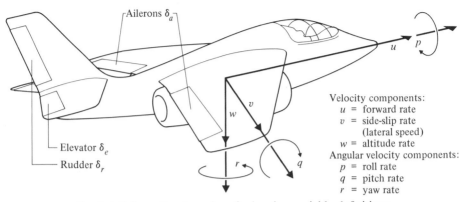

Fig. 9.1 Schematic of an aircraft showing variable definitions.

Velocity components:
u = forward rate
v = side-slip rate
    (lateral speed)
w = altitude rate
Angular velocity components:
p = roll rate
q = pitch rate
r = yaw rate

plant description which is substantially simpler for design purposes and yet yields no significant degradation from an analysis based on the full multivariable system. The ideas are best illustrated by examples.

The linearized equations of motion of an aircraft (Fig. 9.1) are of eighth order but are almost always separated into two fourth-order sets representing longitudinal motion (w, u, q) and lateral motion (p, r, v). The elevator control surfaces affect longitudinal motion while the aileron and rudder primarily affect lateral motion. Although there is a small amount of coupling of lateral motion into longitudinal motion, this is ignored with no serious consequences when the control or "stability-augmentation" systems are designed independently for the two fourth-order systems.

The aircraft lateral equations are also multivariable and are of the form

$$\mathbf{x}_{k+1} = \mathbf{\Phi}\mathbf{x}_k + \mathbf{\Gamma}\mathbf{u}_k, \tag{9.1}$$

where

$$\mathbf{u} = \begin{bmatrix} \delta_r \\ \delta_a \end{bmatrix}, \qquad \mathbf{x} = \begin{bmatrix} v \\ r \\ \phi_p \\ p \end{bmatrix}, \qquad \text{and} \qquad p = \dot{\phi}_p.$$

A control law of the standard form

$$\begin{bmatrix} \delta_r \\ \delta_a \end{bmatrix} = - \begin{bmatrix} K_{11} & K_{12} & K_{13} & K_{14} \\ K_{21} & K_{22} & K_{23} & K_{24} \end{bmatrix} \begin{bmatrix} v \\ r \\ \phi_p \\ p \end{bmatrix} \tag{9.2}$$

shows that there are eight entries in the gain matrix to be selected, and the specification of four closed loop roots will clearly leave many possible values of **K** which will meet the specifications. This is an example of the incompleteness of the pole-placement approach to multivariable system design.

A decoupling which removes the ambiguity is to restrict the control law to

$$\begin{bmatrix} \delta_r \\ \delta_a \end{bmatrix} = - \begin{bmatrix} K_{11} & K_{12} & 0 & 0 \\ 0 & 0 & K_{23} & K_{24} \end{bmatrix} \begin{bmatrix} v \\ r \\ \phi_p \\ p \end{bmatrix}. \tag{9.3}$$

This makes good physical sense because the rudder primarily yaws the aircraft about a vertical axis (r-motion), thus directly causing sideslip (v) while the ailerons primarily roll the aircraft about an axis through the nose, thus causing changes in ϕ_p and p. Given an achievable set of desired pole locations, there are unique values of the four nonzero components of $\mathbf{K}$; however, the governing equations are not in the same linear form as in (6.28) and therefore may be difficult to solve.

A further decoupling that would permit an easy gain calculation is to assume that Eq. (9.1) is of the form

$$\begin{bmatrix} v \\ r \\ \phi_p \\ p \end{bmatrix}_{k+1} = \begin{bmatrix} \phi_{11} & \phi_{12} & 0 & 0 \\ \phi_{21} & \phi_{22} & 0 & 0 \\ 0 & 0 & \phi_{33} & \phi_{34} \\ 0 & 0 & \phi_{43} & \phi_{44} \end{bmatrix} \begin{bmatrix} v \\ r \\ \phi_p \\ p \end{bmatrix}_k + \begin{bmatrix} \Gamma_{11} & 0 \\ \Gamma_{21} & 0 \\ 0 & \Gamma_{32} \\ 0 & \Gamma_{42} \end{bmatrix} \begin{bmatrix} \delta_r \\ \delta_a \end{bmatrix} \tag{9.4}$$

and that the control law is given by Eq. (9.3).

This makes some physical sense but ignores important coupling between the two aircraft modes. It does, however, decouple the system into second-order systems for which the methods of Chapter 6 can be applied directly to obtain the gains. The resulting closed-loop characteristic roots of the full lateral equations can be checked by calculating the eigenvalues of the closed loop matrix:[1]

$$\mathbf{\Phi}_{\text{closed loop}} = \begin{bmatrix} \phi_{11} - \Gamma_{11}K_{11} & \phi_{12} - \Gamma_{11}K_{12} & \phi_{13} & \phi_{14} \\ \phi_{21} - \Gamma_{21}K_{11} & \phi_{22} - \Gamma_{21}K_{12} & \phi_{23} & \phi_{24} \\ \phi_{31} & \phi_{32} & \phi_{33} - \Gamma_{32}K_{23} & \phi_{34} - \Gamma_{32}K_{24} \\ \phi_{41} & \phi_{42} & \phi_{43} - \Gamma_{42}K_{24} & \phi_{44} - \Gamma_{42}K_{24} \end{bmatrix} \tag{9.5}$$

which results from combining Eqs. (9.1) and (9.3).

If the plant coupling that was ignored in the gain computation is important, the roots obtained from (9.5) will differ from those used to compute the gains using (9.3) and (9.4). In many cases, the method will be accurate enough and one need look no further. In other cases, one could revise the "desired" root locations and iterate until the correct roots from (9.5) are satisfactory or else turn to the methods of optimal control to be described in the following sections.

---

[1] Standard packages are available in most computation centers for the calculation of matrix eigenvalues.

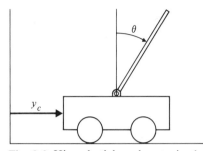

Fig. 9.2 Hinged stick and motorized cart.

The same ideas apply to estimation. Take for example the inverted pendulum on a motorized cart (Fig. 9.2). The equations of motion can be written as

$$\begin{bmatrix} \mathbf{x}_c \\ \mathbf{x}_s \end{bmatrix}_{k+1} = \begin{bmatrix} \boldsymbol{\phi}_{cc} & \boldsymbol{\phi}_{cs} \\ \boldsymbol{\phi}_{sc} & \boldsymbol{\phi}_{ss} \end{bmatrix} \begin{bmatrix} \mathbf{x}_c \\ \mathbf{x}_s \end{bmatrix} + \begin{bmatrix} \boldsymbol{\Gamma}_c \\ \boldsymbol{\Gamma}_s \end{bmatrix} u, \tag{9.6}$$

where

$$\mathbf{x}_c = \begin{bmatrix} y_c \\ \dot{y}_c \end{bmatrix}, \quad \text{cart position and velocity,}$$

$$\mathbf{x}_s = \begin{bmatrix} \theta \\ \dot{\theta} \end{bmatrix}, \quad \text{stick angle and angular rate,}$$

and the available measurements are

$$\begin{bmatrix} y \\ \theta \end{bmatrix} = \begin{bmatrix} 1 & 0 & 0 & 0 \\ 0 & 0 & 1 & 0 \end{bmatrix} \begin{bmatrix} y_c \\ \dot{y}_c \\ \theta \\ \dot{\theta} \end{bmatrix}. \tag{9.7}$$

The stick pictured in Fig. 9.2 is substantially lighter than the cart. This means that stick motion has a small dynamic effect on cart motion, which in turn implies that $\boldsymbol{\phi}_{cs} \simeq \mathbf{0}$. This does not imply that $\boldsymbol{\phi}_{sc} = 0$; in fact, cart motion is the mechanism for influencing stick motion and hence stabilizing it.

An estimator for the system described by Eqs. (9.6) and (9.7) requires the determination of eight elements of an estimator gain matrix, $\mathbf{L}$. Hence specifying the four estimator roots and using the methods of Chapter 6 would not determine this $\mathbf{L}$ uniquely—another example of underspecified gain caused by the multivariable nature of the problem.

But since we can assume that $\boldsymbol{\phi}_{cs} = \mathbf{0}$, the cart equation in (9.6) uncouples from the stick equation, and we simply design an estimator for

$$[\mathbf{x}_c]_{k+1} = \boldsymbol{\phi}_{cc}[\mathbf{x}_c]_k + \boldsymbol{\Gamma}_c u,$$

$$y_c = [1 \quad 0] \begin{bmatrix} y_c \\ \dot{y}_c \end{bmatrix}, \tag{9.8}$$

which can be done with the methods described in Chapter 6.

There is one-way coupling into the stick equation, but this just acts like an additional control input and can be ignored in the calculation of the $\mathbf{L}_s$-matrix using the pole-placement methods of Chapter 6. The final (predictor) estimator would be of the form

$$
\begin{aligned}
[\hat{\mathbf{x}}_c]_{k+1} &= \boldsymbol{\phi}_{cc}[\hat{\mathbf{x}}_c]_k + \boldsymbol{\Gamma}_c u + \mathbf{L}_c(y_c - \hat{y}_c), \\
[\hat{\mathbf{x}}_s]_{k+1} &= \boldsymbol{\phi}_{ss}[\hat{\mathbf{x}}_s]_k + \boldsymbol{\phi}_{sc}[\hat{\mathbf{x}}_c]_k + \boldsymbol{\Gamma}_s u + \mathbf{L}_s(\theta - \hat{\theta}),
\end{aligned}
\tag{9.9}
$$

where $\mathbf{L}_c$ and $\mathbf{L}_s$ are both 2×1 matrices.

Even without the very weak one-way coupling in $\boldsymbol{\phi}_{cs}$ which was obvious for this example, one could go ahead and assume this to be the case, then check the resulting full-system characteristic roots using a method similar to the previous airplane example.

In short, it is very important to use your knowledge of the physical aspects of the system at hand to break the design into simpler and more tractable subsets. With luck, the whole job can be finished this way. At worst, insight will be gained which will aid in the design procedures to follow and in the construction and checkout of the control system.

9.3 OPTIMAL CONTROL

Optimal control methods are attractive because they handle multi-input systems easily and allow the designer quickly to determine many good candidate values of the $\mathbf{K}$-matrix. We will develop the time-varying optimal control solution first and then reduce it to a steady-state solution. This amounts to another method of computing the $\mathbf{K}$-matrix in the control law

$$
u = -\mathbf{K}\mathbf{x}
\tag{6.19}
$$

discussed in Chapter 6 and in the examples of Section 9.2 [Eqs. (9.2) and (9.3)].

9.3.1 Time-Varying Optimal Solution

Given a discrete plant

$$
\mathbf{x}_{k+1} = \boldsymbol{\Phi}\mathbf{x}_k + \boldsymbol{\Gamma}\mathbf{u}_k,
\tag{9.10}
$$

we wish to pick $\mathbf{u}_k$ so that a "cost function"

$$
\mathscr{J} = \frac{1}{2} \sum_{k=0}^{N} [\mathbf{x}_k^T \mathbf{Q}_1 \mathbf{x}_k + \mathbf{u}_k^T \mathbf{Q}_2 \mathbf{u}_k]
\tag{9.11}
$$

is minimized. $\mathbf{Q}_1$ and $\mathbf{Q}_2$ are symmetric weighting matrices to be selected by the designer who bases his choice on the relative importance of the various states and controls. Some weight will almost always be selected for the control ($|\mathbf{Q}_2| \neq 0$); otherwise the solution will include large components in the control gains, and the states would be driven to zero at a ridiculously fast rate which could saturate the actuator device.[2] The $\mathbf{Q}$'s must also be non-negative.[3]

Another way of stating the problem given by Eqs. (9.10) and (9.11) is that we wish to minimize

$$\mathscr{J} = \frac{1}{2} \sum_{k=0}^{N} [\mathbf{x}_k^T \mathbf{Q}_1 \mathbf{x}_k + \mathbf{u}_k^T \mathbf{Q}_2 \mathbf{u}_k] \qquad (9.11)$$

subject to the constraint that

$$-\mathbf{x}_{k+1} + \mathbf{\Phi}\mathbf{x}_k + \mathbf{\Gamma}\mathbf{u}_k = 0, \qquad k = 0, 1, \ldots, N. \qquad (9.10)$$

This is a standard constrained-minima problem which can be solved using the method of Lagrange multipliers. There will be one Lagrange multiplier vector, $\boldsymbol{\lambda}_{k+1}$, for each value of k. The procedure is to rewrite (9.10) and (9.11) as:

$$\mathscr{J}' = \sum_{k=0}^{N} \left[\frac{1}{2}\mathbf{x}_k^T \mathbf{Q}_1 \mathbf{x}_k + \frac{1}{2}\mathbf{u}_k^T \mathbf{Q}_2 \mathbf{u}_k + \boldsymbol{\lambda}_{k+1}^T(-\mathbf{x}_{k+1} + \mathbf{\Phi}\mathbf{x}_k + \mathbf{\Gamma}\mathbf{u}_k) \right] \qquad (9.12)$$

and find the minimum of $\mathscr{J}'$ with respect to $\mathbf{x}_k$, $\mathbf{u}_k$, and $\boldsymbol{\lambda}_k$. The subscript on $\boldsymbol{\lambda}$ is arbitrary conceptually, but we let it be $k + 1$ because this choice will yield a particularly easy form of the equations later on.

Proceeding with the minimization leads to:

$$\frac{\partial \mathscr{J}'}{\partial \mathbf{u}_k} = \mathbf{u}_k^T \mathbf{Q}_2 + \boldsymbol{\lambda}_{k+1}^T \mathbf{\Gamma} = 0, \qquad \text{control equations,} \qquad (9.13)$$

$$\frac{\partial \mathscr{J}'}{\partial \boldsymbol{\lambda}_{k+1}} = -\mathbf{x}_{k+1} + \mathbf{\Phi}\mathbf{x}_k + \mathbf{\Gamma}\mathbf{u}_k = 0, \qquad \text{state equations,} \qquad (9.10)$$

and

$$\frac{\partial \mathscr{J}'}{\partial \mathbf{x}_k} = \mathbf{x}_k^T \mathbf{Q}_1 - \boldsymbol{\lambda}_k^T + \boldsymbol{\lambda}_{k+1}^T \mathbf{\Phi} = 0, \qquad \text{adjoint equations.} \qquad (9.14)$$

The last set of the equations, the adjoint equations, may be written as

$$\boldsymbol{\lambda}_k = \mathbf{\Phi}^T \boldsymbol{\lambda}_{k+1} + \mathbf{Q}_1 \mathbf{x}_k . \qquad (9.14)$$

Combining (9.10), (9.13), and (9.14) gives us a set of coupled difference equations defining the optimal solution of $\mathbf{x}_k$ and $\boldsymbol{\lambda}_k$ and hence $\mathbf{u}_k$, provided the initial (or final) conditions are known. The initial conditions on $\mathbf{x}_k$ must be given; however, usually $\boldsymbol{\lambda}_0$ would not be known, and we are led to the endpoint to establish a final condition. From Eq. (9.11) we see that $\mathbf{u}_N$ will be zero for the minimum $\mathscr{J}$ because

---

[2] If the sampling rate, T, is long, however, a control which moves the state along as rapidly as possible may be feasible. Such controls are called "dead-beat" since they beat the state to a dead stop in at most n steps. They correspond to placement of all poles at $Z = 0$.

[3] Matrix equivalent of a nonnegative number; it ensures that $\mathbf{x}^T \mathbf{Q}_1 \mathbf{x}$ and $\mathbf{u}^T \mathbf{Q}_2 \mathbf{u}$ are nonnegative for all possible $\mathbf{x}$ and $\mathbf{u}$.

$\mathbf{u}_N$ has no effect on $\mathbf{x}_N$ [see Eq. (9.10)]. Thus Eq. (9.13) suggests that $\boldsymbol{\lambda}_{N+1} = 0$, and Eq. (9.14) thus shows that a suitable condition is

$$\boldsymbol{\lambda}_N = \mathbf{Q}_1 \mathbf{x}_N. \tag{9.15}$$

The solution to the optimal control problem is now completely specified. It consists of the two difference equations (9.10) and (9.14) with $\mathbf{u}$ given by (9.13), the final condition on $\boldsymbol{\lambda}$ given by Eq. (9.15), and the initial condition on $\mathbf{x}_0$ assumed given in the problem statement. The solution to this two-point boundary-value problem is not so easy.

One method called the sweep method by Bryson and Ho (1969) is to assume

$$\boldsymbol{\lambda}_k = \mathbf{S}_k \mathbf{x}_k; \tag{9.16}$$

then (9.13) becomes

$$\begin{aligned} \mathbf{Q}_2 \mathbf{u}_k &= -\boldsymbol{\Gamma}^T \mathbf{S}_{k+1} \mathbf{x}_{k+1} \\ &= -\boldsymbol{\Gamma}^T \mathbf{S}_{k+1} (\boldsymbol{\Phi} \mathbf{x}_k + \boldsymbol{\Gamma} \mathbf{u}_k). \end{aligned}$$

Solving for $\mathbf{u}_k$, we obtain

$$\begin{aligned} \mathbf{u}_k &= -(\mathbf{Q}_2 + \boldsymbol{\Gamma}^T \mathbf{S}_{k+1} \boldsymbol{\Gamma})^{-1} \boldsymbol{\Gamma}^T \mathbf{S}_{k+1} \boldsymbol{\Phi} \mathbf{x}_k \\ &= -\mathbf{R}^{-1} \boldsymbol{\Gamma}^T \mathbf{S}_{k+1} \boldsymbol{\Phi} \mathbf{x}_k. \end{aligned} \tag{9.17}$$

In (9.17) we have defined $\mathbf{R} = \mathbf{Q}_2 + \boldsymbol{\Gamma}^T \mathbf{S}_{k+1} \boldsymbol{\Gamma}$ for convenience. If we now substitute (9.16) into (9.14) for $\boldsymbol{\lambda}_k$ and $\boldsymbol{\lambda}_{k+1}$, we eliminate $\boldsymbol{\lambda}$. Then we substitute (9.17) into (9.10) to eliminate $\mathbf{x}_{k+1}$ as follows. From (9.14)

$$\boldsymbol{\lambda}_k = \boldsymbol{\Phi}^T \boldsymbol{\lambda}_{k+1} + \mathbf{Q}_1 \mathbf{x}_k$$

and substituting (9.16), we have

$$\mathbf{S}_k \mathbf{x}_k = \boldsymbol{\Phi}^T \mathbf{S}_{k+1} \mathbf{x}_{k+1} + \mathbf{Q}_1 \mathbf{x}_k.$$

Now we use (9.10) for $\mathbf{x}_{k+1}$,

$$\mathbf{S}_k \mathbf{x}_k = \boldsymbol{\Phi}^T \mathbf{S}_{k+1} (\boldsymbol{\Phi} \mathbf{x}_k + \boldsymbol{\Gamma} \mathbf{u}_k) + \mathbf{Q}_1 \mathbf{x}_k. \tag{9.18}$$

Next we use (9.17) for $\mathbf{u}_k$ in (9.18),

$$\mathbf{S}_k \mathbf{x}_k = \boldsymbol{\Phi}^T \mathbf{S}_{k+1} (\boldsymbol{\Phi} \mathbf{x}_k - \boldsymbol{\Gamma} \mathbf{R}^{-1} \boldsymbol{\Gamma}^T \mathbf{S}_{k+1} \boldsymbol{\Phi} \mathbf{x}_k) + \mathbf{Q}_1 \mathbf{x}_k,$$

and collect all terms on one side:

$$[\mathbf{S}_k - \boldsymbol{\Phi}^T \mathbf{S}_{k+1} \boldsymbol{\Phi} + \boldsymbol{\Phi}^T \mathbf{S}_{k+1} \boldsymbol{\Gamma} \mathbf{R}^{-1} \boldsymbol{\Gamma}^T \mathbf{S}_{k+1} \boldsymbol{\Phi} - \mathbf{Q}_1] \mathbf{x}_k = 0. \tag{9.19}$$

Since Eq. (9.19) must hold for any $\mathbf{x}_k$ whatever, the coefficient matrix must be identically zero, from which follows a backward equation in $\mathbf{S}_k$,

$$\mathbf{S}_k = \boldsymbol{\Phi}^T [\mathbf{S}_{k+1} - \mathbf{S}_{k+1} \boldsymbol{\Gamma} \mathbf{R}^{-1} \boldsymbol{\Gamma}^T \mathbf{S}_{k+1}] \boldsymbol{\Phi} + \mathbf{Q}_1, \tag{9.20}$$

which is often rewritten as

$$\mathbf{S}_k = \boldsymbol{\Phi}^T \mathbf{M}_{k+1} \boldsymbol{\Phi} + \mathbf{Q}_1, \tag{9.21}$$

where

$$\mathbf{M}_{k+1} = \mathbf{S}_{k+1} - \mathbf{S}_{k+1}\mathbf{\Gamma}[\mathbf{Q}_2 + \mathbf{\Gamma}^T\mathbf{S}_{k+1}\mathbf{\Gamma}]^{-1}\mathbf{\Gamma}^T\mathbf{S}_{k+1}. \qquad (9.22)$$

Note that the matrix to be inverted ($\mathbf{R}$) has the same dimension as the number of controls, which is usually less than the number of states.

The boundary condition on the recursion relationship for $\mathbf{S}_{k+1}$ is obtained from (9.15) and (9.16); thus

$$\mathbf{S}_N = \mathbf{Q}_1, \qquad (9.23)$$

and we see that Eqs. (9.21) and (9.22) must be solved backward with the initial condition (9.23). To solve for $\mathbf{u}_k$, we use (9.17) to obtain

$$\mathbf{u}_k = -\mathbf{K}_k\mathbf{x}_k, \qquad (9.24)$$

where

$$\mathbf{K}_k = (\mathbf{Q}_2 + \mathbf{\Gamma}^T\mathbf{S}_{k+1}\mathbf{\Gamma})^{-1}\mathbf{\Gamma}^T\mathbf{S}_{k+1}\mathbf{\Phi} \qquad (9.25)$$

and is the desired "optimal" time-varying feedback gain.

Let us now summarize the entire procedure:

1. Let $\mathbf{S}_N \leftarrow \mathbf{Q}_1$ and $\mathbf{K}_N \leftarrow \mathbf{0}$. $\qquad (9.23)$
2. Let $k \leftarrow N$.
3. Let $\mathbf{M}_k \leftarrow \mathbf{S}_k - \mathbf{S}_k\mathbf{\Gamma}[\mathbf{Q}_2 + \mathbf{\Gamma}^T\mathbf{S}_k\mathbf{\Gamma}]^{-1}\mathbf{\Gamma}^T\mathbf{S}_k$. $\qquad (9.22)$
4. Let $\mathbf{K}_{k-1} \leftarrow (\mathbf{Q}_2 + \mathbf{\Gamma}^T\mathbf{S}_k\mathbf{\Gamma})^{-1}\mathbf{\Gamma}^T\mathbf{S}_k\mathbf{\Phi}$. $\qquad (9.25)$
5. Store $\mathbf{K}_{k-1}$.
6. Let $\mathbf{S}_{k-1} \leftarrow \mathbf{\Phi}^T\mathbf{M}_k\mathbf{\Phi} + \mathbf{Q}_1$. $\qquad (9.21)$
7. Let $k \leftarrow k - 1$.
8. Go to step 3.

For any given initial condition for $\mathbf{x}$, to apply the control, we use the stored gains and

$$\mathbf{x}_{k+1} = \mathbf{\Phi}\mathbf{x}_k + \mathbf{\Gamma}\mathbf{u}_k, \qquad (9.10)$$

where

$$\mathbf{u}_k = -\mathbf{K}_k\mathbf{x}_k. \qquad (9.24)$$

Note that the optimal gain, $\mathbf{K}_k$, changes at each time step but can be precomputed and stored for later use as long as the length, N, of the problem is known; for example, no knowledge of the initial state $\mathbf{x}_0$ is required for computation of the control gain $\mathbf{K}_k$.

As an example of the time-varying nature of the control gains, Eqs. (9.21) through (9.25) were solved for the satellite attitude-control example described in Appendix A.1. The system transfer function is $1/s^2$.

The state weighting matrix was arbitrarily chosen to be

$$Q_1 = \begin{bmatrix} 1 & 0 \\ 0 & 0 \end{bmatrix}, \tag{9.26a}$$

which means that the angle state was weighted but not the angular velocity. The control weighting matrix is a scalar in this case because there is a single control input; three values were selected for evaluation:

$$Q_2 = 0.01, \ 0.1, \ \text{and} \ 1.0. \tag{9.26b}$$

The problem length for purposes of defining $\mathcal{J}$ was chosen to be 51 steps, which, with the sample period of $T = 0.1$ sec, means that the total time was 5.1 sec.

Figure 9.3 contains the resulting gain time histories plotted by the computer. We see from the figure that the problem length only affects the values of K near the end, and in fact, the first portions of all cases show constant values of the

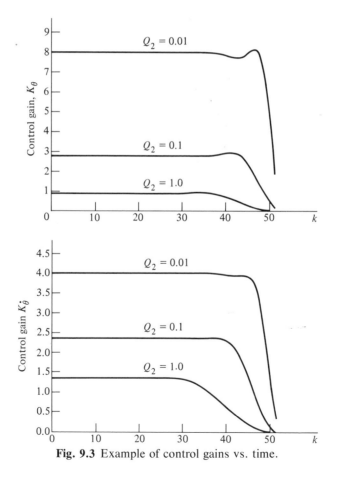

Fig. 9.3 Example of control gains vs. time.

gains. This means that for problems of long duration (long compared to transients in the gains), the optimal controller for much of the time is identical in structure to the constant gain cases discussed in Chapter 6 and Section 9.2 except that the values of the constant gains are based on a cost function rather than root locations.

For multi-input problems, the time-varying gains act exactly like the example above. The next section develops a method to compute the "constant" value of the optimal gains so that they can be used in place of the time-varying values, thus yielding a much simpler implementation and yet one that is almost optimum.

Before we leave the time-varying case it is informative to evaluate the optimal cost function $\mathscr{J}$ in terms of $\boldsymbol{\lambda}$ and $\mathbf{S}$. If we substitute (9.13) and (9.14) for $\boldsymbol{\lambda}_{k+1}^T \boldsymbol{\Gamma}$ and $\boldsymbol{\lambda}_{k+1}^T \boldsymbol{\Phi}$ in (9.12), we find

$$\mathscr{J} = \frac{1}{2} \sum_{k=0}^{N} [\mathbf{x}_k^T \mathbf{Q}_1 \mathbf{x}_k + \mathbf{u}_k^T \mathbf{Q}_2 \mathbf{u}_k - \boldsymbol{\lambda}_{k+1}^T \mathbf{x}_{k+1} + (\boldsymbol{\lambda}_k^T - \mathbf{x}_k^T \mathbf{Q}_1) \mathbf{x}_k + (-\mathbf{u}_k^T \mathbf{Q}_2) \mathbf{u}_k]$$

$$= \frac{1}{2} \sum_{k=0}^{N} (\boldsymbol{\lambda}_k^T \mathbf{x}_k - \boldsymbol{\lambda}_{k+1}^T \mathbf{x}_{k+1})$$

$$= \frac{1}{2} \boldsymbol{\lambda}_0^T \mathbf{x}_0 - \frac{1}{2} \boldsymbol{\lambda}_{N+1}^T \mathbf{x}_{N+1}.$$

However, from (9.15), $\boldsymbol{\lambda}_{N+1} = 0$, and thus, using (9.16), we find

$$\mathscr{J} = \frac{1}{2} \boldsymbol{\lambda}_0^T \mathbf{x}_0$$

$$= \frac{1}{2} \mathbf{x}_0^T \mathbf{S}_0 \mathbf{x}_0. \tag{9.27}$$

Thus we see that having computed $\mathbf{S}$, we can immediately evaluate the cost associated with the control.

9.3.2 Steady-State Optimal Solution

The optimal solution for the time-varying gain, $\mathbf{K}$, is described by Eqs. (9.13), (9.14), and (9.10). We can combine these equations by premultiplying (9.14) by $\boldsymbol{\Phi}^{-T}$ and solving (9.13) for $\mathbf{u}_k$, thus arriving at a set of difference equations in standard form in $\mathbf{x}$ and $\boldsymbol{\lambda}$ if we assume that $\mathbf{Q}_2$ and $\boldsymbol{\Phi}$ are nonsingular. These equations are called *Hamilton's equations,* and their system matrix is the control Hamiltonian matrix:

$$\begin{bmatrix} \mathbf{x} \\ \boldsymbol{\lambda} \end{bmatrix}_{k+1} = \begin{bmatrix} \boldsymbol{\Phi} + \boldsymbol{\Gamma} \mathbf{Q}_2^{-1} \boldsymbol{\Gamma}^T \boldsymbol{\Phi}^{-T} \mathbf{Q}_1 & -\boldsymbol{\Gamma} \mathbf{Q}_2^{-1} \boldsymbol{\Gamma}^T \boldsymbol{\Phi}^{-T} \\ -\boldsymbol{\Phi}^{-T} \mathbf{Q}_1 & \boldsymbol{\Phi}^{-T} \end{bmatrix} \begin{bmatrix} \mathbf{x} \\ \boldsymbol{\lambda} \end{bmatrix}_k. \tag{9.28}$$

The Hamiltonian for the control problem is, therefore,

$$\mathscr{H}_c = \begin{bmatrix} \boldsymbol{\Phi} + \boldsymbol{\Gamma} \mathbf{Q}_2^{-1} \boldsymbol{\Gamma}^T \boldsymbol{\Phi}^{-T} \mathbf{Q}_1 & -\boldsymbol{\Gamma} \mathbf{Q}_2^{-1} \boldsymbol{\Gamma}^T \boldsymbol{\Phi}^{-T} \\ -\boldsymbol{\Phi}^{-T} \mathbf{Q}_1 & \boldsymbol{\Phi}^{-T} \end{bmatrix}. \tag{9.29}$$

Since the Hamiltonian matrix is constant, we can solve (9.28) by first transforming to a new state which has a diagonal system matrix, and from this solution easily obtain the steady-state optimal control. In Appendix 9.1 we prove that the eigenvalues of this matrix are such that the reciprocal of every eigenvalue is also an eigenvalue. Therefore, half the roots of the characteristic equation must be inside the unit circle and half must be outside. In this case, therefore, $\mathcal{H}_c$ can be diagonalized to the form.[4]

$$\mathcal{H}_c^* = \begin{bmatrix} \mathbf{E}^{-1} & 0 \\ 0 & \mathbf{E} \end{bmatrix}, \tag{9.30}$$

where $\mathbf{E}$ is a diagonal matrix of the unstable roots ($|z| > 1$) and $\mathbf{E}^{-1}$ is a diagonal matrix of the stable roots ($|z| < 1$). $\mathcal{H}_c^*$ is obtained by the similarity transformation

$$\mathcal{H}_c^* = \mathbf{W}^{-1}\mathcal{H}_c\mathbf{W}, \tag{9.31}$$

where $\mathbf{W}$ is the matrix of eigenvectors of $\mathcal{H}_c$ and can be written in block form as

$$\mathbf{W} = \begin{bmatrix} \mathbf{X}_I & \mathbf{X}_0 \\ \mathbf{\Lambda}_I & \mathbf{\Lambda}_0 \end{bmatrix}, \tag{9.32}$$

where

$$\begin{bmatrix} \mathbf{X}_0 \\ \mathbf{\Lambda}_0 \end{bmatrix}$$

is the matrix of eigenvectors associated with the eigenvalues (roots) outside the unit circle and

$$\begin{bmatrix} \mathbf{X}_I \\ \mathbf{\Lambda}_I \end{bmatrix}$$

is the matrix of eigenvectors associated with the eigenvalues of $\mathcal{H}_c$ which are inside the unit circle.

This same transformation matrix, $\mathbf{W}$, can be used to transform $\mathbf{x}$ and $\boldsymbol{\lambda}$ to the normal modes of the system, that is,

$$\begin{bmatrix} \mathbf{x}^* \\ \boldsymbol{\lambda}^* \end{bmatrix} = \mathbf{W}^{-1}\begin{bmatrix} \mathbf{x} \\ \boldsymbol{\lambda} \end{bmatrix}, \tag{9.33}$$

where $\mathbf{x}^*$ and $\boldsymbol{\lambda}^*$ are the normal modes. Conversely, we also have

$$\begin{bmatrix} \mathbf{x} \\ \boldsymbol{\lambda} \end{bmatrix} = \mathbf{W}\begin{bmatrix} \mathbf{x}^* \\ \boldsymbol{\lambda}^* \end{bmatrix} = \begin{bmatrix} \mathbf{X}_I & \mathbf{X}_0 \\ \mathbf{\Lambda}_I & \mathbf{\Lambda}_0 \end{bmatrix}\begin{bmatrix} \mathbf{x}^* \\ \boldsymbol{\lambda}^* \end{bmatrix}. \tag{9.34}$$

The solution to the coupled set of difference equations (9.28) can be simply stated in terms of the initial and final conditions and the normal modes, since the

---

[4] In rare cases, $\mathcal{H}_c$ will have repeated roots and cannot be made diagonal by a change of variables. In those cases, a small change in $\mathbf{Q}_1$ or $\mathbf{Q}_2$ will remove the problem or else we must compute the Jordan form for $\mathcal{H}_c^*$. See Strang (1976). Recent methods using the $\mathbf{QZ}$ algorithm are also possible.

solution for the normal modes is given by

$$\begin{bmatrix} \mathbf{x}^* \\ \boldsymbol{\lambda}^* \end{bmatrix}_N = \begin{bmatrix} \mathbf{E}^{-N} & 0 \\ 0 & \mathbf{E}^N \end{bmatrix} \begin{bmatrix} \mathbf{x}^* \\ \boldsymbol{\lambda}^* \end{bmatrix}_0. \tag{9.35}$$

To obtain the steady state, we let N go to infinity; therefore x_N^* goes to zero and, in general, λ_N^* would go to infinity since each element of $\mathbf{E}$ is greater than one. So we see that the only sensible solution for the steady-state ($N \to \infty$) case is for $\lambda_0^* = 0$ and therefore $\lambda_k^* = 0$ for all k.[5]

From (9.34) and (9.35) with $\lambda_k^* \equiv 0$, we have

$$\mathbf{x}_k = \mathbf{X}_I \mathbf{x}_k^* = \mathbf{X}_I \mathbf{E}^{-k} \mathbf{x}_0^*, \tag{9.36}$$

$$\boldsymbol{\lambda}_k = \boldsymbol{\Lambda}_I \mathbf{x}_k^* = \boldsymbol{\Lambda}_I \mathbf{E}^{-k} \mathbf{x}_0^*. \tag{9.37}$$

Therefore (9.36) leads to

$$\mathbf{x}_0^* = \mathbf{E}^k \mathbf{X}_I^{-1} \mathbf{x}_k. \tag{9.38}$$

Thus, from (9.37) and (9.38),

$$\boldsymbol{\lambda}_k = \boldsymbol{\Lambda}_I \mathbf{X}_I^{-1} \mathbf{x}_k = \mathbf{S}_\infty \mathbf{x}_k. \tag{9.39}$$

Equation (9.39) is the same form as our assumption (9.16) for the sweep method, so we conclude that $\mathbf{S}_\infty$ given by (9.39) is the steady-state solution to (9.20) and that the control law for this system corresponding to $\mathscr{J}$ with $N \to \infty$ is

$$\mathbf{u}_k = -\mathbf{K}_\infty \mathbf{x}_k, \tag{9.40}$$

where, from (9.39) and (9.25),

$$\mathbf{K}_\infty = (\mathbf{Q}_2 + \boldsymbol{\Gamma}^T \mathbf{S}_\infty \boldsymbol{\Gamma})^{-1} \boldsymbol{\Gamma}^T \mathbf{S}_\infty \boldsymbol{\Phi}. \tag{9.41}$$

Furthermore, from (9.27), the cost associated with using this control law is

$$\mathscr{J}_\infty = \tfrac{1}{2} \mathbf{x}_0^T \mathbf{S}_\infty \mathbf{x}_0. \tag{9.42}$$

The complete computational procedure is summarized[6] below:

1. Compute eigenvalues of the system matrix $\mathscr{H}_c$ defined by Eq. (9.29).

2. Compute eigenvectors associated with the stable ($|z| < 1$) eigenvalues and call them

$$\begin{bmatrix} \mathbf{X}_I \\ \boldsymbol{\Lambda}_I \end{bmatrix}.$$

3. Compute control gain $\mathbf{K}_\infty$ from Eq. (9.41) with $\mathbf{S}_\infty$ given by (9.39).

---

[5] From (9.34) we see that if λ^* is not zero then the state $\mathbf{x}$ will grow in time and the system will be unstable. However, if the system is controllable we know that a control exists which will make the system stable and give a finite value to $\mathscr{J}$. Since we have the optimal control in (9.28) it must follow that the optimal system is stable and $\lambda^* \equiv 0$ if $\mathbf{Q}_1$ is such that all the states affect $\mathscr{J}$.

[6] It is really necessary only to have a transformation that makes $\mathscr{H}_c$ upper triangular with all stable eigenvalues in the upper left half of the diagonal.

The stable eigenvalues from step 1 are the resulting system closed-loop roots with constant gain $\mathbf{K}_\infty$ from step 3. The reason for this is: Before we let N go to infinity, the time-varying gain system was written in (9.29) as linear, constant-coefficient difference equations in $\mathbf{x}$ and $\boldsymbol{\lambda}$ (9.29). The eigenvalues of this system (of $\mathcal{H}_c$) are equally valid for N small (time-varying gain region) and N very large (constant-gain region). In the constant-gain region, it would be easy to uncouple $\mathbf{x}$ from $\boldsymbol{\lambda}$ by using (9.40), (9.41), and (9.10); and since eigenvalues remain unchanged for any linear combination of states, half of the eigenvalues of the $\mathcal{H}_c$ system must be the eigenvalues of the closed-loop system

$$\mathbf{x}_{k+1} = \boldsymbol{\Phi}\mathbf{x}_k + \boldsymbol{\Gamma}\mathbf{u}_k, \tag{9.10}$$

where

$$\mathbf{u}_k = -\mathbf{K}_\infty \mathbf{x}_k. \tag{6.19}$$

We have already used the notion that the stable roots of $\mathcal{H}_c$ correspond to $\mathbf{x}^*$, that is, (9.35). This is the only reasonable correspondence since the desired solution is that which minimizes the cost function (9.11). It therefore also follows that the stable eigenvalues of the $\mathcal{H}_c$ system correspond to the eigenvalues of the closed-loop system composed of (9.10), (9.40), and (9.42). We can also show that the matrix $\mathbf{X}_I$ of (9.34) and (9.36) is the matrix of eigenvectors of the optimal steady-state closed-loop system.

An approach by Vaughn (1970) similar to the development in this section may be used to obtain a nonrecursive solution to the time-varying case which could be considerably faster to solve than (9.21) through (9.25).

9.4 OPTIMAL ESTIMATION

Optimal estimation methods are attractive because they handle multi-output systems easily and allow a designer quickly to determine many good candidate designs of the L-matrix. We will develop the time-varying optimal estimation solution (commonly known as the "Kalman filter"), then show the correspondence between it and the time-varying optimal control solution, and finally develop the optimal steady-state estimation solution. This amounts to another method of computing the L-matrix in the equations for the current estimator

$$\hat{\mathbf{x}}_n = \bar{\mathbf{x}}_n + \mathbf{L}(y_n - \bar{y}_n), \tag{6.38}$$

where

$$\bar{\mathbf{x}}_{n+1} = \boldsymbol{\Phi}\hat{\mathbf{x}}_n + \boldsymbol{\Gamma}\mathbf{u}_n,$$

discussed in Chapter 6 and in the predictor estimator of the examples in Section 9.2, Eq. (9.9).

9.4.1 The Kalman Filter

Given a discrete plant

$$\mathbf{x}(n + 1) = \mathbf{\Phi}\mathbf{x}(n) + \mathbf{\Gamma}\mathbf{u}(n) + \mathbf{\Gamma}_1\mathbf{w}(n) \tag{9.43}$$

with measurements

$$\mathbf{y}(n) = \mathbf{H}\mathbf{x}(n) + \mathbf{v}(n), \tag{9.44}$$

where the process noise $\mathbf{w}(n)$ and measurement noise $\mathbf{v}(n)$ are random sequences with zero mean, that is,

$$\mathcal{E}\{\mathbf{w}(n)\} = \mathcal{E}\{\mathbf{v}(n)\} = \mathbf{0},$$

have no time correlation or are "white" noise, that is,

$$\mathcal{E}\{\mathbf{w}(i)\mathbf{w}^T(j)\} = \mathcal{E}\{\mathbf{v}(i)\mathbf{v}^T(j)\} = 0 \quad \text{if } i \neq j,$$

and have covariances or "noise levels" defined by

$$\mathcal{E}\{\mathbf{w}(n)\mathbf{w}^T(n)\} \overset{\Delta}{=} \mathbf{R}_w, \qquad \mathcal{E}\{\mathbf{v}(n)\mathbf{v}^T(n)\} \overset{\Delta}{=} \mathbf{R}_v,$$

we wish to pick $\mathbf{L}$ so that the estimate of $\mathbf{x}_k$, given all the data, is optimal.

Temporarily, let us pretend that without using the current measurement $\mathbf{y}(n)$, we already have a prior estimate of the state at the time of a measurement which we will call $\bar{\mathbf{x}}(n)$. The problem therefore is to update this estimate based on the current measurement.

Comparing this problem to the identification problem discussed in Chapter 8, we see that the estimation measurement equation, (9.44), relates the measurements to the quantity to be estimated just as (8.28) does[7]; hence the optimal state estimation solution is basically given by the recursive weighted least-squares solution and is essentially that given by (8.52c) with $\gamma = 1$ and $a = R_v^{-1}$. With modifications for notation, the solution equations are

$$\hat{\mathbf{x}}(n) = \bar{\mathbf{x}}(n) + \mathbf{M}\mathbf{H}^T(\mathbf{H}\mathbf{M}\mathbf{H}^T + \mathbf{R}_v)^{-1}(\mathbf{y}(n) - \mathbf{H}\bar{\mathbf{x}}(n)), \tag{9.45}$$

$$\mathbf{P}(n) = \mathbf{M}(n) - \mathbf{M}(n)\mathbf{H}^T(\mathbf{H}\mathbf{M}\mathbf{H}^T + \mathbf{R}_v)^{-1}\mathbf{H}\mathbf{M}(n). \tag{9.46}$$

The variables in Chapter 8 have been changed as follows to arrive at the above[8]:

$$\hat{\boldsymbol{\theta}}(N + 1) \leftarrow \hat{\mathbf{x}}(n), \qquad \hat{\boldsymbol{\theta}}(N) \leftarrow \bar{\mathbf{x}}(n), \qquad \boldsymbol{\psi}^T \leftarrow \mathbf{H};$$
$$\mathbf{P}(N) \leftarrow \mathbf{M}(n), \qquad \mathbf{P}(N + 1) \leftarrow \mathbf{P}(n), \qquad \mathbf{a}^{-1} \leftarrow \mathbf{R}_v.$$

---

[7] If in (8.28) we let $\boldsymbol{\theta} = \mathbf{x}$, $\boldsymbol{\psi}^T = \mathbf{H}$, and $\mathbf{e} = \mathbf{v}$, then if $N = n$, we have (9.44). The recursive problem is to compute an estimate using $y(n)$ given the estimate $\bar{\mathbf{x}}(n) \sim \hat{\boldsymbol{\theta}}(n - 1)$ which does not include the observation $\mathbf{y}(n)$. The new estimate is $\hat{\mathbf{x}}(n)$, which is equivalent to $\hat{\boldsymbol{\theta}}(n)$ as used in Chapter 8.

[8] The reason for using different symbols such as $\mathbf{x}$ and $\hat{\mathbf{x}}$ and $\mathbf{M}$ and $\mathbf{P}$ here although the same symbols were used in Chapter 8 will be seen shortly. Read the arrow as "is replaced by."

This idea of combining the previous estimate with the current measurement based on the relative accuracy of the two quantities is one of the basic ideas of the Kalman filter. The other idea has to do with using the dynamics of $\bar{\mathbf{x}}(n)$ which, so far, we have not included.

The important distinction in the estimation of the state $\bar{\mathbf{x}}$ that did not come up in the estimation of parameters in Chapter 8 is the notion of the dynamics of the state between $n - 1$ and n. The state estimate at $n - 1$, given data up through $n - 1$, is $\hat{\mathbf{x}}(n - 1)$, whereas we defined $\bar{\mathbf{x}}(n)$ to be the estimate at n given the same data up through $n - 1$. In Chapter 8, these two quantities were identical and both called $\hat{\boldsymbol{\theta}}(n - 1)$ since $\boldsymbol{\theta}$ was presumed to be a constant, but here the estimates differ due to the fact that the state will change according to the system dynamics as time passes. Specifically, the estimate $\bar{\mathbf{x}}(n)$ is given from $\hat{\mathbf{x}}(n - 1)$ by using (9.43) without noise because we know that this is the expected value of $\mathbf{x}(n)$ if $\mathscr{E}\{\mathbf{x}(n - 1)\}$ is $\hat{\mathbf{x}}(n - 1)$ since the expected value of the plant noise, $\mathbf{w}(n - 1)$, is zero. Thus

$$\bar{\mathbf{x}}(n) = \boldsymbol{\Phi}\hat{\mathbf{x}}(n - 1) + \boldsymbol{\Gamma}\mathbf{u}(n - 1). \tag{9.47}$$

The change in estimate from $\hat{\mathbf{x}}(n - 1)$ to $\bar{\mathbf{x}}(n)$ is called a "time update," whereas the change in the estimate from $\bar{\mathbf{x}}(n)$ to $\hat{\mathbf{x}}(n)$ as given by (9.45) is a "measurement update" which occurs at the fixed time n but expresses the improvement in the estimate due to the measurement $\mathbf{y}(n)$. The same kind of time and measurement updates apply to the estimate covariances, $\mathbf{P}$ and $\mathbf{M}$: $\mathbf{P}$ represents the estimate accuracy immediately after a measurement, whereas $\mathbf{M}$ is the propagated value of $\mathbf{P}$ and is valid just before measurements. From (9.43) and (9.47) we see that

$$\mathbf{x}(n + 1) - \bar{\mathbf{x}}(n + 1) = \boldsymbol{\Phi}(\mathbf{x}(n) - \hat{\mathbf{x}}(n)) + \boldsymbol{\Gamma}_1\mathbf{w}(n); \tag{9.48}$$

hence the covariance of the state at time $n + 1$ but *before* taking $\mathbf{y}(n + 1)$ into account is

$$\mathscr{E}[(\mathbf{x}(n + 1) - \bar{\mathbf{x}}(n + 1))(\mathbf{x}(n + 1) - \bar{\mathbf{x}}(n + 1))^T].$$

If the measurement noise, $\mathbf{v}$, and the process noise, $\mathbf{w}$, are uncorrelated so that $\mathbf{x}(n)$ and $\mathbf{w}(n)$ are also uncorrelated, the cross product terms vanish and we get

$$\mathscr{E}[\boldsymbol{\Phi}(\mathbf{x}(n) - \hat{\mathbf{x}}(n))(\mathbf{x}(n) - \hat{\mathbf{x}}(n))^T\boldsymbol{\Phi}^T + \boldsymbol{\Gamma}_1\mathbf{w}(n)\mathbf{w}^T(n)\boldsymbol{\Gamma}_1^T]. \tag{9.49}$$

We define

$$\mathbf{M} = \mathscr{E}\{(\mathbf{x} - \bar{\mathbf{x}})(\mathbf{x} - \bar{\mathbf{x}})^T\}, \qquad \mathbf{P} = \mathscr{E}\{(\mathbf{x} - \hat{\mathbf{x}})(\mathbf{x} - \hat{\mathbf{x}})^T\}, \qquad \mathbf{R}_w = \mathscr{E}\{\mathbf{w}\mathbf{w}^T\},$$

and (9.49) reduces to the time update

$$\mathbf{M}(n + 1) = \boldsymbol{\Phi}\mathbf{P}(n)\boldsymbol{\Phi}^T + \boldsymbol{\Gamma}_1\mathbf{R}_w\boldsymbol{\Gamma}_1^T. \tag{9.50}$$

The measurement update is given by the proper substitutions into (8.52a) and (8.52b), which include $\mathbf{P}(N)$ replaced by $\mathbf{M}(n + 1)$ and $\mathbf{P}(N + 1)$ replaced by $\mathbf{P}(n + 1)$.

This completes the required relations for the optimal, time-varying gain, state estimation or Kalman filter. A summary of the required relations is:

At the measurement time (measurement update)

$$\hat{\mathbf{x}}(n) = \bar{\mathbf{x}}(n) + \mathbf{MH}^T(\mathbf{HMH}^T + \mathbf{R}_v)^{-1}(\mathbf{y}(n) - \mathbf{H}\bar{\mathbf{x}}(n)), \tag{9.45}$$

$$\mathbf{P}(n) = \mathbf{M}(n) - \mathbf{M}(n)\mathbf{H}^T(\mathbf{HM}(n)\mathbf{H}^T + \mathbf{R}_v)^{-1}\mathbf{HM}(n). \tag{9.46}$$

Between measurements (time update):

$$\bar{\mathbf{x}}(n + 1) = \mathbf{\Phi}\hat{\mathbf{x}}(n) + \mathbf{\Gamma}\mathbf{u}(n), \tag{9.47}$$

$$\mathbf{M}(n + 1) = \mathbf{\Phi}\mathbf{P}(n)\mathbf{\Phi}^T + \mathbf{\Gamma}_1\mathbf{R}_w\mathbf{\Gamma}_1^T. \tag{9.50}$$

$\mathbf{M}(0) = \mathscr{E}\{\bar{\mathbf{x}}(0)\bar{\mathbf{x}}^T(0)\}$ is given in the problem statement.

As an example of the time-varying nature of the estimator gain,

$$\mathbf{L}(n) = \mathbf{MH}^T(\mathbf{HMH}^T + \mathbf{R}_v)^{-1}, \tag{9.51}$$

(9.45) through (9.50) will be solved for the satellite example (Appendix A.1). Since only θ is directly sensed, we have from (6.3)

$$H = [1 \quad 0],$$

and we assume that the measurement noise covariance is

$$R_v = 0.1 \text{ deg}^2.$$

The white disturbance noise is assumed to be acting as a torque; therefore the disturbance-input distribution matrix is taken to be

$$\mathbf{\Gamma}_1 = \begin{bmatrix} 0 \\ 1 \end{bmatrix},$$

and we will assume several values for the variance magnitude of this disturbance:

$$\mathbf{R}_w = 0.001, 0.01, 0.1 \ (\text{deg}^2/\text{sec}^4).$$

Given an actual design problem, one can often assign a meaningful value to $\mathbf{R}_v$ which is based on the sensor accuracy. The same cannot be said for $\mathbf{R}_w$. The assumption of white process noise is usually a mathematical artifice which is used because of the ease of solving the resulting optimization problem. Physically, $\mathbf{R}_w$ is crudely accounting for unknown disturbances, whether they be steps, white noise, or somewhere inbetween and for imperfections in the plant model. If $\mathbf{R}_w$ was chosen to be zero, as one might be tempted to do for the multitude of design problems that do not appear to have any *white* process noise, the estimator gain would eventually go to zero. This is so because the optimal thing to do in the ideal situation of no disturbances and a perfect plant model is to estimate open loop and forget the noisy measurements. In practice, this will not work because there are always some disturbances, and the plant model is never perfect. We therefore are often forced to pick values of $\mathbf{R}_w$ "out of a hat" in the design process and to settle on an acceptable one based on the quality of the estimation that results in subsequent simulations. Initial conditions for the recursion relation for $\mathbf{M}$ and $\mathbf{P}$ are also required. Physically, they represent the *a priori* estimate of the accuracy of $\overline{\mathbf{x}}(0)$, which in turn represents the *a priori* estimate of the state. In lieu of any better information, one could logically postulate that the components of $\overline{\mathbf{x}}(0)$ contained in $\mathbf{y}$ be equal to the first measurement, and the remaining components equal to zero.

Continuing on with the satellite example, we see that Fig. 9.4 shows the time history of the $\mathbf{L}$'s for the values of the $\mathbf{R}$'s given above. We see from the figure that after an initial settling time, the gains essentially reach a steady state. The subject of the next section is a method to compute the value of these steady-state gains without going through the computation of (9.46) and (9.50) over many time steps until they produce a constant solution.

In obtaining the values of $\mathbf{L}$ in Fig. 9.4, it was necessary to solve for the time history of $\mathbf{P}$. Since $\mathbf{P}$ is the covariance of the estimation errors, we might be tempted to use $\mathbf{P}$ as an indicator of estimation accuracy. However, it is only a

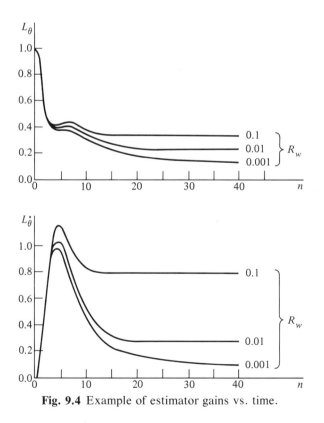

Fig. 9.4 Example of estimator gains vs. time.

meaningful accuracy predictor when we choose the values of $\mathbf{R}_v$ and $\mathbf{R}_w$ based on some knowledge of processes which are approximately white.

The intent of this section and example is to give some insight into the nature of the solution so as to motivate and provide a basis for the following section. Readers interested in the application of Kalman filters are encouraged to review works devoted to that subject, such as Bryson and Ho (1969), Athans (1971), and Anderson and Moore (1979).

9.4.2 Steady-State Optimal Estimation

The equations to be solved that determine the estimator gain have been given in the last section:

$$\mathbf{P}(n) = \mathbf{M}(n) - \mathbf{M}(n)\mathbf{H}^T(\mathbf{H}\mathbf{M}(n)\mathbf{H}^T + \mathbf{R}_v)^{-1}\mathbf{H}\mathbf{M}(n), \tag{9.46}$$

$$\mathbf{M}(n + 1) = \mathbf{\Phi}\mathbf{P}(n)\mathbf{\Phi}^T + \mathbf{\Gamma}_1\mathbf{R}_w\mathbf{\Gamma}_1^T. \tag{9.50}$$

Comparing (9.46) and (9.50) to the optimal control recursion relationships,

$$\mathbf{M}(k) = \mathbf{S}(k) - \mathbf{S}(k)\mathbf{\Gamma}[\mathbf{Q}_2 + \mathbf{\Gamma}^T\mathbf{S}(k)\mathbf{\Gamma}]^{-1}\mathbf{\Gamma}^T\mathbf{S}(k), \tag{9.22}$$

$$\mathbf{S}(k) = \mathbf{\Phi}^T\mathbf{M}(k + 1)\mathbf{\Phi} + \mathbf{Q}_1, \tag{9.20}$$

we see that they are precisely of the same form, except that (9.50) goes forward instead of backward as (9.20) does. Therefore, we can simply change variables and directly use the steady-state solution of the control problem as the desired steady-state solution to the estimation problem, even though the equations are solved in opposite directions in time.

Table 9.1 lists the correspondences that result by direct comparison of the control and estimation recursion relations: (9.22) with (9.46) and (9.20) with (9.50).

By analogy with the control problem, Eqs. (9.46) and (9.50) must have arisen from two coupled equations with the same form as (9.28). Using the correspondences in Table 9.1, the system matrix, (9.28), becomes the estimation Hamiltonian:

$$\mathcal{H}_e = \begin{bmatrix} \mathbf{\Phi}^T + \mathbf{H}^T\mathbf{R}_v^{-1}\mathbf{H}\mathbf{\Phi}^{-1}\mathbf{\Gamma}_1\mathbf{R}_w\mathbf{\Gamma}_1^T & -\mathbf{H}^T\mathbf{R}_v^{-1}\mathbf{H}\mathbf{\Phi}^{-1} \\ -\mathbf{\Phi}^{-1}\mathbf{\Gamma}_1\mathbf{R}_w\mathbf{\Gamma}_1^T & \mathbf{\Phi}^{-1} \end{bmatrix}. \tag{9.52}$$

Table 9.1 Control and Estimation Duality

Control	Estimation
$\mathbf{\Phi}$	$\mathbf{\Phi}^T$
$\mathbf{M}$	$\mathbf{P}$
$\mathbf{S}$	$\mathbf{M}$
$\mathbf{Q}_1$	$\mathbf{\Gamma}_1\mathbf{R}_w\mathbf{\Gamma}_1^T$
$\mathbf{\Gamma}$	$\mathbf{H}^T$
$\mathbf{Q}_2$	R_v

Therefore, the steady-state value of $\mathbf{M}$ is deduced by comparison with (9.39) and is

$$\mathbf{M}_\infty = [\Lambda_I \mathbf{X}_I^{-1}], \qquad (9.53)$$

where $\begin{bmatrix} \mathbf{X}_I \\ \Lambda_I \end{bmatrix}$ are the eigenvectors of (9.52) associated with the stable eigenvalues. Hence the steady Kalman-filter gain is[9]

$$\mathbf{L}_\infty = \mathbf{M}_\infty \mathbf{H}^T (\mathbf{H} \mathbf{M}_\infty \mathbf{H}^T + \mathbf{R}_v)^{-1}. \qquad (9.54)$$

9.5 EXAMPLE OF MULTIVARIABLE CONTROL

As an illustration of a multivariable control using optimal control techniques, we will consider control of the paper-machine head box described briefly in Appendix A.5. The continuous equations of motion are given by

$$\dot{\mathbf{x}} = \begin{bmatrix} -0.2 & 0.1 & 1 \\ -0.05 & 0 & 0 \\ 0 & 0 & -1 \end{bmatrix} \mathbf{x} + \begin{bmatrix} 0 & 1 \\ 0 & 0.7 \\ 1 & 0 \end{bmatrix} \mathbf{u}. \qquad (9.55)$$

We assume the designer has the following information about the expected responses of this system:

1. The maximum sampling frequency is 5 Hz ($T = 0.2$).
2. The settling time to demands on x_1 ($=$ total head) should be less than 1.6 sec (8 periods).
3. The settling time to demands on x_2 ($=$ liquid level) should be less than 5 sec (25 periods).
4. The units on the states have been selected so that the maximum permissible deviation on each x_i is 1.0.
5. The units on control have been selected so that the maximum permissible deviation on u_1 (air control) is 1 unit and that of u_2 (stock control) is 10 units.

9.5.1 Selection of Loss Matrices $\mathbf{Q}_1$ and $\mathbf{Q}_2$

As we mentioned in Section 9.3, the selection of $\mathbf{Q}_1$ and $\mathbf{Q}_2$ is only weakly connected to the performance specifications and a certain amount of trial and error is usually required with an interactive computer program before a satisfactory design results. There are, however, a few guidelines which can be employed. For example, Bryson and Ho (1969) and Kwakernaak and Sivan (1972) suggest essentially the same approach. This is to take $\mathbf{Q}_1 = \mathbf{H}^T \mathbf{Q}_1 \mathbf{H}$ so that the states enter the loss via the important outputs (which may lead to $\mathbf{H} = \mathbf{I}$ if all states are to be kept under close regulation) and to select $\overline{\mathbf{Q}}_1$ and $\mathbf{Q}_2$ to be diagonal with entries so se-

---

[9] In the steady state, the filter has constant coefficients and, for the assumed model, is the same as the Wiener filter.

lected that a fixed percentage change of each variable makes an equal contribution to the loss.[10] For example, suppose we have three outputs with maximum deviations m_1, m_2, and m_3. The loss, for diagonal $\overline{\mathbf{Q}}_1$, is

$$\bar{q}_{11}y_1^2 + \bar{q}_{22}y_2^2 + \bar{q}_{33}y_3^2.$$

The rule is that if $y_1 = \alpha m_1$, $y_2 = \alpha m_2$, and $y_3 = \alpha m_3$, then

$$\bar{q}_{11}y_1^2 = \bar{q}_{22}y_2^2 = \bar{q}_{33}y_3^2;$$

thus

$$\bar{q}_{11}\alpha^2 m_1^2 = \bar{q}_{22}\alpha^2 m_2^2 = \bar{q}_{33}\alpha^2 m_3^2.$$

A satisfactory solution for elements of $\overline{\mathbf{Q}}_1$ is then

$$\overline{Q}_{1,11} = 1/m_1^2, \qquad \overline{Q}_{1,22} = 1/m_2^2, \qquad \overline{Q}_{1,33} = 1/m_3^2. \tag{9.56}$$

Similarly for $\mathbf{Q}_2$ we select a matrix with diagonal elements

$$Q_{2,11} = 1/u_{1\,\text{max}}^2, \qquad Q_{2,22} = 1/u_{2\,\text{max}}^2. \tag{9.57}$$

There remains a scalar ratio between the state and the control terms which we will call ρ. Thus the total loss is

$$\ell = \rho\mathbf{x}^T\mathbf{H}^T\overline{\mathbf{Q}}_1\mathbf{H}\mathbf{x} + \mathbf{u}^T\overline{\mathbf{Q}}_2\mathbf{u}, \tag{9.58}$$

where $\overline{\mathbf{Q}}_1$ and $\overline{\mathbf{Q}}_2$ are given by (9.56) and (9.57) and ρ is to be selected by trial and error. A computer-interactive procedure that allows examination of root locations and transient response for selected values of $\mathbf{Q}_1$ and $\mathbf{Q}_2$ expedites this process considerably. For the example problem described by (9.55), $m_1 = m_2 = m_3 = 1$, $u_{1\,\text{max}} = 1$, and $u_{2\,\text{max}} = 10$.

The designer may introduce another degree of freedom into this problem by requiring that all the closed-loop poles be inside a circle of radius $1/\alpha$, where $\alpha \geq 1$. If we do this, then the magnitude of every transient in the closed loop will decay at least as fast as $1/\alpha^k$ which forms pincers around the transients and allows a degree of direct control over the settling time. We can introduce this effect in the following way.

Suppose that as a modification to the performance criterion of (9.11), we consider

$$\mathcal{I}_\alpha = \sum_{k=0}^{\infty} [\mathbf{x}^T\mathbf{Q}_1\mathbf{x} + \mathbf{u}^T\mathbf{Q}_2\mathbf{u}]\alpha^{2k}. \tag{9.59}$$

---

[10] Kwakernaak and Sivan suggest using a percentage change from nominal values; Bryson and Ho use percentage change from the maximum values.

We can distribute the scalar term α^{2k} in (9.59) as $\alpha^k \alpha^k$ and write it as

$$\mathcal{J}_\alpha = \sum_{k=0}^{\infty} [(\alpha^k \mathbf{x})^T \mathbf{Q}_1(\alpha^k \mathbf{x}) + (\alpha^k \mathbf{u})^T \mathbf{Q}_2(\alpha^k \mathbf{u})] \tag{9.60}$$

$$= \sum_{k=0}^{\infty} [\mathbf{z}^T \mathbf{Q}_1 \mathbf{z} + \mathbf{v}^T \mathbf{Q}_2 \mathbf{v}], \tag{9.61}$$

where

$$\mathbf{z}(k) = \alpha^k \mathbf{x}(k), \qquad \mathbf{v}(k) = \alpha^k \mathbf{u}(k). \tag{9.62}$$

The equations in $\mathbf{z}$ and $\mathbf{v}$ are readily found. Consider

$$\mathbf{z}(k + 1) = \alpha^{k+1} \mathbf{x}(k + 1).$$

From (9.10) we have the state equations for $\mathbf{x}(k + 1)$, so that

$$\mathbf{z}(k + 1) = \alpha^{k+1}[\mathbf{\Phi}\mathbf{x}(k) + \mathbf{\Gamma}\mathbf{u}(k)].$$

If we multiply through by the α^{k+1}-term, we can write this as

$$\mathbf{z}(k + 1) = \alpha\mathbf{\Phi}(\alpha^k \mathbf{x}(k)) + \alpha\mathbf{\Gamma}(\alpha^k \mathbf{u}(k)),$$

but from the definitions in (9.62), this is the same as

$$\mathbf{z}(k + 1) = \alpha\mathbf{\Phi}\mathbf{z}(k) + \alpha\mathbf{\Gamma}\mathbf{v}(k). \tag{9.63}$$

The performance function (9.61) and the equations of motion (9.63) define a new problem in optimal control for which the solution is a control law

$$\mathbf{v} = -\mathbf{K}\mathbf{z},$$

which, if we work backward, is

$$\alpha^k \mathbf{u}(k) = -\mathbf{K}(\alpha^k \mathbf{x}(k))$$

or

$$\mathbf{u}(k) = -\mathbf{K}\mathbf{x}(k). \tag{9.64}$$

We conclude from all this that if we use the control law (9.64) in the state equations (9.10), then a trajectory results which is optimal for the performance $\mathcal{J}_\alpha$ given by (9.60). Furthermore, the state trajectory satisfies (9.62), where $\mathbf{z}(k)$ is a stable vector so that $\mathbf{x}(k)$ must decay *at least* as fast as $1/\alpha^k$ or else $\mathbf{z}(k)$ could not be guaranteed to be stable.

To apply the pincers we need to relate the settling time to the value of α. Suppose we define settling time of x_j as that time t_s such that if $x_j(0) = 1$ and all other states are zero at $k = 0$, then the transients in x_j are less than 0.05 (5% of the maximum) for all times greater than t_s. If we approximate the transient in x_j as

$$x_j(k) \approx x_j(0)(1/\alpha)^k,$$

then when $kT = t_s$, we must have

$$x_j(kT) \le 0.05x_j(0) \le x_j(0)(1/\alpha)^k,$$

which will be satisfied if α is such that

$$(1/\alpha)^k \le 0.05 = \tfrac{1}{20},$$

or

$$\alpha > 20^{1/k} = 20^{T/t_s}. \tag{9.65}$$

For our example case, we have asked that t_s be 1.6 sec for x_1 and 5 sec for x_2. If, for purposes of illustration, we select the more stringent of these, and set $t_s = 1.6$ for which $t_s/T = 8$, then (9.65) requires

$$\alpha > 20^{1/8} = 1.454.$$

From this discussion we conclude that reasonable initial loss matrices are

$$\mathbf{Q}_1 = \begin{bmatrix} 1 & 0 & 0 \\ 0 & 1 & 0 \\ 0 & 0 & 0 \end{bmatrix}, \qquad \mathbf{Q}_2 = \begin{bmatrix} 1 & 0 \\ 0 & 0.01 \end{bmatrix}, \qquad \alpha = 1.46.$$

The design gives a control gain

$$\mathbf{K} = \begin{bmatrix} 5.85 & -8.41 & 3.55 \\ 0.986 & 5.22 & -0.24 \end{bmatrix}$$

and closed-loop poles at

$$z_i = 0.0638, 0.4836, 0.5597,$$

which are well within $1/1.46 = 0.685$. The transient responses to unit-state initial conditions are shown in Fig. 9.5. Note that the scale of 9.5(c) is magnified by approximately 8.

Examination of these results shows that the requirement on settling time has been met since the states settle to within 5% by eight sample periods. However, the control effort on u_1 is rather large, since the elements of the first row of $\mathbf{K}$ add (in magnitude) to 17.81. This is large compared to the value of 1 assumed to be the maximum limit on u_1 for unit values of the states. To correct this situation, the designer can relax the demands on response time a bit, perhaps to $\alpha = 1.3$ ($k \cong 11$), and raise substantially the cost on u_1 to 100. If this fails to satisfy, reducing the cost on x_2 will possibly relieve the demand on u_1 somewhat. Little is to be gained by reciting results of these moves here; a great deal will be learned by the student who has access to a computer program and simulation facility which will permit a certain amount of interactive learning!

The key computer aid to the solution of steady-state control and estimation problems as described in this chapter is an algorithm to solve for eigenvalues and eigenvectors, such as the QR algorithm. In addition to this facility, one needs the

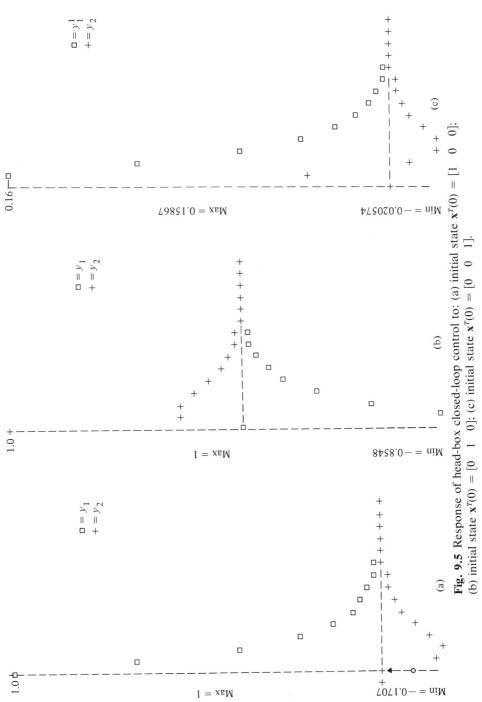

Fig. 9.5 Response of head-box closed-loop control to: (a) initial state $\mathbf{x}^T(0) = [1 \quad 0 \quad 0]$; (b) initial state $\mathbf{x}^T(0) = [0 \quad 1 \quad 0]$; (c) initial state $\mathbf{x}^T(0) = [0 \quad 0 \quad 1]$.

ability to manipulate matrices so as to form the control and estimation Hamiltonian matrices and to form the control law and estimate law, **K** and **L**. Finally, as in Chapter 6, we need a program to take these matrix gains, form the closed-loop system matrix, and plot transient responses.

9.6 SUMMARY

In this chapter we have introduced the ideas of optimal control as a means of designing multivariable control systems. The time-varying optimal control solution to both the control problem and the estimation problem was obtained. In both cases, the steady-state optimal solution was also obtained and found to be much more useful for implementation purposes.

The optimization procedures were found to provide a technique for which many good design candidates could be generated which could then be evaluated to ascertain whether they met all the design goals. The procedures did not eliminate trial-and-error methods; rather they transferred iteration on classical compensation parameters to iteration on optimal cost-function parameters.

Appendix to Chapter 9

Reciprocal Root Property and the Symmetrical Root Locus

If we take the z-transforms of (9.10), (9.13), and (9.14), we obtain the equations for the steady-state optimal control

$$z\mathbf{X}(z) = \boldsymbol{\Phi}\mathbf{X}(z) + \boldsymbol{\Gamma}\mathbf{U}(z), \qquad (9.66a)$$

$$\mathbf{Q}_2\mathbf{U}(z) = -z\boldsymbol{\Gamma}^T\boldsymbol{\Lambda}(z), \qquad (9.66b)$$

$$\boldsymbol{\Lambda}(z) = \mathbf{Q}_1\mathbf{X}(z) + z\boldsymbol{\Phi}^T\boldsymbol{\Lambda}(z). \qquad (9.66c)$$

If we substitute (9.66b) into (9.66a) and write the remaining two equations in terms of the variables $\mathbf{X}(z)$ and $z\boldsymbol{\Lambda}(z)$, we find, in matrix form,

$$\begin{bmatrix} z\mathbf{I} - \boldsymbol{\Phi} & \boldsymbol{\Gamma}\mathbf{Q}_2^{-1}\boldsymbol{\Gamma}^T \\ -\mathbf{Q}_1 & z^{-1}\mathbf{I} - \boldsymbol{\Phi}^T \end{bmatrix} \begin{bmatrix} \mathbf{X}(z) \\ z\boldsymbol{\Lambda}(z) \end{bmatrix} = [\mathbf{0}]. \qquad (9.67)$$

Thus the poles of the Hamiltonian system are those values of z for which

$$\det \begin{bmatrix} z\mathbf{I} - \boldsymbol{\Phi} & \boldsymbol{\Gamma}\mathbf{Q}_2^{-1}\boldsymbol{\Gamma}^T \\ -\mathbf{Q}_1 & z^{-1}\mathbf{I} - \boldsymbol{\Phi}^T \end{bmatrix} = 0.$$

If we now reduce the term $-\mathbf{Q}_1$ to zero by adding $\mathbf{Q}_1(z\mathbf{I} - \boldsymbol{\Phi})^{-1}$ times the first rows to the second rows, we have

$$\det \begin{bmatrix} z\mathbf{I} - \boldsymbol{\Phi} & \boldsymbol{\Gamma}\mathbf{Q}_2^{-1}\boldsymbol{\Gamma}^T \\ \mathbf{0} & z^{-1}\mathbf{I} - \boldsymbol{\Phi}^T + \mathbf{Q}_1(z\mathbf{I} - \boldsymbol{\Phi})^{-1}\boldsymbol{\Gamma}\mathbf{Q}_2^{-1}\boldsymbol{\Gamma}^T \end{bmatrix} = 0.$$

Since this matrix is blockwise triangular, we have

$$\det(z\mathbf{I} - \mathbf{\Phi})\det\{z^{-1}\mathbf{I} - \mathbf{\Phi}^T + \mathbf{Q}_1(z\mathbf{I} - \mathbf{\Phi})^{-1}\mathbf{\Gamma}\mathbf{Q}_2^{-1}\mathbf{\Gamma}^T\}.$$

Now we factor the term $z^{-1}\mathbf{I} - \mathbf{\Phi}^T$ from the second term to find

$$\det(z\mathbf{I} - \mathbf{\Phi})\det\{(z^{-1}\mathbf{I} - \mathbf{\Phi}^T)\{\mathbf{I} + (z^{-1}\mathbf{I} - \mathbf{\Phi}^T)^{-1}\mathbf{Q}_1(z\mathbf{I} - \mathbf{\Phi})^{-1}\mathbf{\Gamma}\mathbf{Q}_2^{-1}\mathbf{\Gamma}^T\}\}.$$

To simplify the notation, we note that $\det(z\mathbf{I} - \mathbf{\Phi}) = a(z)$, the plant characteristic polynomial, and $\det(z^{-1}\mathbf{I} - \mathbf{\Phi}^T) = a(z^{-1})$. Thus, using the fact that $\det \mathbf{AB} = \det \mathbf{A} \det \mathbf{B}$, we find that the Hamiltonian characteristic equation is

$$a(z)a(z^{-1})\det\{\mathbf{I} + \rho(z^{-1}\mathbf{I} - \mathbf{\Phi}^T)^{-1}\mathbf{H}^T\mathbf{H}(z\mathbf{I} - \mathbf{\Phi})^{-1}\overline{\mathbf{\Gamma}}\overline{\mathbf{\Gamma}}^T\} = 0. \qquad (9.68)$$

In (9.68), we factored $\mathbf{Q}_1$ and $\mathbf{Q}_2$ so that $\mathbf{Q}_1 = \rho\mathbf{H}^T\mathbf{H}$ and $\mathbf{\Gamma}\mathbf{Q}_2^{-1}\mathbf{\Gamma}^T = \overline{\mathbf{\Gamma}}\overline{\mathbf{\Gamma}}^T$. If $\mathbf{Q}_1$ and $\mathbf{Q}_2$ are diagonal, the factoring is obvious; otherwise it may be done without loss of generality. Now we use the result (C.3) from Appendix C for the determinant of a sum of $\mathbf{I}$ and a matrix product $\mathbf{AB}$, choosing $\mathbf{A} = (z^{-1}\mathbf{I} - \mathbf{\Phi}^T)^{-1}\mathbf{H}^T$ to write

$$a(z)a(z^{-1})\det[1 + \rho\mathbf{H}(z\mathbf{I} - \mathbf{\Phi})^{-1}\overline{\mathbf{\Gamma}}\overline{\mathbf{\Gamma}}^T(z^{-1}\mathbf{I} - \mathbf{\Phi}^T)^{-1}\mathbf{H}^T] = 0. \qquad (9.69)$$

If we replace z by z^{-1} in (9.69), the result is unchanged since $\det A^T = \det A$. Therefore, if z_i is a characteristic root of the optimal system, so is the reciprocal z_i^{-1}.

An interesting special case of (9.69) occurs in the single-input case, in which the loss function is $\ell = u^2 + \rho y^2$ and $y = \mathbf{H}x$. If this is so, then $\mathbf{Q}_1 = \rho\mathbf{H}^T\mathbf{H}$ and the characteristic equation is given by (9.69). However, may we now interpret $\mathbf{H}(z\mathbf{I} - \mathbf{\Phi})^{-1}\mathbf{\Gamma}$ properly as the plant transfer function $G(z)$. The characteristic equation for optimal control is thus

$$1 + \rho G(z^{-1})G(z) = 0. \qquad (9.70)$$

Equation (9.70) is an equation in root-locus form with respect to ρ, the parameter which reflects the relative weighting on output error y and control u. If ρ is small, the roots are near the poles of the plant [or the stable reflections of the poles if $G(z)$ is unstable], and as ρ gets large, the roots go toward the zeros of $G(z^{-1})G(z)$ which are inside the unit circle. This last observation suggests yet another way to select the loss matrices: select an output signal $y = \overline{\mathbf{H}}x$ so that the *zeros* of this (perhaps artificial) plant are near where you want the closed-loop poles to be. Then, as ρ is made large, you may be sure that the optimal control will move the poles in those directions. A sophisticated discussion of this technique is given in Harvey and Stein (1978).

PROBLEMS AND EXERCISES

9.1 a) Derive Eqs. (9.28) and hence verify the control Hamiltonian as given by (9.29).

b) Demonstrate that if $\mathbf{W}$ is a transformation which brings $\mathscr{H}_c$ to diagonal form as given by (9.30), then the first n columns of $\mathbf{W}$ are eigenvectors of $\mathscr{H}_c$ associated with stable eigenvalues.

c) Demonstrate that if the optimal steady-state control, $\mathbf{K}_\infty$, given by (9.40) is used in the plant equation (9.10), then the matrix in the upper left corner of $\mathbf{W}$, which is $\mathbf{X}_I$, has columns which are eigenvectors of the closed-loop system.

9.2 *Reciprocal root locus.*

a) Compute the closed-loop pole locations of the satellite design problem, using the steady-state gains given by Fig. 9.3.

b) Take the discrete transfer function from u to θ for the satellite design problem (Appendix A) and form the reciprocal root locus as described in Appendix 9.1. Plot the locus and locate the values of ρ corresponding to the selections of Q_2 given in (9.26b).

c) Show that for a system with one control it is always possible to construct a reciprocal root locus corresponding to the optimal steady-state control. [*Hint*: Show that if $\mathbf{Q}_1 = \mathbf{HH}^T$ and $\mathbf{G}(z)$ is a column-matrix transfer function given by $\mathbf{H}(z\mathbf{I} - \mathbf{\Phi})^{-1}\mathbf{\Gamma}$, then the roots of the optimal control are given by $1 + \rho\mathbf{G}^T(z^{-1})\mathbf{G}(z) = 0$, which is a scalar reciprocal root-locus problem, which can be put in the form $1 + \rho G_1(z^{-1})G_1(z) = 0$. Use (9.68).]

d) Give the equivalent scalar plant transfer function $G_1(z)$, if we have the satellite control problem and use $\mathbf{Q}_1 = \mathbf{I}$, the 2×2 identity. Draw the reciprocal root locus for this case.

9.3 Compute the location of the closed-loop poles of the optimal filter for the satellite problem for each of the values of R_w given by the asymptotes to the curves of Fig. 9.4.

9.4 a) Design the antenna of Appendix A to have optimal control corresponding to the loss function $\ell(y, u) = u^2 + \rho y^2$ for $\rho = 0.01, 0.1$, and 1. Plot the step response of the closed loop for initial errors in y and $\dot{y}$ in each case. Which value of ρ most nearly meets the step-response specification given in Chapter 5?

b) Draw the reciprocal loot locus corresponding to the design of part (a).

c) Design an optimal steady-state filter for the antenna. Assume that the receiver noise has a covariance $R_v = 10^{-6}$ rad^2 and that the wind gust noise, w_a, is white, and we want three cases corresponding to $R_w = 10^{-2}, 10^{-4}$, and 10^{-6}.

d) Plot the step response of the complete system with control law corresponding to $\rho = 0.1$ and filter corresponding to $R_w = 10^{-4}$.

9.5 The lateral equations of motion for the DC-8 in cruise are:

$$\begin{bmatrix} \dot{v} \\ \dot{r} \\ \dot{\phi}_p \\ \dot{p} \end{bmatrix} = \begin{bmatrix} -0.0868 & -1 & -0.0391 & 0 \\ 2.14 & -0.228 & 0 & -0.0204 \\ 0 & 0 & 0 & 1 \\ -4.41 & 0.334 & 0 & -1.181 \end{bmatrix} \begin{bmatrix} v \\ r \\ \phi_p \\ p \end{bmatrix} + \begin{bmatrix} 0.0222 & 0 \\ -1.165 & -0.0652 \\ 0 & 0 \\ 0.549 & -2.11 \end{bmatrix} \begin{bmatrix} \delta_r \\ \delta_a \end{bmatrix}.$$

a) Design two second-order controllers by ignoring the cross-coupling terms. Pick roots at $s = -1 \pm j1.5$ rad/sec for the yaw mode (v and r) and $s = -2, -0.1$ for the roll mode (ϕ_p and p). Use a sample period of $T = 0.1$ sec.

b) Determine the root locations that result when the cross-coupling terms are included. (You will need a computer program for finding the eigenvalues of a 4×4 matrix for this step.)

c) Assuming one can measure r and ϕ_p, design two second-order estimators by ignoring the cross-coupling terms. Place all poles at $s = -2$ rad/sec.

d) Determine the root locations that result from (c) when the cross-coupling terms are included.

9.6 The equations of motion for a stick balancer (Fig. 9.2) are given by

$$
\begin{bmatrix} \dot{\theta} \\ \dot{\omega} \\ \dot{x} \\ \dot{v} \end{bmatrix} = \begin{bmatrix} 0 & 1 & 0 & 0 \\ 21 & 0 & 0 & 0.8 \\ 0 & 0 & 0 & 1 \\ 0 & 0 & 0 & -0.4 \end{bmatrix} \begin{bmatrix} \theta \\ \omega \\ x \\ v \end{bmatrix} + \begin{bmatrix} 0 \\ -2 \\ 0 \\ 1 \end{bmatrix} u
$$

$$
\begin{bmatrix} y_1 \\ y_2 \end{bmatrix} = \begin{bmatrix} 1 & 0 & 0 & 0 \\ 0 & 0 & 1 & 0 \end{bmatrix} \begin{bmatrix} \theta \\ \omega \\ x \\ v \end{bmatrix}
$$

Design two second-order estimators, one for the stick (θ, ω) and one for the cart (x, v). Verify that the one-way coupling (cart to stick only) causes the full fourth-order estimator to have the same roots as the separate second-order designs.

10 / Sample Rate Selection

10.1 INTRODUCTION

The selection of the best sample rate for a digital control system is a compromise among many factors. The basic motivation to lower the sample rate (f_s) is cost. A decrease in sample rate means more time is available for the control calculations, hence slower computers are possible for a given control function or more control capability is available for a given computer. Either result lowers the cost per function. For systems with A/D converters, less demand on conversion speed will also lower cost. These economic arguments indicate that the best engineering choice is the slowest possible sample rate which meets all performance specifications.

Factors which provide a lower limit to the acceptable sample rate are: (1) tracking effectiveness as measured either by closed-loop bandwidth or by time response requirements, such as rise time and settling time; (2) regulation effectiveness as measured by the error response to random plant disturbances; (3) sensitivity to plant parameter variations; and (4) error due to the measurement noise which leaks through the analog prefilters. A fictitious limit occurs when one designs by digital filtering techniques (as in Section 5.3). The inherent approximation in the method may give rise to system instabilities as the sample rate is lowered, and this can lead the designer to conclude that a lower limit on f_s has been reached when in fact the proper conclusion is that the approximations are invalid. The solution is not to sample faster but to switch to one of the exact discrete design methods covered in Sections 5.4, 5.5, 5.6, and Chapter 6.

10.2 TRACKING EFFECTIVENESS IN TERMS OF BANDWIDTH, TIME RESPONSE, AND ROUGHNESS

An absolute lower bound to the sample rate is set by a specification to track certain command or reference input signals, a bound which has a theoretical basis in the sampling theorem discussed in Section 4.3. Assuming we can represent our

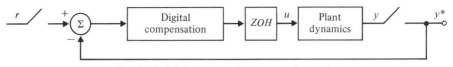

Fig. 10.1 Digital control system schematic.

digital control system by a single loop structure as depicted in Fig. 10.1, we can specify the tracking performance in terms of the frequency response from r to y. The sampling theorem states that in order to reconstruct an unknown band-limited continuous signal from samples of that signal, one must use a sample rate which is at least twice as high as the highest frequency contained in the unknown signal. This theorem applies to a feedback controller such as the one illustrated in Fig. 10.1 because r is an unknown signal which must be reconstructed by the plant if the system output y is to follow r, and we can model the reference signals as having spectral content up to a given frequency or closed-loop "bandwidth," f_c. The sample rate, therefore, must be at least twice the required closed-loop bandwidth of the system.

It is important to note the distinction between the closed-loop bandwidth, f_c, and the highest frequencies in the open-loop plant dynamics, f_p, since these two frequencies can be quite different. For example, closed-loop bandwidths could be an order of magnitude *less* than open-loop modes or resonances for some vehicle control problems. We can extract information concerning the state of the plant resonances for purposes of control from sampling the output y in Fig. 10.1 without satisfying the sampling theorem because some *a priori* knowledge (albeit imprecise) is available concerning these dynamics, and the system is *not* required to track these frequencies. This *a priori* knowledge of the dynamic model of the plant can be included in the compensation in the form of a notch filter if we use the classical design, or as part of the observer dynamics if we use state-space methods. The questions that immediately arise, however, are how accurately these high-frequency dynamics must be known and whether this accuracy is a function of the sampling rate. These are important questions of system sensitivity and are treated in Section 10.4 as a separate issue affecting the selection of sampling rates.

The "closed-loop bandwidth" limitation provides the fundamental lower bound on the sample rate. In practice, however, the theoretical lower bound of sampling at twice the bandwidth of the reference input signal would not be judged sufficient in terms of desired time responses because the control system must be causal and the intersample behavior is constrained by the plant dynamics. For a system with a rise time on the order of 1 sec and a required closed-loop bandwidth on the order of 0.5 Hz, it is not unreasonable to insist on a sample rate of 2 to 10 Hz, which is a factor of 4 to 20 times the closed-loop bandwidth. This is so in order to: (1) reduce the delay between a command change and the system response to the command change, and (2) smooth the system output response to the control inputs which are applied to the plant through a ZOH.[1] For systems with human input commands, the time delay alone suggests that the sample period be kept to a small fraction of the rise time, such as 10%; that is, a 10-Hz sample frequency should be used if a 1-second rise time is required.

As an example of these concepts, we consider the stability augmentation for aircraft pitch dynamics.[2] The equations of motion are given by

$$\begin{bmatrix} \dot{q} \\ \dot{\alpha} \end{bmatrix} = \begin{bmatrix} M_q & M_\alpha \\ 1 & Z_\alpha \end{bmatrix} \begin{bmatrix} q \\ \alpha \end{bmatrix} + \begin{bmatrix} M_\delta \\ Z_{\delta_e} \end{bmatrix} \delta_e, \tag{10.1}$$

where the aircraft stability derivatives (M's and Z's) vary considerably over the aircraft operating envelope. The aircraft state vector consists of q, the pitch rate, and α, the angle of attack. The control is through the elevator control surface, δ_e, which is assumed to act instantaneously, although in practice there would be a lag due to the actuator response. The purpose of the stability augmentation system is to create closed-loop aircraft responses that are considered desirable by pilots and are less variable over the many operating regions than the uncontrolled vehicle responses would be.

The digital controller consists of a sensor for q (a rate gyro), an estimator to reconstruct α, and state feedback from q and $\hat{\alpha}$ to generate the control δ_e. Suppose we use the methods of Chapter 6 and select the closed loop roots to provide a natural frequency of about $f_c = 0.5$ Hz with a damping ratio $\zeta \cong 0.7$.

Responses to step δ_e commands for designs with different sampling rates are shown in Fig. 10.2 for the case where the open-loop natural frequency (set by the M's and Z's) is about 2 Hz. This example shows that at a sample rate of 10 times the bandwidth (Fig. 10.2b), both the input delay *and* the lack of smooth response are likely to be limiting factors on sample rate selection.

The same aircraft under different flight conditions where the Z's and M's of (10.1) result in slower open-loop natural frequencies (~ 0.1 Hz) exhibits much smoother behavior (Fig. 10.3) but the input data could still be a limiting factor.

For regulator control systems where the response to command inputs is unimportant, the command delay would not be a factor in f_s selection. Furthermore, in systems where no person or instrument is adversely affected by a jerky response (for example, in active wing flutter control or a computer storage-disk tracking arm), the smoothness would not be a critical factor in selecting f_s.

Hence, at this point, we are left with the conclusion that the sample rate must be at least twice the desired closed-loop bandwidth and possibly as much as 20 times the system closed-loop bandwidth, depending on the particular performance requirements of the system and the relation between the tracking signal and the plant dynamics.

---

[1] Zero-order hold.

[2] This example and those in 10.3 and 10.4 are by Katz (1975).

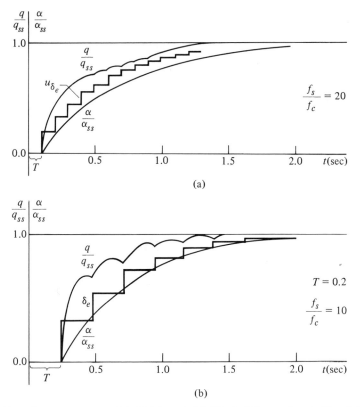

Fig. 10.2 Aircraft response to stick (δ_e) input: (a) Mach 1.2, sea level, $T = 0.1$ sec.   (b) Mach 1.2, sea level, $T = 0.2$ sec.

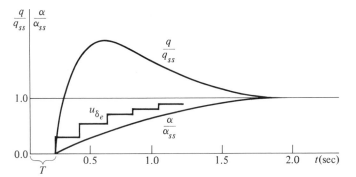

Fig. 10.3 Aircraft response to stick input, including pilot delay, T_2 (Mach 0.2, sea level, $T = 0.2$ sec).

10.3 DISTURBANCE REJECTION

Disturbance rejection is an important aspect of any control system if not the most important one. Disturbances enter a system with various characteristics ranging from steps to white noise. For purposes of sample-rate determination, the higher frequency random disturbances are the most influential.

The ability of the control system to reject disturbances with a good continuous controller represents a lower bound on the error response than can be hoped for when implementing the controller digitally. In fact, some degradation over the continuous design must occur because the sampled values are slightly out of date at all times except at the very moment of sampling. However, if the sample rate is very fast compared to the frequencies contained in the noisy disturbance, no appreciable loss should be expected from the digital system compared with the continuous controller. At the other extreme, if the sample time is very long compared with the characteristic frequencies of the noise, the response of the system due to noise is essentially the same as could result if there were no control at all. The selection of a sample rate will place the response somewhere in between these two extremes, and thus the impact of sample rate on the disturbance rejection of the system may be very influential to the designer in selecting the sample rate.

An example of the concepts of disturbance rejection should help illustrate the ideas. Again, we use the aircraft pitch dynamics expressed by Eq. (10.1) with the addition of wind gust noise entering as a vertical force in the $\dot{\alpha}$ equation. We assume that the wind noise is equivalent to a white signal passed through a first-order filter with time constant τ_w and standard deviation σ_w. Figure 10.4 shows the variance of the pitch rate in response to the wind gusts as a function of sample time for specific values of τ_s and τ_w and the aircraft in the condition used for Fig. 10.2. The free response of the aircraft shows the limit without any control ($T \rightarrow \infty$). From the figure we see that a sample rate of T_2 for which f_s is 20 Hz

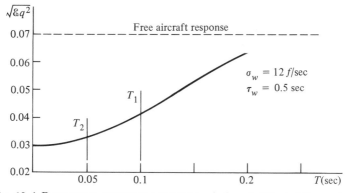

Fig. 10.4 Root-mean-square response to wind gust (Mach 1.2, sea level).

$(40 \times f_c)$ yields almost as much noise attenuation as can be obtained with a continuous controller, and that sample rates any slower will degrade that performance. At a sample rate of 5 Hz $(10 \times f_c)$, no appreciable attenuation is being accomplished by the controller. These results are highly dependent on the noise correlation time, τ_w; for example, for smaller τ_w, the rise in the curve would occur at shorter sample times, and conversely, longer τ_w's would result in the rise occurring at longer sample times.

10.4 SENSITIVITY TO PARAMETER VARIATIONS

Any control design relies to some extent on a knowledge of the parameters representing plant dynamics. Discrete systems exhibit an increasing sensitivity to parameter errors for a decreasing f_s when the sample interval becomes comparable to the period of any of the open-loop vehicle dynamics. The determination of the degree of sensitivity as a function of sample rate can be carried out relatively easily as we will demonstrate using again the example of stability augmentation for aircraft pitch motion described by Eq. 10.1.

In Chapter 6 we assumed that the actual plant and the model of the plant used in the estimator were precisely the same. In any real implementation these two would be slightly different.

Let us suppose the plant is described as

$$\begin{aligned} \mathbf{x}_{n+1} &= \mathbf{\Phi}_p\mathbf{x}_n + \mathbf{\Gamma}_p\mathbf{u}_n, \\ \mathbf{y}_n &= \mathbf{H}\mathbf{x}_n, \end{aligned} \tag{10.2}$$

and the estimator controller as

$$\begin{aligned} \hat{\mathbf{x}}_n &= \bar{\mathbf{x}}_n + \mathbf{L}(\mathbf{y} - \mathbf{H}\bar{\mathbf{x}}_n), \\ \bar{\mathbf{x}}_{n+1} &= \mathbf{\Phi}_e\hat{\mathbf{x}}_n + \mathbf{\Gamma}_e\mathbf{u}_n, \\ \mathbf{u} &= -\mathbf{K}\hat{\mathbf{x}}. \end{aligned} \tag{10.3}$$

In the ideal case, $\mathbf{\Phi}_p = \mathbf{\Phi}_e(=\mathbf{\Phi})$ and $\mathbf{\Gamma}_p = \mathbf{\Gamma}_e(=\mathbf{\Gamma})$, and the system closed-loop roots are given by the controller and the estimator roots designed separately; i.e., they are the roots of the characteristic equation (see also Eq. 6.54)

$$|z\mathbf{I} - [\mathbf{\Phi} - \mathbf{\Gamma}\mathbf{K}]||z\mathbf{I} - (\mathbf{\Phi} - \mathbf{L}\mathbf{H}\mathbf{\Phi})| = 0. \tag{10.4}$$

If we allow $\mathbf{\Phi}_e \neq \mathbf{\Phi}_p$ and $\mathbf{\Gamma}_e \neq \mathbf{\Gamma}_p$, the roots do not separate, and the system characteristic equation is obtained from the full $2n \times 2n$ determinant that results from Eqs. (10.2) and (10.3):

$$\begin{vmatrix} \mathbf{\Phi}_p - \mathbf{I}z & -\mathbf{\Gamma}_p\mathbf{K} \\ \mathbf{L}\mathbf{H}\mathbf{\Phi}_p & (\mathbf{I} - \mathbf{L}\mathbf{H})(\mathbf{\Phi}_e - \mathbf{\Gamma}_e\mathbf{K}) - \mathbf{L}\mathbf{H}\mathbf{\Gamma}_p\mathbf{K} - \mathbf{I}z \end{vmatrix} = 0. \tag{10.5}$$

Either the pole-placement approach of Chapter 6 or the steady-state optimal discrete design method of Chapter 9 could be used to arrive at the **K**- and **L**-matrices. In both cases, the root sensitivity is obtained by assuming some error in $\mathbf{\Phi}$ or $\mathbf{\Gamma}$ (thus $\mathbf{\Phi}_e \neq \mathbf{\Phi}_p$ and $\mathbf{\Gamma}_e \neq \mathbf{\Gamma}_p$) and comparing the resulting roots from Eq. (10.5) with the ideal case of Eq. (10.4).

If a system has been designed using the methods of Chapter 5, root sensitivity is obtained by repeating the closed-loop root analysis with a perturbed plant, or if one parameter is particularly troublesome, an analysis of a root locus versus that parameter may be worthwhile.

For an example, suppose the equation of the aircraft pitch stability model (Eq. 10.1) is augmented with equations of a first bending mode which are given by

$$\dot{x}_3 = \omega_b x_4,$$
$$\dot{x}_4 = -\omega_b x_3 - 2\zeta_b \omega_b x_4 + \omega_b K_1 Z_{\delta_e} \delta e + K_2 \omega_b Z_\alpha \dot{\alpha}, \qquad (10.6)$$

where

ζ_b = bending-mode damping,
ω_b = bending-mode frequency,
K_1, K_2 = quantities depending on the specific aircraft shape and mass distribution.

For systems with all plant dynamics in the vicinity of the closed-loop bandwidth or slower, root-location changes due to parameter errors will not likely be constraining factors unless the parameter error was quite large. The effect of the bending mode will illustrate a situation where sensitivity is likely to be a constraint on sample-rate selection.

In the flight condition chosen for analysis (zero altitude, Mach 1.2), the open-loop rigid-body poles are located at $s = -2 \pm j13.5$ rad/sec, while the bending mode is lightly damped ($\zeta_b = 0.01$) with a natural frequency of 25 rad/sec (4 Hz). The closed-loop poles of the rigid body were relocated by the optimal discrete synthesis described in Chapter 9 to $s = -16 \pm j10$ with essentially no change in the bending-mode root locations. The optimal compensator (controller plus estimator) generates a very deep and narrow notch filter which filters out the unwanted bending frequencies. The width of the notch filter is directly related to the low damping of the bending mode and the noise properties of the system. Furthermore, the optimal estimator gains for the bending mode are very low, causing low damping in this estimation error mode. If the bending frequency of the vehicle varies from the assumed frequency, the components of the incoming signal to the estimator due to the bending miss the notch and are transmitted as a positive feedback to the elevator. This is demonstrated in Fig. 10.5 as a function of sample rate $\omega_s = 2\pi f_s$. Note the insensitivity to sample rate when knowledge of the bending mode is perfect (no error in ω_b) and the strong influence of sample rate for the case where the bending mode has a 10% error. For this example, the bending modes were unstable ($\zeta_b < 0$) for all sample rates with a 10% ω_b error, in-

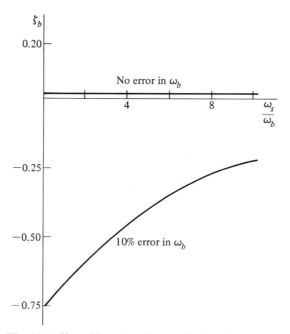

Fig. 10.5 Closed-loop bending-mode damping vs. ω_s.

dicating a very sensitive system and totally unsatisfactory design if ω_b is subject to change.

The sensitivity of the design to changes in ω_b can be reduced by increasing the width of the notch in the controller frequency response through variations of the weighting factors $\mathbf{Q}_1$ and $\mathbf{Q}_2$ which apply to the bending modes, and by introducing additional disturbance noise in the bending-mode state of Eq. (10.6). Figure 10.6

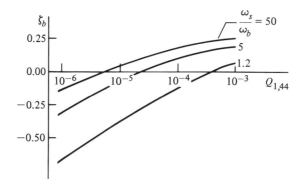

Fig. 10.6 Closed-loop bending-mode damping vs. $Q_{1,44}$, weighting factor (10% ω_b error).

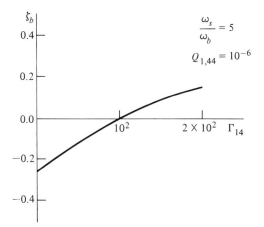

Fig. 10.7 Closed-loop bending-mode damping vs. noise magnitude, Γ_{14} (10% error).

shows the effect on ζ_b due to the bending-mode weighting factor $Q_{1,44}$ and demonstrates a substantial reduction in sensitivity. Note that it is possible to have $\zeta_b > 0.05$ with a 10% ω_b error and a sample rate only 20% faster than the bending mode. Figure 10.7 depicts a similar trend for the parameter Γ_{14} representing the magnitude of the disturbance noise on the bending-mode state. Ultimately a final design could consist of selecting the combination of Q_1 and Γ_{14} and possibly other currently unidentified constants which yield the "best" performance in terms of sensitivity, noise response, and rigid-body response.

In summary, the example becomes increasingly sensitive to bending-mode frequency errors as the sample rate decreases; however, for perfect plant-parameter information, there is no effect of sample rate on bending-mode damping. Designs were demonstrated that yielded good bending-mode damping with a 10% ω_b error and a sample rate 20% faster than the bending mode. Since a typical rise-time requirement for this type of autopilot would be on the order of 1 sec, an adequate closed-loop response would require a sample rate on the order of 10 Hz. The sensitivity example was based on one bending mode at 4 Hz, and in this case, sensitivity considerations did not impose a higher sample-rate requirement than that required for tracking. However, sensitivity of the system to off-nominal vehicle parameters must be evaluated, and especially in slow sample-rate cases, the system should be specifically designed to minimize the sensitivity.

10.5 EFFECT OF PREFILTER DESIGN

Digital control systems with analog sensors typically include an analog prefilter between the sensor and the sampler or A/D converter as an antialiasing device.

The prefilters are low pass, and the simplest transfer function is

$$G_p(s) = \frac{a}{s + a},$$ (10.7)

so that the noise above the prefilter breakpoint (a in Eq. 10.7) is attenuated. The design goal is to provide enough attenuation at half the sample rate ($\omega_s/2$) so that the noise above $\omega_s/2$, when aliased into lower frequencies by the sampler, will not be detrimental to the control-system performance.

A conservative design procedure is to select the breakpoint and ω_s sufficiently higher than the system bandwidth so that the phase lag from the prefilter does not significantly alter the system stability; thus the prefilter can be ignored in the basic control-system design. Furthermore, for a good reduction in the high-frequency noise at $\omega_s/2$, the sample rate should be selected about 5 or 10 times higher than the prefilter breakpoint. The implication of this prefilter design procedure is that sample rates need to be on the order of 20 to 100 times faster than the system bandwidth. If carried out in this way, the prefilter design procedure is likely to provide the lower bound on the selection of the sample rate.

An alternative design procedure is to allow significant phase lag from the prefilter at the system bandwidth and thus require that the control design be carried out with the analog prefilter characteristics included. This procedure allows us to use sample rates as low as 5 to 10 times the system bandwidth, but at the expense of increased complexity in the design procedure. In addition, some cases may require increased complexity in the control implementation to maintain sufficient stability in the presence of prefilter phase lag; that is, the existence of the prefilter may, itself, lead to a more complex digital control algorithm.

To illustrate the effect of prefilter characteristics, we present the results of two cases.[3] The first is an elementary example of a plant described by a single integrator and a zero-order hold. The second is the stability augmentation control for short-period motion of an aircraft as described by (10.1).

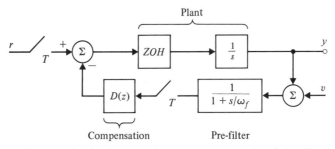

Fig. 10.8 Block diagram of a simple control system for the study of the effects of a prefilter bandwidth and sampling rate on measurement noise rejection.

---

[3] These examples are by Peled (1978).

The block diagram used in the first example is shown in Fig. 10.8. The compensation $D(z)$ was designed to make the step response from r to y follow an exponential with a 1-sec time constant as described by a plant bandwidth $\omega_p = 1$, using either a constant, $D(z) = K$, or a first-order controller, $D(z) = K(z - \alpha)/(z - \beta)$. With the design fixed for the best step response for a given sampling rate, the response to measurement noise v was computed for various values of sampling rate (specified by $\omega_s = 2\pi/T$) and prefilter bandwidth given by ω_f. The results of both continuous $[y(t)]$ and discrete $[y(kT)]$ responses are plotted in Fig. 10.9, where two trends are apparent. For large ω_s/ω_p, which is fast sampling, the noise response is small and the differences in the choice of prefilter bandwidth are also small. However, as ω_s/ω_p is reduced, the effect of the prefilter bandwidth is more dramatic. If ω_f is very small ($0.2 \le \omega_f \le 2$), the rise in noise response with sample period is modest. However, if a wide bandwidth prefilter is used ($\omega_f \ge 5$), the noise response rises by a factor of 3 over the fast sampling value. Similar results are shown for the first-order compensator. The difference between the two cases — parts (a) and (b) of Fig. 10.9 (p. 286) — is in the quality of the step responses, the first-order compensator yielding a much better transient at the expense of some increase in measurement noise response.

As a second example, the response of the pitch rate for the short-period motion of the aircraft described by (10.1) to measurement noise is plotted in Fig. 10.10 (p. 287). The rise in noise response for a wide-band prefilter is shown in curve e. An interesting phenomenon is shown by curves b, c, and d for low sampling rates. As ω_s/ω_p approaches 2, the feedforward part of the control is able to achieve the best step response, and the feedback is essentially turned off, so the measurement noise does not appear at the output with the resulting very low value for σ_q.

Two main conclusions can be drawn from these examples. The first is that prefilters for removing measurement noise are effective and can be rationally designed. The second is that the bandwidth of the prefilter should be selected primarily for the reduction of sensor noise effects, and values as low as the system closed-loop bandwidth ($\omega_f/\omega_p = 1$) should be considered. The influence of the sampling-rate selection on measurement noise and prefilter characteristics is not pronounced, even for ω_s-values as low as twice the system closed-loop bandwidth.

10.6 SUMMARY

In this chapter we have presented the four major considerations which influence the selection of sampling rate. The fundamental compromise is between system cost, which goes up as we are forced to sample faster, and system performance, which is degraded if we are forced to sample more slowly. The ability of the control system to track command inputs is influenced by the sampling rate, being limited in absolute terms by the sampling theorem to twice the highest frequency ex-

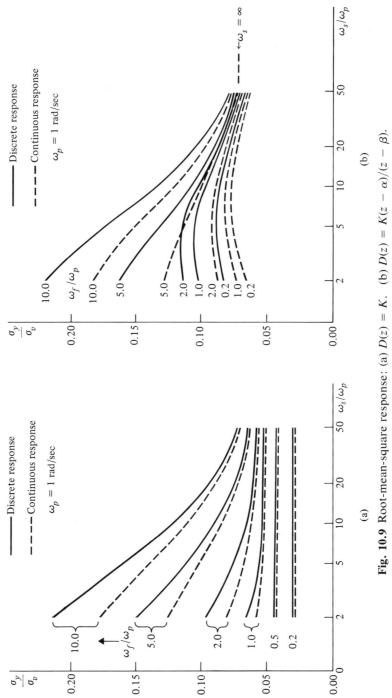

Fig. 10.9 Root-mean-square response: (a) $D(z) = K$. (b) $D(z) = K(z - \alpha)/(z - \beta)$.

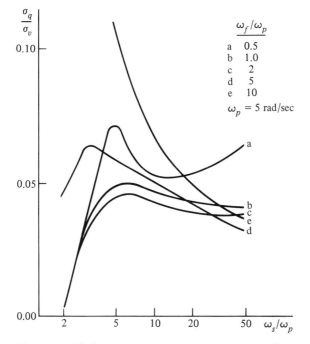

Fig. 10.10 Pitch rate response to measurement noise.

pected to be tracked, and in terms of tracking error by system "roughness" and response delay to the range of 10 to 20 times the expected closed-loop system bandwidth. The ability of the system to regulate its error against plant disturbances is measured mainly by high-frequency disturbances and typically shows an *s*-curve shape of variation in error power versus sample period. For long periods, the response is as if the loop were opened, and for short periods (high-frequency sampling) the response is as if the system were continuous. The knee of the curve typically occurs for sampling frequencies between 5 and 20 times the bandwidth of the open-loop noise response of the plant.

Parameters in the plant are typically subject to variation, and these variations can influence the selection of sampling rate. Thus, for example, it often happens in vehicle control that a high-frequency vibration or bending mode must be made stable by the digital controller. Analysis of such a case shows a more sensitive system as sample rate is reduced; design methods, including an increase in the equivalent plant noise, were able to effect a satisfactory response for sampling rates in the range of the bending-mode frequency.

Finally, the effects of prefilter or anti-aliasing filters on response to sensor noise were analyzed. The result of this study shows that prefilters can be designed with bandwidths comparable to the system closed-loop bandwidth if the dynamic effects of the filter are properly accounted for in the digital compensation design.

Sample rate selection is a complicated matter involving compromise among many factors, but effective analysis tools are available to guide the designer in making the most cost-effective choice.

PROBLEMS AND EXERCISES

10.1 Consider the satellite design problem of Appendix A.

a) Show that there is no lower limit on sampling rate necessary to achieve stability.

b) What is the limiting value, in terms of sampling period T, on rise time t_r and settling time t_s as T gets larger and larger? [*Hint:* Consider the direct design method of Ragazzini (Section 5.6) and place the poles at $z = 0$.]

c) For the design mentioned in the hint to part (b), plot the error-frequency response magnitude and mark the highest frequency for which the steady-state error magnitude is less than 5% of the input magnitude. What is this frequency as a fraction of the sampling frequency?

10.2 For the satellite design problem of Chapter 6, we derived a control law

$$K = (0.1/T^2 \quad 0.35/T)$$

in Section 6.3. Use the results of Appendix D, Eq. (D.28) or (D.35) to compute the rms error to a discrete torque error with spectrum $R_w = 1$. Plot the error for $T = 0.5, 0.2, 0.1, 0.05, 0.02, 0.01$ sec and comment on the selection of T for disturbance rejection in this case.

10.3 In Appendix A, Section A.4, we discuss a plant consisting of two coupled masses. Assume $M = 20$, $m = 1$, $k = 144$, $b = 0.5$, and $T = 0.15$ sec.

a) Design an optimal control for this system with quadratic loss $\ell = y^2 + u^2$. Plot the resulting step response.

b) Design an optimal filter based on measurements of $d(kT)$ with $R_v = 0.1$ and a disturbing noise in the form of a force on the main mass M with $R_w = 0.001$.

c) Combine the control and estimation into a controller and plot the frequency response from d to u.

d) Simulate the system designed above and, leaving the controller coefficients unchanged, vary the value of the damping coefficient b and note the sensitivity of the design to this parameter.

e) Repeat steps (a) to (d) for $T = 0.25$.

10.4 Find compensation for a plant given by

$$G(s) = \frac{0.8}{s(s + 0.8)}.$$

Use s-plane design techniques followed by the pole-zero mapping methods of Chapter 3 to obtain a discrete equivalent. Select compensation so that the resulting undamped natural frequency (ω_n) is 4 rad/sec and the damping ratio (ζ) is 0.5. Do the design for two sample periods: 100 ms and 600 ms, then repeat, using an exact z-plane analysis of the resulting discrete system, and examine the resulting root locations.   What do you conclude concerning the 600-ms sample period with these design techniques?

10.5 The satellite attitude control transfer function (Appendix A) is

$$G_1(z) = \frac{T^2(z + 1)}{2(z - 1)^2}.$$

Determine the lead network and gain that yield dominant poles at $\zeta = 0.5$ and $\omega_n = 2$ rad/sec for $f_s = 1, 2$, and 4 Hz.   Sketch the control time history for the first 5 sec and compare for the different sample rates.

10.6 For each design in Problem 10.5, increase the gain by 20% and determine the change in the system damping.

10.7 Consider a plant consisting of a diverging exponential, that is,

$$\frac{x(s)}{u(s)} = \frac{a}{s - a}.$$

Controlled discretely with a ZOH, this yields a difference equation, namely

$$x_{k+1} = e^{aT}x_k + [e^{aT} - 1]u_k.$$

Assume proportional feedback,

$$u_k = -Kx_k,$$

and compute the gain K which yields a z-plane root at $z = e^{-bT}$.   Assume $a = 1$, $b = 2$, and do the problem for $T = 0.1, 1.0, 2, 5$ sec.   Is there an upper limit on the sample period that will stabilize this system?   Compute the percent error in K that will result in an unstable system for $T = 2$ and 5 sec.   Do you judge that the $T = 5$ sec case is practical?   Why?

Appendix A / Examples

A.1 SINGLE-AXIS
SATELLITE ATTITUDE CONTROL

Satellites often require attitude control for proper orientation of antennas and sensors with respect to the earth. Figure A.1 shows a communications satellite with a three-axis attitude-control system. To gain insight into the three-axis problem we often consider one axis at a time. Figure A.2 depicts this case where motion is only allowed about an axis perpendicular to the page. The equations of motion of the system are given by

$$I\ddot{\theta} = M_C + M_D, \tag{A.1}$$

where I is the moment of inertia of the satellite about its mass center, M_C is the control torque applied by the thrustors, M_D are the disturbance torques, and θ is the angle of the satellite axis with respect to an ''inertial'' reference. The inertial reference must have no angular acceleration. Normalizing, we define

$$u = M_C/I, \qquad w_d = M_D/I \tag{A.2}$$

and obtain

$$\ddot{\theta} = u + w_d. \tag{A.3}$$

Taking the Laplace transform

$$\theta(s) = \frac{1}{s^2}\left[u(s) + w_d(s)\right], \tag{A.4}$$

which becomes, with no disturbance,

$$\frac{\theta(s)}{u(s)} = \frac{1}{s^2} = G_1(s). \tag{A.5}$$

Fig. A.1 Communications satellite. (Courtesy Ford Aerospace and Communications Corporation)

In the discrete case with u being applied through a zero-order hold, we can use the methods of Chapter 2 to obtain the discrete transfer function

$$G_1(z) = \frac{\theta(z)}{u(z)} = \frac{T^2}{2} \frac{(z + 1)}{(z - 1)^2}.$$
(A.6)

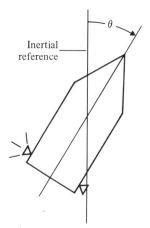

Inertial reference

θ

Fig. A.2 Satellite control schematic.

A.2 A SERVOMECHANISM FOR ANTENNA AZIMUTH CONTROL

It is desired to control the elevation of an antenna designed to track a satellite (Fig. A.3). The antenna and drive parts have a moment of inertia J and damping B arising to some extent from bearing and aerodynamic friction, but mostly from the back emf of the DC-drive motor (Fig. A.4). The equations of motion are

$$J\ddot{\theta} + B\dot{\theta} = T_c + T_d, \tag{A.7}$$

where T_c is the net torque from the drive motor and T_d is the disturbance torque due to wind. If we define

$$B/J = a, \qquad u = T_c/B, \qquad w_d = T_d/B,$$

the equations reduce to

$$\frac{1}{a}\ddot{\theta} + \dot{\theta} = u + w_d. \tag{A.8}$$

Transformed, they are

$$\theta(s) = \frac{1}{s\left(\dfrac{s}{a} + 1\right)}\left[u(s) + w_d(s)\right] \tag{A.9}$$

Fig. A.3 Satellite tracking antenna. (Courtesy Ford Aerospace and Communications Corporation)

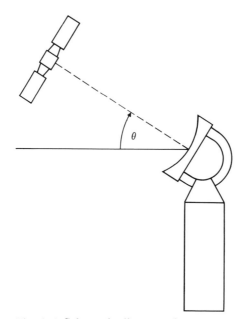

Fig. A.4 Schematic diagram of antenna.

or, with no disturbances,

$$\frac{\theta(s)}{u(s)} = \frac{1}{s\left(\dfrac{s}{a} + 1\right)} = G_2(s). \tag{A.10}$$

The discrete case with $u(n)$ applied through a zero-order hold yields

$$G_2(z) = \frac{\theta(z)}{u(z)} = K \frac{(z + b)}{(z - 1)(z - e^{-aT})}, \tag{A.11}$$

where

$$K = \frac{(aT - 1 + e^{-aT})}{a}, \qquad b = \frac{1 - e^{-aT} - aTe^{-aT}}{aT - 1 + e^{-aT}}.$$

The equation given by (A.7) is only a linear approximation to the true motion, of course. In reality there are many nonlinear effects which must be considered in the final design. To give some idea of these considerations, we present the major features of the angle, θ_s, of the satellite to be followed and of the servomechanism which must be designed to achieve the tracking.

The general shape of $\theta_s(t)$ and its velocity and acceleration are sketched in Fig. A.5. The peak values of velocity and acceleration depend critically on the altitude of the satellite above the earth and on the elevation above the horizon of the

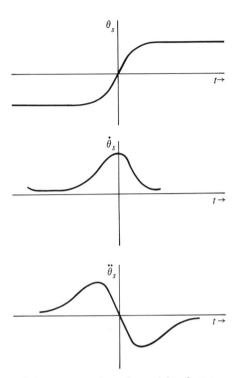

Fig. A.5 General shape of the command angle and its first two derivatives which the tracking antenna must follow.

orbit seen from the antenna.   For our problem, we have taken the velocity of 0.01 rad/sec, which is relatively slow.   An orbit of 200 mi above the earth which passes at an elevation above the antenna of 86° requires an azimuth rate of 0.34 rad/sec (about 20°/sec).   For the purpose of setting the tracking accuracy and gains, a reasonable assumption is that the antenna should be capable of following the peak velocity in the steady state with acceptable error.

The size of the allowable tracking error is determined by the antenna properties.   The purpose, of course, is to permit acceptable communication signals to be received by the antenna electronics.   For this, we must consider the dependence of signal amplitude on pointing error.   A sketch of a typical pattern is shown in Fig. A.6.   The beam width $\Delta\theta$ is the range of tracking error permissible if acceptable communications are to be achieved.   Typical values of the beam width may vary from a few degrees (0.1 rad) to less than one degree (0.01 rad).   The total error will be composed mainly of tracking errors due to θ_s motion, wind-gust errors, and random errors caused by noise in the measurement of satellite position.   A reasonable allowance is to permit tracking errors of 10% of the beam width and an equal contribution from wind gusts.   For a beam width of 0.1 rad,

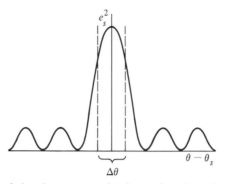

Fig. A.6 Plot of signal power received as a function of tracking error.

then, we allow a tracking error of 0.01 rad, from which it follows that $K_v = 1$ (if $\dot{\theta}_{max} = 0.01$ rad/sec). For a 60-ft tracking antenna, a typical beam width is 0.01 rad, and if 0.34 rad/sec must be followed, the velocity constant for a 10% error (0.001 rad) is $K_v = 0.34/(0.001) = 340$. For a velocity constant of this magnitude, a Type II system with two poles at $z = 1$ is probably required.

In addition to considerations of tracking error, the designer must take into account the fact that the drive motor has finite torque and power capability. In selecting the drive motor, the total torque must be capable of meeting the acceleration demands of tracking, overcoming the wind torque, and overcoming the static friction of the drive-train gears and antenna mount. The power of the motor must be able to supply this torque at the maximum tracking velocity.

A.3 TANK FLUID TEMPERATURE CONTROL

The temperature of a tank of fluid with a constant flow rate in and out is to be controlled by adjusting the temperature of the incoming fluid. The temperature of the incoming fluid is controlled by a mixing valve that adjusts the relative amounts of hot and cold supplies of the fluid (Fig. A.7). Because of the distance between the

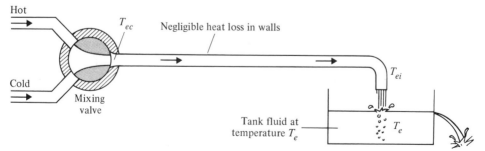

Fig. A.7 Tank temperature control.

valve and the point of discharge into the tank, there is a time delay between the application of a change in the mixing valve and the discharge of the flow with the changed temperature into the tank.

The differential equation governing the tank temperature is

$$\dot{T}_e = \frac{1}{cM}(q_{in} - q_{out}), \tag{A.12}$$

where

$$T_e = \text{tank temperature,}$$
$$c = \text{specific heat of the fluid,}$$
$$M = \text{fluid mass contained in the tank,}$$
$$q_{in} = c\dot{m}_{in}T_{ei},$$
$$q_{out} = c\dot{m}_{out}T_e,$$
$$\dot{m} = \text{mass flow rate } (= \dot{m}_{in} = \dot{m}_{out}),$$
$$T_{ei} = \text{temperature of fluid entering tank,}$$

but the temperature at the input to the tank at time t is the control temperature τ_d sec in the past which may be expressed as

$$T_{ei}(t) = T_{ec}(t - \tau_d), \tag{A.13}$$

where

$\tau_d = $ delay time,
$T_{ec} = $ temperature of fluid immediately after the control valve and directly controllable by the valve.

Combining constants, we obtain

$$\dot{T}_e(t) + aT_e(t) = aT_{ec}(t - \tau_d), \tag{A.14}$$

where

$$a = \dot{m}/M$$

which, transformed, becomes

$$\frac{T_e(s)}{T_{ec}(s)} = \frac{e^{-\tau_d s}}{s/a + 1} = G_3(s). \tag{A.15}$$

To form the discrete transfer function of G_3 preceded by a zero-order hold, we must compute

$$G_3(z) = Z\left\{\frac{1 - e^{-Ts}}{s}\frac{e^{-\tau_d s}}{s/a + 1}\right\}.$$

We assume that $\tau_d = lT - mT$, $0 \le m < 1$.   Then

$$G_3(z) = Z\left\{\frac{1 - e^{-Ts}}{s}\frac{e^{-lTs}\ e^{mTs}}{s/a + 1}\right\}$$

$$= (1 - z^{-1})z^{-l}Z\left\{\frac{e^{mTs}}{s(s/a + 1)}\right\}$$

$$= (1 - z^{-1})z^{-l}Z\left\{\frac{e^{mTs}}{s} - \frac{e^{mTs}}{s + a}\right\}$$

$$= \frac{z - 1}{z}\frac{1}{z^l} Z\{1(t + mT) - e^{-a(t+mT)}1(t + mT)\}$$

$$= \frac{z - 1}{z}\frac{1}{z^l}\left\{\frac{z}{z - 1} - \frac{e^{-amT} z}{z - e^{-aT}}\right\}$$

$$= \frac{1}{z^l}\frac{(1 - e^{-amT}) z + e^{-amT} - e^{-aT}}{z - e^{-aT}}$$

$$= \frac{1 - e^{-amT}}{z^l}\frac{z + \alpha}{z - e^{-aT}};$$

$$\alpha = \frac{e^{-amT} - e^{-aT}}{1 - e^{-amT}}. \tag{A.16}$$

From (A.16) it is easy to see that the zero location, $-\alpha$, varies from $\alpha = \infty$ at $m = 0$ to $\alpha = 0$ as $m \to 1$ and that $G_3(1) = 1.0$ for all a, m, and l.

For the specific values of $\tau_d = 1.5$, $T = 1$, $a = 1$ (A.16) reduces to ($l = 2$, $m = \frac{1}{2}$)

$$G_3(z) = \frac{z + 0.6065}{z^2(z - 0.3679)}.$$

A.4 CONTROL THROUGH A FLEXIBLE STRUCTURE

Many controlled systems have some structural flexibility in some portion of the system.   The spacecraft (Fig. A.1) may not be perfectly rigid, the tracker (Fig. A.3) may have some flexibility between the angle observed by the antenna and the angle of the base, and almost any mechanical system would exhibit some degree of structural flexibility.

Conceptually, these systems are equivalent to the double mass-spring device shown in Fig. A.8.   The equations of motion are

$$\begin{aligned}M\ddot{y} + (\dot{y} - \dot{d})b + (y - d)k &= F, \\ m\ddot{d} + (\dot{d} - \dot{y})b + (d - y)k &= 0,\end{aligned} \tag{A.17}$$

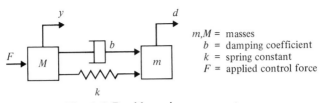

Fig. A.8 Double spring-mass system.

which, when transformed, become,

$$\begin{bmatrix} Ms^2 + bs + k & -(bs + k) \\ -(bs + k) & ms^2 + bs + k \end{bmatrix} \begin{bmatrix} y(s) \\ d(s) \end{bmatrix} = \begin{bmatrix} 1 \\ 0 \end{bmatrix} F(s). \tag{A.18}$$

Two transfer functions are of interest: one between F and d, the other between F and y. The first represents the case where there is structural flexibility between the sensor and the actuator. The second is the case where the sensor is placed close to the actuator, but there exists a mechanical oscillation elsewhere in the system which is coupled to the mass to which the actuator and sensor are attached.

Using (A.18) we can obtain both transfer functions:

$$\frac{d(s)}{F(s)} = \frac{1}{M} \frac{\left(\dfrac{b}{m} s + \dfrac{k}{m}\right)}{s^2 \left[s^2 + \left(1 + \dfrac{m}{M}\right)\left(\dfrac{b}{m} s + \dfrac{k}{m}\right)\right]}, \tag{A.19}$$

which becomes for the typical case with very low damping:

$$\frac{d(s)}{F(s)} \simeq \frac{1}{M} \frac{k/M}{s^2 \left[s^2 + \left(1 + \dfrac{m}{M}\right)\left(\dfrac{b}{m} s + \dfrac{k}{m}\right)\right]} = G_4(s) \tag{A.20}$$

and

$$\frac{y(s)}{F(s)} = \frac{1}{M} \frac{\left(s^2 + \dfrac{b}{m} s + \dfrac{k}{m}\right)}{s^2 \left[s^2 + \left(1 + \dfrac{m}{M}\right)\left(\dfrac{b}{m} s + \dfrac{k}{m}\right)\right]} = G_5(s). \tag{A.21}$$

For $m/M \ll 1$, Eq. (A.21) indicates that the system dynamics are essentially $1/s^2$, but there are poles and zeros at the structural bending-mode frequency that almost cancel each other. A small value of m/M in this simplified example represents the fact that the flexibility is not a dominant response to the control input F and $1/s^2$ is a good model for many such problems.

Equation (A.20) always has the $1/s^2$ ("rigid-body") poles plus the resonance

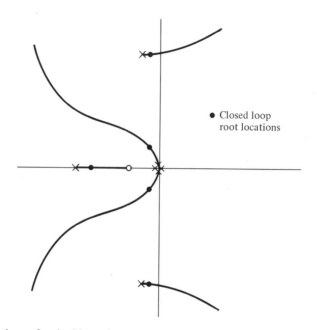

● Closed loop
root locations

Fig. A.9 Root locus for double spring-mass system with flexibility between sensor and actuator.

poles without the neighboring zeros of (A.21) which make a control designer's job less difficult. In fact, if a lead compensation is used to stabilize (A.20), the structural mode is destabilized, as shown in the root locus of Fig. A.9, and depending on the feedback gain and original structural damping, the structural mode could well be driven unstable.

A.5 CONTROL OF A PRESSURIZED FLOW BOX FOR A PAPER MACHINE

This example is multivariable since it has two control variables and two measurable states. The basic function of the system is to keep the paper stock flow out the bottom opening (slice) (Fig. A.10) at a consistent rate. The differential equations of motion are[1]

$$
\begin{bmatrix} \dot{H} \\ \dot{h} \\ \dot{u}_a \end{bmatrix} = \begin{bmatrix} -0.2 & +0.1 & +1 \\ -0.05 & 0 & 0 \\ 0 & 0 & -1 \end{bmatrix} \begin{bmatrix} H \\ h \\ u_a \end{bmatrix} + \begin{bmatrix} 0 & 1 \\ 0 & 0.7 \\ 1 & 0 \end{bmatrix} \begin{bmatrix} u_c \\ u_s \end{bmatrix}, \qquad (A.22)
$$

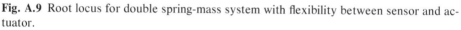

---

[1] These numbers are mainly contrived and may not correspond to any real head box, alive or dead.

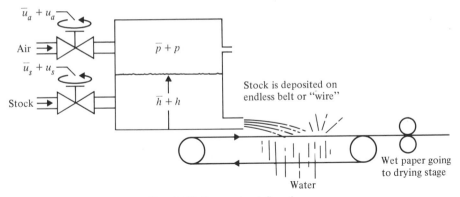

Fig. A.10 Pressurized flow box.

where

$$H = h + p/\rho_s g = \text{total head perturbation,}$$
$$h = \text{stock level perturbation,}$$
$$u_a = \text{perturbation in air-valve opening,}$$
$$u_c = \text{command value to air valve,}$$
$$u_s = \text{perturbation in stock valve opening,}$$
$$\bar{(\ )} = \text{denotes mean value.}$$

Note from Eq. (A.22) that there is two-way coupling between H and h and that u_s affects both states: therefore, it is difficult to decouple the system so as to influence either H or h alone with a single control. The third equation in (A.22) represents dynamics of the air-valve actuator which is such that u_a does not respond instantaneously to an air command u_c. The transfer functions between u_c and u_s and H and h are

$$G_6(s) = \frac{1}{\Delta(s)} \begin{bmatrix} s & (s + 1)(s + 0.07) \\ -0.05 & 0.7(s + 1)(s + 0.13) \end{bmatrix},$$

where $\Delta(s) = (s + 1)(s + 0.1707)(s + 0.02929)$.

Appendix B

B.1 PROPERTIES OF z-TRANSFORMS

Let $\mathscr{F}_i(s)$ be the Laplace transform of $f_i(t)$ and $F_i(z)$ be the z-transform of $f_i(kT)$.

Number	Laplace transform	Samples	z-Transform	Comment
—	$\mathscr{F}_i(s)$	$f_i(kT)$	$F_i(z)$	
1	$\alpha\mathscr{F}_1(s) + \beta\mathscr{F}_2(s)$	$\alpha f_1(kT) + \beta f_2(kT)$	$\alpha F_1(z) + \beta F_2(z)$	The z-transform is linear
2	$\mathscr{F}_1(e^{Ts})\mathscr{F}_2(s)$	$\displaystyle\sum_{\ell=-\infty}^{\infty} f_1(\ell T)f_2(kT - \ell T)$	$F_1(z)F_2(z)$	Discrete convolution corresponds to product of z-transforms
3	$e^{+nTs}\mathscr{F}(s)$	$f(kT + nT)$	$z^n F(z)$	Shift in time
4	$\mathscr{F}(s + a)$	$e^{-akT}f(kT)$	$F(e^{aT}z)$	Shift in frequency
5	—	$\displaystyle\lim_{k\to\infty} f(kT)$	$\displaystyle\lim_{z\to1}(z - 1)F(z)$	If all poles of $(z - 1)F(z)$ are inside the unit circle and $F(z)$ converges for $1 \le |z|$
6	$\mathscr{F}(s/\omega_n)$	$f(\omega_n kT)$	$F(z; \omega_n T)$	Time and frequency scaling
7	—	$f_1(kT)f_2(kT)$	$\displaystyle\frac{1}{2\pi i}\oint_{C_3} F_1(\zeta)F_2(z/\zeta)\frac{d\zeta}{\zeta}$	Time product
8	$\mathscr{F}_3(s) = \mathscr{F}_1(s)\mathscr{F}_2(s)$	$\displaystyle\int_{-\infty}^{\infty} f_1(\tau)f_2(kT - \tau)d\tau$	$F_3(z)$	Continuous convolution does *not* correspond to product of z-transforms

B.2 TABLE OF z-TRANSFORMS

$\mathscr{F}(s)$ is the Laplace transform of $f(t)$ and $F(z)$ is the z-transform of $f(nT)$. Unless otherwise noted, $f(t) = 0$, $t < 0$, and the region of convergence of $F(z)$ is outside a circle $r < |z|$ such that all poles of $F(z)$ are inside r.

Number	$\mathscr{F}(s)$	$f(nT)$	$F(z)$
1	—	$1,\ n = 0;\ 0,\ n \neq 0$	1
2	—	$1,\ n = k;\ 0,\ n \neq k$	z^{-k}
3	$\dfrac{1}{s}$	$1(nT)$	$\dfrac{z}{z-1}$
4	$\dfrac{1}{s^2}$	nT	$\dfrac{Tz}{(z-1)^2}$
5	$\dfrac{1}{s^3}$	$\dfrac{1}{2!}(nT)^2$	$\dfrac{T^2}{2}\dfrac{z(z+1)}{(z-1)^3}$
6	$\dfrac{1}{s^4}$	$\dfrac{1}{3!}(nT)^3$	$\dfrac{T^3}{6}\dfrac{z(z^2+4z+1)}{(z-1)^4}$
7	$\dfrac{1}{s^m}$	$\lim\limits_{a\to 0}\dfrac{(-1)^{m-1}}{(m-1)!}\dfrac{\partial^{m-1}}{\partial a^{m-1}}\,e^{-anT}$	$\lim\limits_{a\to 0}\dfrac{(-1)^{m-1}}{(m-1)!}\dfrac{\partial^{m-1}}{\partial a^{m-1}}\dfrac{z}{z-e^{-aT}}$
8	$\dfrac{1}{s+a}$	e^{-anT}	$\dfrac{z}{z-e^{-aT}}$
9	$\dfrac{1}{(s+a)^2}$	nTe^{-anT}	$\dfrac{Tze^{-aT}}{(z-e^{-aT})^2}$
10	$\dfrac{1}{(s+a)^3}$	$\dfrac{1}{2}(nT)^2e^{-anT}$	$\dfrac{T^2}{2}\,e^{-aT}\dfrac{z(z+e^{-aT})}{(z-e^{-aT})^3}$
11	$\dfrac{1}{(s+a)^m}$	$\dfrac{(-1)^{m-1}}{(m-1)!}\dfrac{\partial^{m-1}}{\partial a^{m-1}}(e^{-anT})$	$\dfrac{(-1)^{m-1}}{(m-1)!}\dfrac{\partial^{m-1}}{\partial a^{m-1}}\dfrac{z}{z-e^{-aT}}$
12	$\dfrac{a}{s(s+a)}$	$1 - e^{-anT}$	$\dfrac{z(1-e^{-aT})}{(z-1)(z-e^{-aT})}$

13	$\dfrac{a}{s^2(s+a)}$	$\dfrac{1}{a}(anT - 1 + e^{-anT})$
14	$\dfrac{b-a}{(s+a)(s+b)}$	$(e^{-anT} - e^{-bnT})$
15	$\dfrac{s}{(s+a)^2}$	$(1 - anT)e^{-anT}$
16	$\dfrac{a^2}{s(s+a)^2}$	$1 - e^{-anT}(1 + anT)$
17	$\dfrac{(b-a)s}{(s+a)(s+b)}$	$be^{-bnT} - ae^{-anT}$
18	$\dfrac{a}{s^2+a^2}$	$\sin anT$
19	$\dfrac{s}{s^2+a^2}$	$\cos anT$
20	$\dfrac{s+a}{(s+a)^2+b^2}$	$e^{-anT}\cos bnT$
21	$\dfrac{b}{(s+a)^2+b^2}$	$e^{-anT}\sin bnT$
22	$\dfrac{a^2+b^2}{s((s+a)^2+b^2)}$	$1 - e^{-anT}\left(\cos bnT + \dfrac{a}{b}\sin bnT\right)$

z-transforms:

13. $\dfrac{z[(aT - 1 + e^{-aT})z + (1 - e^{-aT} - aTe^{-aT})]}{a(z - 1)^2(z - e^{-aT})}$

14. $\dfrac{(e^{-aT} - e^{-bT})z}{(z - e^{-aT})(z - e^{-bT})}$

15. $\dfrac{z[z - e^{-aT}(1 + aT)]}{(z - e^{-aT})^2}$

16. $\dfrac{z[z(1 - e^{-aT} - aTe^{-aT}) + e^{-2aT} - e^{-aT} + aTe^{-aT}]}{(z - 1)(z - e^{-aT})^2}$

17. $\dfrac{z[z(b - a) - (be^{-aT} - ae^{-bT})]}{(z - e^{-aT})(z - e^{-bT})}$

18. $\dfrac{z\sin aT}{z^2 - (2\cos aT)z + 1}$

19. $\dfrac{z(z - \cos aT)}{z^2 - (2\cos aT)z + 1}$

20. $\dfrac{z(z - e^{-aT}\cos bT)}{z^2 - 2e^{-aT}(\cos bT)z + e^{-2aT}}$

21. $\dfrac{z\,e^{-aT}\sin bT}{z^2 - 2e^{-aT}(\cos bT)z + e^{-2aT}}$

22. $\dfrac{z(Az + B)}{(z - 1)(z^2 - 2e^{-aT}(\cos bT)z + e^{-2aT})}$

$A = 1 - e^{-aT}\cos bT - \dfrac{a}{b}e^{-aT}\sin bT$

$B = e^{-2aT} + \dfrac{a}{b}e^{-aT}\sin bT - e^{-aT}\cos bT$

B.3 w-PLANE TRANSFER FUNCTIONS

A general formula for computing $G(w)$ given that

$$G(z) = \frac{K \prod_{i=1}^{m} (z + a_i)}{(z - 1)^{\ell} \prod_{j=1}^{n} (z + b_j)},$$ (B.1)

can be written

$$K \frac{\prod_{i=1}^{m} (1 + a_i) \left(1 - \dfrac{w}{2/T}\right)^{\ell - m + n} \prod_{i=1}^{m} \left(1 + \dfrac{w}{(2/T)[(1 + a_i)(1 - a_i)]}\right)}{\prod_{j=1}^{n} (1 + b_j) T^{\ell} w^{\ell} \prod_{j=1}^{n} \left(1 + \dfrac{w}{(2/T)[(1 + b_j)(1 - b_j)]}\right)}.$$ (B.2)

A more general treatment of w-transforms is found in Saucedo and Schiring (1968).

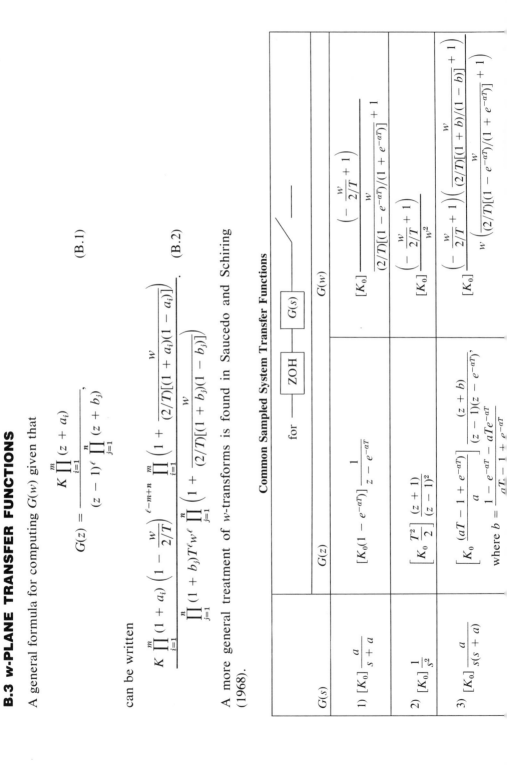

Common Sampled System Transfer Functions

for ZOH — $G(s)$

$G(s)$	$G(z)$	$G(w)$
1) $[K_0]\dfrac{a}{s+a}$	$[K_0(1 - e^{-aT})]\dfrac{1}{z - e^{-aT}}$	$[K_0]\dfrac{\left(-\dfrac{w}{2/T} + 1\right)}{\dfrac{w}{(2/T)[(1 - e^{-aT})/(1 + e^{-aT})]} + 1}$
2) $[K_0]\dfrac{1}{s^2}$	$\left[K_0\dfrac{T^2}{2}\right]\dfrac{(z + 1)}{(z - 1)^2}$	$[K_0]\dfrac{\left(-\dfrac{w}{2/T} + 1\right)}{w^2}$
3) $[K_0]\dfrac{a}{s(s + a)}$	$\left[K_0\dfrac{(aT - 1 + e^{-aT})}{a}\right]\dfrac{(z + b)}{(z - 1)(z - e^{-aT})},$ where $b = \dfrac{1 - e^{-aT} - aTe^{-aT}}{aT - 1 + e^{-aT}}$	$[K_0]\dfrac{\left(-\dfrac{w}{2/T} + 1\right)\left(\dfrac{w}{(2/T)[(1 + b)/(1 - b)]} + 1\right)}{w\left(\dfrac{w}{(2/T)[(1 - e^{-aT})/(1 + e^{-aT})]} + 1\right)}$

Appendix C / A Few Results
from Matrix Analysis

Although we assume the reader has some acquaintance with linear equations and determinants, there are a few results of a more advanced character which even elementary control-system theory requires, and these are collected here for reference in the text. For further study, a good choice is Strang (1976).

C. 1 DETERMINANTS
AND THE MATRIX INVERSE

The determinant of a product of two square matrices is the product of their determinants.

$$\det \mathbf{AB} = \det \mathbf{A} \det \mathbf{B} \qquad (C.1)$$

If a matrix is diagonal, then the determinant is the product of the elements on the diagonal.

If the matrix is partitioned with square elements on the main diagonal, then an extension of this result applies, namely

$$\det \begin{bmatrix} \mathbf{A} & \mathbf{0} \\ \mathbf{B} & \mathbf{C} \end{bmatrix} = \det \mathbf{A} \det \mathbf{C} \qquad \text{if} \quad \mathbf{A} \text{ and } \mathbf{C} \text{ are square matrices.} \quad (C.2)$$

Suppose $\mathbf{A}$ is a matrix of dimensions $m \times n$ and $\mathbf{B}$ is of dimension $n \times m$. Let $\mathbf{I}_m$ and $\mathbf{I}_n$ be the identity matrices of size $m \times m$ and $n \times n$, respectively. Then

$$\det[\mathbf{I}_n + \mathbf{BA}] = \det[\mathbf{I}_m + \mathbf{AB}]. \qquad (C.3)$$

To show this result, we consider the determinant of the matrix product

$$\det \begin{bmatrix} \mathbf{I}_m & \mathbf{0} \\ \mathbf{B} & \mathbf{I}_n \end{bmatrix} \begin{bmatrix} \mathbf{I}_m & \mathbf{A} \\ -\mathbf{B} & \mathbf{I}_n \end{bmatrix} = \det \begin{bmatrix} \mathbf{I}_m & \mathbf{A} \\ 0 & \mathbf{I}_n + \mathbf{BA} \end{bmatrix} = \det[\mathbf{I}_n + \mathbf{BA}].$$

But this is also equal to

$$\det \begin{bmatrix} \mathbf{I}_m & -\mathbf{A} \\ \mathbf{0} & \mathbf{I}_n \end{bmatrix} \begin{bmatrix} \mathbf{I}_m & \mathbf{A} \\ -\mathbf{B} & \mathbf{I}_n \end{bmatrix} = \det \begin{bmatrix} \mathbf{I}_m + \mathbf{AB} & \mathbf{0} \\ -\mathbf{B} & \mathbf{I}_n \end{bmatrix} = \det[\mathbf{I}_m + \mathbf{AB}],$$

and therefore these two determinants are equal to each other, which is (C.3).

If the determinant of a matrix $\mathbf{A}$ is not zero, then we can define a related matrix $\mathbf{A}^{-1}$ called "$\mathbf{A}$ inverse" which has the property that

$$\mathbf{AA}^{-1} = \mathbf{A}^{-1}\mathbf{A} = \mathbf{I}. \tag{C.4}$$

According to property (C.1) we have

$$\det \mathbf{AA}^{-1} = \det \mathbf{A} \cdot \det \mathbf{A}^{-1} = 1,$$

or

$$\det \mathbf{A}^{-1} = \frac{1}{\det \mathbf{A}}.$$

It can be shown that there is an $n \times n$ matrix called the *adjugate of* $\mathbf{A}$ with elements composed of sums of products of the elements of $\mathbf{A}^1$ and having the property that

$$\mathbf{A} \cdot \operatorname{adj} \mathbf{A} = \det \mathbf{A} \cdot \mathbf{I}. \tag{C.5}$$

Thus, if the determinant of $\mathbf{A}$ is not zero, the inverse of $\mathbf{A}$ is given by

$$\mathbf{A}^{-1} = \frac{\operatorname{adj} \mathbf{A}}{\det \mathbf{A}}.$$

A famous and useful formula for the inverse of a combination of matrices has come to be called the *matrix inversion lemma* in the control literature. It arises in the development of recursive algorithms for estimation as found in Chapter 8. The formula is as follows: If $\det \mathbf{A}$, $\det \mathbf{C}$, and $\det(\mathbf{A} + \mathbf{BCD})$ are different from zero, then we have the matrix inversion lemma:

$$(\mathbf{A} + \mathbf{BCD})^{-1} = \mathbf{A}^{-1} - \mathbf{A}^{-1}\mathbf{B}(\mathbf{C}^{-1} + \mathbf{DA}^{-1}\mathbf{B})^{-1}\mathbf{DA}^{-1}. \tag{C.6}$$

The truth of (C.6) is readily confirmed if we multiply both sides by $\mathbf{A} + \mathbf{BCD}$ to obtain

$$\mathbf{I} = \mathbf{I} + \mathbf{BCDA}^{-1} - \mathbf{B}(\mathbf{C}^{-1} + \mathbf{DA}^{-1}\mathbf{B})^{-1}\mathbf{DA}^{-1} - \mathbf{BCDA}^{-1}\mathbf{B}(\mathbf{C}^{-1} + \mathbf{DA}^{-1}\mathbf{B})^{-1}\mathbf{DA}^{-1}$$
$$= \mathbf{I} + \mathbf{BCDA}^{-1} - [\mathbf{B} + \mathbf{BCDA}^{-1}\mathbf{B}][\mathbf{C}^{-1} + \mathbf{DA}^{-1}\mathbf{B})^{-1}\mathbf{DA}^{-1}.$$

If we subtract $\mathbf{I}$ from both sides and factor $\mathbf{BC}$ from the left on the third term, we find

$$\mathbf{0} = \mathbf{BCDA}^{-1} - \mathbf{BC}[\mathbf{C}^{-1} + \mathbf{DA}^{-1}\mathbf{B}][\mathbf{C}^{-1} + \mathbf{DA}^{-1}\mathbf{B}]^{-1}\mathbf{DA}^{-1},$$

---

[1] If $\mathbf{A}^{ij}$ is the $n - 1 \times n - 1$ matrix (minor) found by deleting row i and column j from $\mathbf{A}$, then the entry in row i and column j of the adj $\mathbf{A}$ is $(-1)^{i+j}\det \mathbf{A}^{ji}$.

which is

$$0 = 0 \quad \text{which was to be demonstrated.}$$

C.2 EIGENVALUES AND EIGENVECTORS

We consider the discrete dynamic system

$$\mathbf{x}_{k+1} = \mathbf{\Phi}\mathbf{x}_k, \tag{C.7}$$

where, for purposes of illustration, we will let

$$\mathbf{\Phi} = \begin{bmatrix} \frac{5}{6} & -\frac{1}{6} \\ 1 & 0 \end{bmatrix}. \tag{C.8}$$

If we assume that it is possible for this system to have a motion given by a geometric series such as z^k, we can assume that there is a vector $\mathbf{v}$ so that $\mathbf{x}_k$ can be written

$$\mathbf{x}_k = \mathbf{v}z^k. \tag{C.9}$$

Substituting (C.9) into (C.7), we must find the vector $\mathbf{v}$ and the number z such that

$$\mathbf{v}z^{k+1} = \mathbf{\Phi}\mathbf{v}z^k,$$

or, multiplying by z^{-k} yields

$$\mathbf{v}z = \mathbf{\Phi}\mathbf{v}. \tag{C.10}$$

If we collect both the terms of (C.10) on the left, we find

$$(z\mathbf{I} - \mathbf{\Phi})\mathbf{v} = \mathbf{0}. \tag{C.11}$$

These linear equations have a solution for a nontrivial $\mathbf{v}$ if and only if the determinant of the coefficient matrix is zero. This determinant is a polynomial of degree n in z ($\mathbf{\Phi}$ is an $n \times n$ matrix) called the *characteristic polynomial* of $\mathbf{\Phi}$, and values of z for which the characteristic polynomial is zero are roots of the characteristic equation and are called *eigenvalues* of $\mathbf{\Phi}$. For example, for the matrix given in (C.8) the characteristic polynomial is

$$\det \left\{ \begin{bmatrix} z & 0 \\ 0 & z \end{bmatrix} - \begin{bmatrix} \frac{5}{6} & -\frac{1}{6} \\ 1 & 0 \end{bmatrix} \right\}.$$

Adding the two matrices, we find

$$\det \left\{ \begin{matrix} z - \frac{5}{6} & +\frac{1}{6} \\ -1 & z \end{matrix} \right\},$$

which can be evaluated to give

$$z(z - \tfrac{5}{6}) + \tfrac{1}{6} = (z - \tfrac{1}{2})(z - \tfrac{1}{3}). \tag{C.12}$$

Thus the characteristic roots of this $\mathbf{\Phi}$ are $\tfrac{1}{2}$ and $\tfrac{1}{3}$. Associated with these characteristic roots are solutions to (C.11) for vectors $\mathbf{v}$, called the *characteristic* or *eigenvectors.* If we let $z = \frac{1}{2}$, then (C.11) requires

$$\left\{ \begin{bmatrix} \frac{1}{2} & 0 \\ 0 & \frac{1}{2} \end{bmatrix} - \begin{bmatrix} \frac{5}{6} & -\frac{1}{6} \\ 1 & 0 \end{bmatrix} \right\} \begin{bmatrix} v_{11} \\ v_{21} \end{bmatrix} = \begin{bmatrix} 0 \\ 0 \end{bmatrix}. \tag{C.13}$$

Adding the matrices, we find that these equations become

$$\begin{bmatrix} -\frac{1}{3} & \frac{1}{6} \\ -1 & \frac{1}{2} \end{bmatrix} \begin{bmatrix} v_{11} \\ v_{21} \end{bmatrix} = \begin{bmatrix} 0 \\ 0 \end{bmatrix}. \tag{C.14}$$

Equations (C.14) are satisfied by any v_{11} and v_{21} such that

$$v_{21} = 2v_{11},$$

from which we conclude that the eigenvector corresponding to $z_1 = \frac{1}{2}$ is given by

$$\mathbf{v}_1 = \begin{bmatrix} a \\ 2a \end{bmatrix}. \tag{C.15}$$

We can arbitrarily select the scale factor "a" in (C.15). Some prefer to make the length[2] of eigenvectors equal to one. Here we make the largest component of $\mathbf{v}$ have unit magnitude. Thus the scaled $\mathbf{v}_1$ is

$$\mathbf{v}_1 = \begin{bmatrix} \frac{1}{2} \\ 1 \end{bmatrix}. \tag{C.16}$$

In similar fashion, the eigenvector $\mathbf{v}_2$ associated with $z_2 = \frac{1}{3}$ can be computed to be

$$\mathbf{v}_2 = \begin{bmatrix} \frac{1}{3} \\ 1 \end{bmatrix}.$$

Note that even if all elements of $\boldsymbol{\Phi}$ are real, it is possible for characteristic values and characteristic vectors to be complex.

C.3 SIMILARITY TRANSFORMATIONS

If we make a change of variables in (C.7) according to $\mathbf{x} = \mathbf{T}\boldsymbol{\xi}$, where $\mathbf{T}$ is an $n \times n$ matrix, then we start with the equations

$$\mathbf{x}_{k+1} = \boldsymbol{\Phi}\mathbf{x}_k,$$

and, substituting for $\mathbf{x}$, we have

$$\mathbf{T}\boldsymbol{\xi}_{k+1} = \boldsymbol{\Phi}\mathbf{T}\boldsymbol{\xi}_k.$$

Then, if we multiply on the left by $\mathbf{T}^{-1}$, we get the equation in $\boldsymbol{\xi}$,

$$\boldsymbol{\xi}_{k+1} = \mathbf{T}^{-1}\boldsymbol{\Phi}\mathbf{T}\boldsymbol{\xi}_k. \tag{C.17}$$

---

[2] Usually we define the length of a vector as the square root of the sum of squares of its components or, if $\|\mathbf{v}\|$ is the symbol for length, then $\|\mathbf{v}\|^2 = \mathbf{v}^T\mathbf{v}$. If $\mathbf{v}$ is complex, as will happen if z_i is complex, then we must take a conjugate, and we define $\|\mathbf{v}\|^2 = (\mathbf{v}^*)^T\mathbf{v}$, where $\mathbf{v}^*$ is the complex conjugate of $\mathbf{v}$.

If we define the new system matrix as $\mathbf{\Psi}$, then the new states satisfy the equations

$$\boldsymbol{\xi}_{k+1} = \mathbf{\Psi}\boldsymbol{\xi}_k,$$

where

$$\mathbf{\Psi} = \mathbf{T}^{-1}\mathbf{\Phi}\mathbf{T}. \tag{C.18}$$

If we now seek the characteristic polynomial of $\mathbf{\Psi}$, we find

$$\det[z\mathbf{I} - \mathbf{\Psi}] = \det[z\mathbf{I} - \mathbf{T}^{-1}\mathbf{\Phi}\mathbf{T}].$$

Since $\mathbf{T}^{-1}\mathbf{T} = \mathbf{I}$, we can write this polynomial as

$$\det[z\mathbf{T}^{-1}\mathbf{T} - \mathbf{T}^{-1}\mathbf{\Phi}\mathbf{T}],$$

and the $\mathbf{T}^{-1}$ and $\mathbf{T}$ can be factored out on the left and right to give

$$\det[\mathbf{T}^{-1}[z\mathbf{I} - \mathbf{\Phi}]\mathbf{T}].$$

Now, using property (C.1) for the determinant, we compute

$$\det \mathbf{T}^{-1} \cdot \det[z\mathbf{I} - \mathbf{\Phi}] \cdot \det \mathbf{T},$$

which, by the equation following (C.4), gives us the final result

$$\det[z\mathbf{I} - \mathbf{\Psi}] = \det[z\mathbf{I} - \mathbf{\Phi}]. \tag{C.19}$$

From (C.19) we see that $\mathbf{\Psi}$ and $\mathbf{\Phi}$ have the same characteristic polynomials. The matrices are said to be "similar," and the transformation (C.18) is a similarity transformation.

A case of a similarity transformation of particular interest is one for which the resulting matrix $\mathbf{\Psi}$ is diagonal. As an attempt to find such a matrix, suppose we *assume* that $\mathbf{\Psi}$ is diagonal and write the transformation $\mathbf{T}$ in terms of its columns, $\mathbf{t}_i$. Then (C.18) may be expressed as

$$\mathbf{T}\mathbf{\Psi} = \mathbf{\Phi}\mathbf{T},$$
$$[\mathbf{t}_1\ \mathbf{t}_2\ \cdots\ \mathbf{t}_n]\mathbf{\Psi} = \mathbf{\Phi}[\mathbf{t}_1\ \mathbf{t}_2\ \cdots\ \mathbf{t}_n]$$
$$= [\mathbf{\Phi}\mathbf{t}_1\ \mathbf{\Phi}\mathbf{t}_2\ \cdots\ \mathbf{\Phi}\mathbf{t}_n]. \tag{C.20}$$

If we *assume* that $\mathbf{\Psi}$ is diagonal with elements $\lambda_1, \lambda_2, \ldots, \lambda_n$, then (C.20) can be written as

$$[\mathbf{t}_1\ \mathbf{t}_2\ \cdots\ \mathbf{t}_n]\begin{bmatrix} \lambda_1 & 0 & \cdots & 0 \\ 0 & \lambda_2 & \cdots & 0 \\ 0 & & & \\ \vdots & & \ddots & \\ \vdots & & & \lambda_n \end{bmatrix} = [\mathbf{\Phi}\mathbf{t}_1\ \ \mathbf{\Phi}\mathbf{t}_2\ \ \cdots\ \ \mathbf{\Phi}\mathbf{t}_n].$$

Multiplying the matrices on the left, we find

$$[\lambda_1 \mathbf{t}_1 \; \lambda_2 \mathbf{t}_2 \; . \; . \; . \; \lambda_n \mathbf{t}_n] = [\mathbf{\Phi t}_1 \; . \; . \; . \; \mathbf{\Phi t}_n]. \tag{C.21}$$

Since the two sides of (C.21) are equal, they must match up column by column, and we can write the equation for column j as

$$\lambda_j \mathbf{t}_j = \mathbf{\Phi t}_j. \tag{C.22}$$

Comparing (C.22) with (C.10), we see that $\mathbf{t}_j$ is an eigenvector of $\mathbf{\Phi}$ and λ_j is an eigenvalue. We conclude that if the transformation $\mathbf{T}$ converts $\mathbf{\Phi}$ into a diagonal matrix $\mathbf{\Psi}$, then the columns of $\mathbf{T}$ must be eigenvectors of $\mathbf{\Phi}$ and the diagonal elements of $\mathbf{\Psi}$ are the eigenvalues of $\mathbf{\Phi}$ [which are also the eigenvalues of $\mathbf{\Psi}$, by (C.19)]. It turns out that if the eigenvalues of $\mathbf{\Phi}$ are distinct, then there are exactly n eigenvectors and they are independent; i.e., we can construct a nonsingular transformation $\mathbf{T}$ from the n eigenvectors.

In the example given above, we would have

$$\mathbf{T} = \begin{bmatrix} \frac{1}{2} & \frac{1}{3} \\ 1 & 1 \end{bmatrix},$$

for which

$$\mathbf{T}^{-1} = \begin{bmatrix} 6 & -2 \\ -6 & 3 \end{bmatrix},$$

and the new diagonal system matrix is

$$\begin{aligned}
\mathbf{T}^{-1}\mathbf{\Phi T} &= \begin{bmatrix} 6 & -2 \\ -6 & 3 \end{bmatrix} \begin{bmatrix} \frac{5}{6} & -\frac{1}{6} \\ 1 & 0 \end{bmatrix} \begin{bmatrix} \frac{1}{2} & \frac{1}{3} \\ 1 & 1 \end{bmatrix} \\
&= \begin{bmatrix} 3 & -1 \\ -2 & 1 \end{bmatrix} \begin{bmatrix} \frac{1}{2} & \frac{1}{3} \\ 1 & 1 \end{bmatrix} \\
&= \begin{bmatrix} \frac{1}{2} & 0 \\ 0 & \frac{1}{3} \end{bmatrix}
\end{aligned}$$

as advertised!

If the elements of $\mathbf{\Phi}$ are real and an eigenvalue is complex, say $\lambda_1 = \alpha + j\beta$, then the conjugate, $\lambda_1^* = \alpha - j\beta$, is also an eigenvalue since the characteristic polynomial has real coefficients. In such a case, the respective eigenvectors will be conjugate. If $\mathbf{v}_1 = \mathbf{r} + j\mathbf{i}$, then $\mathbf{v}_2 = \mathbf{v}_1^* = \mathbf{r} - j\mathbf{i}$, where $\mathbf{r}$ and $\mathbf{i}$ are matrices of real elements representing the real and imaginary parts of the eigenvectors. In such cases, it is common practice to use the real matrices $\mathbf{r}$ and $-\mathbf{i}$ as columns of the transformation matrix $\mathbf{T}$ rather than go through the complex arithmetic required to deal directly with $\mathbf{v}_1$ and $\mathbf{v}_1^*$. The resulting transformed equations are not diagonal but rather the corresponding variables appear in the coupled equations

$$\dot{\eta} = \alpha\eta - \beta\nu, \qquad \dot{\nu} = \beta\eta + \alpha\nu. \tag{C.23}$$

C.4 THE CAYLEY-HAMILTON THEOREM

A very useful property of a matrix $\mathbf{\Phi}$ follows from consideration of the inverse of $z\mathbf{I} - \mathbf{\Phi}$. As we saw in (C.5), we can write

$$(z\mathbf{I} - \mathbf{\Phi})\mathrm{adj}(z\mathbf{I} - \mathbf{\Phi}) = \mathbf{I}\det(z\mathbf{I} - \mathbf{\Phi}). \tag{C.24}$$

The coefficient of $\mathbf{I}$ on the right-hand side of (C.24) is the characteristic polynomial of $\mathbf{\Phi}$ which we can write as

$$a(z) = z^n + a_1 z^{n-1} + a_2 z^{n-2} + \cdots + a_n.$$

The adjugate of $z\mathbf{I} - \mathbf{\Phi}$, on the other hand, is a *matrix* of polynomials in z, found from the determinants of the minors of $z\mathbf{I} - \mathbf{\Phi}$. If we collect the constant matrix coefficients of the powers of z, it is clear that we can write

$$\mathrm{adj}(z\mathbf{I} - \mathbf{\Phi}) = \mathbf{B}_1 z^{n-1} + \mathbf{B}_2 z^{n-2} + \cdots + \mathbf{B}_n)$$

and (C.24) becomes a polynomial equation with matrix coefficients. Written out, it is

$$[z\mathbf{I} - \mathbf{\Phi}][\mathbf{B}_1 z^{n-1} + \mathbf{B}_2 z^{n-2} + \cdots + \mathbf{B}_n] = z^n\mathbf{I} + a_1\mathbf{I}z^{n-1} + \cdots + a_n\mathbf{I}. \tag{C.25}$$

If we now multiply the two matrices on the left and equate coefficients of equal powers of z, we find

$$\mathbf{B}_1 = \mathbf{I},$$
$$\mathbf{B}_2 = \mathbf{\Phi}\mathbf{B}_1 + a_1\mathbf{I} = \mathbf{\Phi} + a_1\mathbf{I},$$
$$\mathbf{B}_3 = \mathbf{\Phi}\mathbf{B}_2 + a_2\mathbf{I} = \mathbf{\Phi}^2 + a_1\mathbf{\Phi} + a_2\mathbf{I},$$
$$\vdots$$
$$\mathbf{B}_n = \mathbf{\Phi}\mathbf{B}_{n-1} + a_{n-1}\mathbf{I} = \mathbf{\Phi}^{n-1} + a_1\mathbf{\Phi}^{n-2} + \cdots + a_{n-1}\mathbf{I},$$
$$0 = \mathbf{\Phi}\mathbf{B}_n + a_n\mathbf{I} = \mathbf{\Phi}^n + a_1\mathbf{\Phi}^{n-1} + a_2\mathbf{\Phi}^{n-1} + \cdots + a_n\mathbf{I}. \tag{C.26}$$

Equation (C.26) is a statement that the matrix obtained when matrix $\mathbf{\Phi}$ is substituted for z in the characteristic polynomial, $a(z)$, is exactly zero! In other words, we have the Cayley-Hamilton theorem according to which

$$a(\mathbf{\Phi}) = 0. \tag{C.27}$$

Appendix D / Summary
Of Facts from the
Theory of Probability
and Stochastic Processes[1]

D.1 RANDOM VARIABLES

We begin with a space of experiments, Ω, whose outcomes are called ω and depend on chance. Over the space Ω and its subsets Ω_i we define a probability function P which assigns a positive number between 0 and 1 to each countable combination of subsets in Ω to which an outcome (or "event") may belong. P has the properties that the probability of *some* outcome is certain in which case it is assigned the value 1,

$$P(\Omega) = 1,$$

and the probability of an outcome which may result from events Ω_i Ω_j which have no common points (the intersection of Ω_i and Ω_j is empty) is the sum of the probabilities of Ω_i and Ω_j,

$$P\{\Omega_i \cup \Omega_j\} = P(\Omega_i) + P(\Omega_j).$$

In addition to the function P we define a *random variable* $x(\omega)$ which maps Ω into the real line such that to each outcome ω in Ω we associate a value x, and the probability that a chance experiment maps into a value x which is less than or equal to the constant a is

$$\Pr(x \leq a) = F_x(a).[2] \tag{D.1}$$

The function $F_x(\xi)$ is called the *distribution function of the random variable.* If F_x is a smooth function,[3] we define its derivative $f_x(\xi)$ as the *density function*

---

[1] See Parzen (1962).
[2] $\Pr\{\cdot\}$ is meant to be read "the probability that $\{\cdot\}$."
[3] Using the impulse, we can include simple discontinuities as must be done if a specific value of x has nonzero probability.

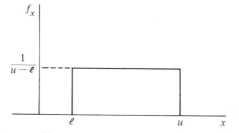

Fig. D.1 Sketch of the uniform density function.

which has the property

$$\Pr\{a \leq x \leq b\} = \int_a^b f_x(\xi)d\xi. \tag{D.2}$$

Since the whole space Ω maps into the line somewhere, we have

$$\int_{-\infty}^{\infty} f_x(\xi)d\xi = 1. \tag{D.3}$$

Two common density functions which we shall have reason to use are *the uniform density* and *the normal* or *Gaussian density*. A random variable having a uniform density has zero probability of having any value outside a finite range between lower limit ℓ and upper limit u, $\ell \leq x \leq u$, and f_x is constant inside this range. Because of (D.3), the constant is $1/(u - \ell)$. A sketch of the uniform density is given in Fig. D.1.

The normal density function is given by the equation

$$f_x(\xi) = \frac{1}{\sqrt{2\pi}\,\sigma_x} \exp\left(-\frac{1}{2}\frac{(\xi - \mu_x)^2}{\sigma_x^2}\right), \tag{D.4}$$

and shown in the sketch in Fig. D.2.

The importance of the normal density derives mainly from the following facts:

1. The distribution of a random variable based on events which themselves consist of a sum of a large number of independent[4] random events are accurately approximated by the normal law. Such distributions describe electrical noise caused by thermal motions of a large number of particles as in a resistor, for example.

2. If two random variables have (jointly) normal distributions, then their sum also has a normal distribution. (As an extension of this second point, if the input to a linear system is normal, then the distribution of its output is also normal.)

---

[4] The technical definition of "independent" will be given shortly.

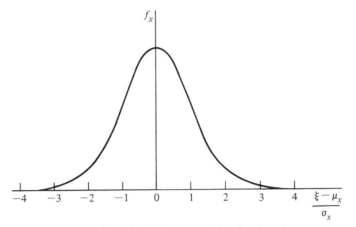

Fig. D.2 Sketch of the normal density function.

D.2 EXPECTATION

By the very nature of variables whose values are dependent on chance we cannot discuss a formula for the calculation of values of x. To describe a random variable, we instead discuss average values such as the arithmetic mean or the average power. Such concepts are contained in the idea of *expectation*. The expected value of a function g of a random variable whose density is f_x is defined as

$$\mathscr{E}\{g(x)\} = \int_{-\infty}^{\infty} g(\xi)f_x(\xi)d\xi$$
$$\overset{\Delta}{=} \overline{g(x)}. \tag{D.5}$$

Important special cases are the mean, variance, and mean square. If $g(x) = x$, then we have the mean, namely

$$\mathscr{E}(x) = \int_{-\infty}^{\infty} \xi f_x(\xi)d\xi \overset{\Delta}{=} \bar{x} = \mu_x. \tag{D.6}$$

For the uniform distribution, the mean is given by

$$\bar{x} = \int_{-\infty}^{\infty} \xi f_x(\xi)d\xi$$
$$= \int_{\ell}^{u} \left[\frac{1}{u - \ell}\right] d\xi = \frac{1}{u - \ell} \left[\frac{\xi^2}{2}\bigg|_{\ell}^{u}\right]$$
$$= \frac{1}{u - \ell} \left[\frac{u^2}{2} - \frac{\ell^2}{2}\right] = \frac{u + \ell}{2}. \tag{D.7}$$

Since the probability density function has the intuitive properties of a histogram of relative frequency of occurrence of a particular x, we see that the mean, like an arithmetic mean, is a weighted average of the random variable values.

If $g(x)$ in (D.5) is $g = (x - \bar{x})^2$, then the expected value is the average of the square of the variation of the variable from the mean, and this number is called the *variance of x*, written var x. For the uniform density we compute

$$\mathcal{E}(x - \bar{x})^2 \overset{\Delta}{=} \text{var } x$$

$$= \int_{-\infty}^{\infty} (\xi - \bar{x})^2 f_x(\xi) d\xi$$

$$= \int_{\ell}^{u} (\xi - \bar{x})^2 \frac{1}{u - \ell} d\xi$$

$$= \frac{1}{u - \ell} \left[\frac{(\xi - \bar{x})^3}{3} \Big|_{\ell}^{u} \right]$$

$$= \frac{1}{u - \ell} \left\{ \frac{(u - \bar{x})^3}{3} - \frac{(\ell - \bar{x})^3}{3} \right\}.$$

Substituting (D.7) for $\bar{x}$, we find

$$\mathcal{E}(x - \bar{x})^2 = \frac{1}{3(u - \ell)} \left\{ \left(\frac{u - \ell}{2} \right)^3 - \left(\frac{\ell - u}{2} \right)^3 \right\}$$

and, simplifying, we get

$$\text{var } x = \frac{(u - \ell)^2}{12}. \tag{D.8}$$

The square root of the variance is called the *standard deviation* and given the symbol σ. Thus, for the uniform density,

$$\sigma_x^2 = \frac{(u - \ell)^2}{12}. \tag{D.9}$$

If we let $g(x)$ be simply x^2, then we compute by (D.5) the mean-square value $\overline{x^2}$. However, consider an expanded expression for the variance, and let the mean value of x be μ:

$$\text{var } x = \int_{-\infty}^{\infty} (\xi - \mu)^2 f_x(\xi) d\xi$$

$$= \int_{-\infty}^{\infty} (\xi^2 - 2\xi\mu + \mu^2) f_x(\xi) d\xi$$

$$= \overline{x^2} - 2\mu^2 + \mu^2$$

$$= \overline{x^2} - \mu^2.$$

Thus we find

$$\overline{x^2} = \text{var } x + \mu^2.$$

For the uniform density, the mean-square value is, therefore,

$$\overline{x^2} = \frac{(u - \ell)^2}{12} + \left(\frac{u + \ell}{2}\right)^2.$$

The integration is more complicated in the case of the normal density, but the definitions are the same, namely, if x has the normal density given by (D.4), then

$$\bar{x} = \int_{-\infty}^{\infty} \xi f_x(\xi) d\xi \stackrel{\Delta}{=} \mu_x \tag{D.10a}$$

and

$$\mathcal{E}(x - \mu_x)^2 = \int_{-\infty}^{\infty} (\xi - \mu_x)^2 f_x(\xi) d\xi$$

$$\text{var } x = \sigma_x^2. \tag{D.10b}$$

Note that the mean and standard deviation already appear as parameters of the density given by (D.4). Because these two parameters completely describe the normal density, it is standard to say that a random variable with normal density having mean μ and standard deviation σ is distributed according to the $N(\mu, \sigma)$ law.

D.3 MORE THAN ONE RANDOM VARIABLE

Frequently the random experiment has an outcome which is mapped into several random variables, say $x_1, x_2, \ldots, x_n$ or, organized as a column matrix,

$$\mathbf{x} = (x_1 \, x_2 \cdots x_n)^T.$$

For this collection we define the probability that

$$x_1 \leq a_1, x_2 \leq a_2, \ldots, x_n \leq a_n$$

as

$$\Pr\{\mathbf{x} \leq \mathbf{a}\}$$

and let this be the distribution $F_\mathbf{x}(\boldsymbol{\xi})$ with vector argument. The corresponding density $f_\mathbf{x}(\boldsymbol{\xi})$ is a function such that the probability that $\mathbf{x}$ is in a box with sides a_i and b_i is given by

$$\Pr\{\mathbf{a} < \mathbf{x} \leq \mathbf{b}\} = \int_{a_1}^{b_1} \cdots \int_{a_n}^{b_n} f_\mathbf{x}(\xi_1, \ldots, \xi_n) d\xi_1 \cdots d\xi_n. \tag{D.11}$$

The mean of the vector is a vector of the means

$$\mathcal{E}\{\mathbf{x}\} = \boldsymbol{\mu}_\mathbf{x} = (\mu_{x_1} \mu_{x_2} \cdots \mu_{x_n})^T, \tag{D.12}$$

and likewise the deviation of the random variables from the mean-value vector is measured by a matrix of terms called the *covariance matrix*, cov $\mathbf{x}$, defined as

$$\text{cov } \mathbf{x} = \mathcal{E}\{(\mathbf{x} - \boldsymbol{\mu}_\mathbf{x})(\mathbf{x} - \boldsymbol{\mu}_\mathbf{x})^T\} \triangleq \mathbf{R}_{\mathbf{xx}} \tag{D.13}$$

Thus cov $\mathbf{x}$ is a matrix with the element in row i and column j given by

$$\mathcal{E}\{(x_i - \mu_{x_i})(x_j - \mu_{x_j})\}.$$

Often the symbol $\mathbf{R}_\mathbf{x}$ will be used for the covariance matrix of a random vector $\mathbf{x}$. In an obvious way we can extend (D.13) to include the case of different random vectors $\mathbf{x}$ and $\mathbf{y}$ with mean vectors $\boldsymbol{\mu}_\mathbf{x}$ and $\boldsymbol{\mu}_\mathbf{y}$. We define the covariance of $\mathbf{x}$ and $\mathbf{y}$ as

$$\text{cov}(\mathbf{x},\mathbf{y}) = \mathcal{E}\{(\mathbf{x} - \boldsymbol{\mu}_\mathbf{x})(\mathbf{y} - \boldsymbol{\mu}_\mathbf{y})^T\} = \mathbf{R}_{\mathbf{xy}}. \tag{D.14}$$

Consider now the special case of two random variables x and y having distribution $F_{x,y}(\xi,\eta)$ and (joint) density $f_{x,y}(\xi,\eta)$. Since the probability that y has any value is the integral from $-\infty$ to $+\infty$ over η, the density of x is given by

$$f_x(\xi) = \int_{-\infty}^{\infty} f_{x,y}(\xi,\eta)d\eta. \tag{D.15}$$

If $f_{x,y}$ is such that

$$f_{x,y}(\xi,\eta) = f_x(\xi)f_y(\eta), \tag{D.16}$$

we say that x and y are *independent*.

It often happens that we know a particular value of one random variable, such as the output y of a dynamic system, and we wish to estimate the value of a related variable, such as the state x. A useful function for this situation is the conditional density, defined as the density of x given that the value of y is η and is defined as

$$f_{x|y}(\xi|\eta) = \frac{f_{xy}(\xi,\eta)}{f_y(\eta)}. \tag{D.17}$$

Note that, from (D.15) and (D.16), if x and y are independent, then

$$f_{x|y}(\xi|\eta) = f_x(\xi). \tag{D.18}$$

The most important multivariable probability density is the normal or gaussian law given by

$$f_\mathbf{x}(\boldsymbol{\xi}) = [(2\pi)^n|\mathbf{R}_\mathbf{x}|]^{-1/2} \exp[-\tfrac{1}{2}(\boldsymbol{\xi} - \boldsymbol{\mu}_\mathbf{x})^T \mathbf{R}_\mathbf{x}^{-1}(\boldsymbol{\xi} - \boldsymbol{\mu}_\mathbf{x})], \tag{D.19}$$

where $|\mathbf{R}_\mathbf{x}|$ is the determinant of the matrix $\mathbf{R}_\mathbf{x}$.

For the multivariable normal law, we can compute the mean,

$$\mathscr{E}\{\mathbf{x}\} = \boldsymbol{\mu}_\mathbf{x},$$

and the covariance matrix,

$$\mathscr{E}\{(\mathbf{x} - \boldsymbol{\mu}_\mathbf{x})(\mathbf{x} - \boldsymbol{\mu}_\mathbf{x})^T\} = \mathbf{R}_\mathbf{x}.$$

Like the scalar normal density, the multivariable law is described entirely by the two parameters $\boldsymbol{\mu}$ and $\mathbf{R}$, the difference being that the multivariable case is described by matrix parameters rather than scalar parameters. In (D.19) we require the inverse of $\mathbf{R}_\mathbf{x}$ and have thus implicitly assumed that this covariance matrix is nonsingular. [See Parzen (1962) for a discussion of the case when $\mathbf{R}_\mathbf{x}$ is singular.]

D.4 STOCHASTIC PROCESSES

In a study of dynamic systems, it is natural to have random variables which evolve in time much as the states and control inputs evolve. However, with random time variables it is not possible to compute z-transforms in the usual way, and furthermore, since specific values of the variables have little value, we need formulas to describe how the means and covariances evolve in time. A random variable which evolves in time is called a *stochastic process,* and here we consider only discrete time.

Suppose we deal first with a stochastic process $w(n)$, where w is a scalar and distributed according to the density $f_w(\xi;n)$. Note that the density function depends on the time of occurrence of the random variable. If a variable has statistical properties (such as f_w) which are independent of the origin of time, then we say the process is *stationary.* Considering values of the process at distinct times, we have separate random variables and we define the covariance of the process w as

$$R_w(j,k) = \mathscr{E}(w(j) - \bar{w}(j))(w(k) - \bar{w}(k)). \tag{D.20}$$

If the process is stationary, then the covariance in (D.20) depends only on the magnitude of the difference in observation times, $k - j$, and we often will write $R_w(j,k) = R_w(k - j)$ and drop the second argument. Since a stochastic process is both random and time dependent, we can imagine averages which are computed over the time variable as well as by the expectation. For example, for a stationary process $w(n)$ we can define the mean as

$$\bar{w}(k) \stackrel{\Delta}{=} \lim_{N \to \infty} \frac{1}{2N + 1} \sum_{n=-N}^{N} w(n + k), \tag{D.21}$$

and the second-order mean or autocorrelation

$$\overline{(w(j) - \overline{w})(w(k) - \overline{w}(k))}$$
$$= \lim_{N \to \infty} \frac{1}{2N + 1} \sum_{n=-N}^{N} \{(w(n + j) - \overline{w}(j))(w(n + k - \overline{w}(k)))\}. \quad \text{(D.22)}$$

For a stationary process, the time average in (D.21) is usually equal to the distribution average, and likewise the second-order average in (D.22) is the same as the covariance in (D.20). Processes for which time averages give the same limits as distribution averages are called *ergodic*.

A very useful aid to understanding the properties of stationary stochastic processes is found by considering the response of a linear stationary system to a stationary input process. Suppose we let the input be w, a stationary scalar process with zero mean and covariance $R_w(j)$, and suppose we take the output to be $y(k)$. We let the unit-pulse response from w to y be $h(j)$. Thus from standard analysis (see Chapter 2), we have

$$y(j) = \sum_{k=-\infty}^{\infty} h(k)w(j - k), \quad \text{(D.23)}$$

and the covariance of $y(j)$ with $y(j + \ell)$ is

$$R_y(\ell) = \mathscr{E}y(j + \ell)y(j)$$
$$= \mathscr{E}\left\{ \sum_{k=-\infty}^{\infty} h(k)w(j + \ell - k) \right\}\left\{ \sum_{n=-\infty}^{\infty} h(n)w(j - n) \right\}. \quad \text{(D.24)}$$

Since the system unit pulse response, $h(k)$, is not random, both $h(k)$ and $h(n)$ may be removed from the integral implied by the $\mathscr{E}$ operation with the result

$$R_y(\ell) = \sum_{k=-\infty}^{\infty} h(k) \sum_{n=-\infty}^{\infty} h(n)\mathscr{E}w(j + \ell - k)w(j - n). \quad \text{(D.25)}$$

The expectation in (D.25) is now recognized as $R_w(\ell - k + n)$, and substituting this expression in (D.25), we find

$$R_y(\ell) = \sum_{k=-\infty}^{\infty} h(k) \sum_{n=-\infty}^{\infty} h(n)R_w(\ell - k + n). \quad \text{(D.26)}$$

Equation (D.26) is not especially enlightening, but the z-transform of it is. We proceed with several simple steps as follows:

$$\mathscr{z}\{R_y(\ell)\} = \sum_{-\infty}^{\infty} R_y(\ell)z^{-\ell}$$
$$= \sum_{\ell=-\infty}^{\infty} \sum_{k=-\infty}^{\infty} h(k) \sum_{n=-\infty}^{\infty} h(n)R_w(\ell - k + n)z^{-\ell}.$$

Exchanging the order, since $h(k)$ and $h(n)$ do not depend on ℓ, we have

$$\mathfrak{z}\{R_y(\ell)\} = \sum_{k=-\infty}^{\infty} h(k) \sum_{n=-\infty}^{\infty} h(n) \sum_{\ell=-\infty}^{\infty} R_w(\ell - k + n)z^{-\ell}.$$

Now we let $m = \ell - k + n$ in the last sum, leading to

$$\mathfrak{z}\{R_y(\ell)\} = \sum_{k=-\infty}^{\infty} h(k) \sum_{n=-\infty}^{\infty} h(n) \sum_{m=-\infty}^{\infty} R_w(m)z^{-(m+k-n)}.$$

Finally we use the fact that $z^{-(m+k-n)} = z^{-m}z^{-k}z^{n}$ and distribute these terms to the corresponding sums with the result

$$\mathfrak{z}\{R_y(\ell)\} = \sum_{k=-\infty}^{\infty} h(k)z^{-k} \sum_{n=-\infty}^{\infty} h(n)z^{n} \sum_{m=-\infty}^{\infty} R_w(m)z^{-m}$$

$$\triangleq \mathcal{S}_y(z). \tag{D.27}$$

For reasons soon to be clear, we call the z-transform of R_y the *spectrum of y* and use the symbol $\mathcal{S}_y(z)$, and similarly for w and $\mathcal{S}_w$. With these symbols and recognition that the z-transform of the unit-pulse response is the system-transfer function, $H(z)$, (D.27) becomes

$$\mathcal{S}_y(z) = H(z)H(z^{-1})\mathcal{S}_w(z). \tag{D.28}$$

To give an interpretation of (D.28) we make two observations. First note that $R_y(0) = \mathcal{E}(y^2)$ is the mean-square value or power in the y-process. By the inverse transform integral, we have

$$\overline{y^2} = R_y(0)$$

$$= \frac{1}{2\pi j} \oint \mathcal{S}_y(z) \frac{dz}{z}$$

$$= \frac{1}{2\pi j} \oint H(z)H(z^{-1})\mathcal{S}_w(z) \frac{dz}{z}. \tag{D.29}$$

Now, as a second step, we suppose that $H(z)$ is the transfer function of a very narrow bandpass filter centered at ω_0, so that $H(z)H(z^{-1})$ is $|H(e^{j\omega_0 T})|^2$ and is nearly zero except at ω_0. Then the integral in (D.29) may be approximated by assuming that $\mathcal{S}_w(z)$ is nearly constant at the value $\mathcal{S}_w(e^{j\omega_0 T})$, where $|H|^2$ is nonzero and may thus be removed from the integral. The result is

$$\overline{y^2} = \mathcal{S}_w(e^{j\omega_0 T}) \frac{1}{2\pi j} \oint H(z)H(z^{-1}) \frac{dz}{z}$$

$$= \mathcal{S}_w(e^{j\omega_0 T})K. \tag{D.30}$$

In (D.30) we have defined the integral as a constant dependent on the exact area of the narrow band characteristic of $H(z)$. But now we can give good intuitive meaning to (D.30). The mean square of the output of a very narrow band filter is proportional to the S_w. If S_w is constant for all z, we say the process is white (after the spectrum of white light which has equal intensity at all frequencies). Hence we call S_w *the power spectral density* of the w-process. Equation (D.28) is the fundamental formula for transform analysis of linear constant systems with stochastic inputs.

An alternative to transform analysis and (D.28) is transient analysis via the state variable formulation described in Chapter 6. In this case, we take the system equations to be

$$\mathbf{x}(k + 1) = \mathbf{\Phi}\mathbf{x}(k) + \mathbf{\Gamma}_1\mathbf{w}(k),$$
$$\mathbf{y}(k) = \mathbf{H}\mathbf{x}(k). \tag{D.31}$$

We assume that the system starts at $k = 0$ with the initial value

$$\mathcal{E}\mathbf{x}(0) = \mathbf{0},$$
$$\mathcal{E}\mathbf{x}(0)\mathbf{x}^T(0) = \mathbf{R}_{\mathbf{xx}}(0;0)$$
$$\overset{\Delta}{=} \mathbf{P}_{\mathbf{x}}(0),$$

and that $\mathbf{w}(k)$ is a stationary process with covariance

$$\mathcal{E}\mathbf{w}(k)\mathbf{w}^T(k + j) = 0 \qquad (j \neq 0)$$
$$= \mathbf{R}_{\mathbf{w}} \qquad (j = 0). \tag{D.32}$$

Note that by (D.27), $S_{\mathbf{w}}(z) = \mathbf{R}_{\mathbf{w}}$, a constant, and hence $\mathbf{w}$ is a white process. With these conditions, we can compute the evolution of $\mathbf{P}_{\mathbf{x}}$, the autocovariance of the state at time k. Thus

$$\mathcal{E}\mathbf{x}(k)\mathbf{x}^T(k) = \mathbf{R}_{\mathbf{xx}}(k;k)$$
$$\overset{\Delta}{=} \mathbf{P}_{\mathbf{x}}(k),$$

and

$$\mathcal{E}\mathbf{x}(k + 1)\mathbf{x}^T(k + 1) \overset{\Delta}{=} \mathbf{P}_{\mathbf{x}}(k + 1)$$
$$= \mathcal{E}\{(\mathbf{\Phi}\mathbf{x}(k) + \mathbf{\Gamma}_1\mathbf{w}(k))(\mathbf{\Phi}\mathbf{x}(k) + \mathbf{\Gamma}_1\mathbf{w}(k))^T\}$$
$$= \mathbf{\Phi}\mathcal{E}\mathbf{x}(k)\mathbf{x}^T(k)\mathbf{\Phi}^T + \mathbf{\Phi}\mathcal{E}x(k)\mathbf{w}^T(k)\mathbf{\Gamma}_1^T$$
$$+ \mathbf{\Gamma}_1\mathcal{E}\mathbf{w}(k)\mathbf{x}^T(k)\mathbf{\Phi}^T + \mathbf{\Gamma}_1\mathcal{E}\mathbf{w}(k)\mathbf{w}^T(k)\mathbf{\Gamma}_1. \tag{D.33}$$

Since the center two terms in (D.33) are zero by (D.32)[5] we reduce (D.33) to

$$\mathbf{P}_{\mathbf{x}}(k + 1) = \mathbf{\Phi}\mathbf{P}_{\mathbf{x}}(k)\mathbf{\Phi}^T + \mathbf{\Gamma}_1\mathbf{R}_{\mathbf{w}}\mathbf{\Gamma}_1^T,$$
$$\mathbf{P}_{\mathbf{x}}(0) = \text{a given matrix initial condition}. \tag{D.34}$$

Equation (D.34) is the fundamental equation for the time-domain analysis of discrete systems with stochastic inputs. Note that (D.34) represents a nonstationary

---

[5] $\mathbf{x}(k)$ is a combination of $w(0)$, $w(1)$, . . . , $w(k - 1)$, all of which are uncorrelated with $w(k)$.

situation since the covariance of $\mathbf{x}(k)$ depends on the time k of the occurrence of $\mathbf{x}$. However, if all the characteristic roots of $\boldsymbol{\Phi}$ are inside the unit circle, then the effects of the initial condition, $\mathbf{P}_x(0)$, gradually diminish, and $\mathbf{P}_x$ approaches a stationary value. This value is given by the solution to the (Lyapunov) equation

$$\mathbf{P}_x = \boldsymbol{\Phi}\mathbf{P}_x\boldsymbol{\Phi}^T + \boldsymbol{\Gamma}_1\mathbf{R}_w\boldsymbol{\Gamma}_1^T. \tag{D.35}$$

References

Ackermann, J., "Der Entwurf Linearer Regelungssysteme im Zustandsraum," *Regelungstechnik und Prozessdatenverarbeitung,* **7,** 297–300, 1972.

Anderson, B. D. O., and J. B. Moore, *Optimal Filtering,* Prentice Hall, Englewood Cliffs, N.J., 1979.

Ash, R. H., and G. R. Ash, "Numerical Computation of Root Loci Using the Newton-Raphson Technique," *IEEE Trans. on Aut. Contr.,* **AC-13,** 5, 576–582, 1968.

Åström, K. J., *Introduction to Stochastic Control Theory,* Academic Press, New York, 1970.

Åström, K. J., and T. Bohlin, "Numerical Identification of Linear Dynamical Systems from Normal Operating Records," *Theory of Self-Adaptive Control Systems,* P. Hammond, ed., Plenum Press, New York City, N.Y., 1966.

Åström, K. J., and P. E. Eykhoff, "System Identification—A Survey," *Automatica,* **7,** 123–162, 1971.

Athans, M., "Special Issue on the Linear-Quadratic-Gaussian Estimation and Control Problem," *IEEE Trans. on Aut. Contr.* **AC-16,** 6, 1971.

Bertram, J. E., "The Effect of Quantization in Sampled-Feedback Systems," *Trans. AIEE,* **77,** pt. 2, 177–182, 1958.

Blackman, R. B., *Linear Data-Smoothing and Prediction in Theory and Practice,* Addison-Wesley, Reading, Mass., 1965.

Bode, H., *Network Analysis and Feedback Amplifier Design,* D. Van Nostrand, New York City, N.Y., 1945.

Bracewell, R. N., *The Fourier Transform and Its Applications,* 2nd ed., McGraw-Hill, New York City, N.Y., 1978.

Bryson, A. E., and Y. C. Ho, *Applied Optimal Control,* Blaisdell, Waltham, Mass., 1969.

Butterworth, S. "On the Theory of Filter Amplifiers," *Wireless Engineering,* **7,** 536–541, 1930.

Cannon, R. H., Jr., *Dynamics of Physical Systems,* McGraw-Hill, New York City, N.Y., 1967.

Clark, R. N., *Introduction to Automatic Control Systems,* John Wiley, New York City, N.Y., 1962.

Daniels, R. W., *Approximation Methods for Electronic Filter Design,* McGraw-Hill, New York City, N.Y., 1974.

D'Azzo, J. J., and C. H. Houpis, *Linear Control Systems Analysis and Design,* McGraw-Hill, New York City, N.Y., 1975.

Dorf, R. C., *Modern Control Systems,* Addison-Wesley, Reading, Mass., 1980.

Evans, W. R., "Control System Synthesis by Root Locus Method," *AIEE Trans.,* **69,** pt. 2, 66–69, 1950.

Fortmann, T. E., and K. L. Hitz, *An Introduction to Linear Control Systems,* Marcel Dekker, New York City, N.Y., 1977.

Golub, G., "Numerical Methods for Solving Linear Least Squares Problems," *Numer. Math.* **7,** 206–216, 1965.

Gopinath, B., "On the Control of Linear Multiple Input-Output Systems," *Bell Sys. Tech. J.,* **50** March 1971.

Graham, D., and R. C. Lathrop, "The Synthesis of Optimum Response: Criteria and Standard Forms," *Trans. AIEE,* **72,** pt. 2, 273–288, 1953.

Hamming, R., *Numerical Methods for Scientists and Engineers,* McGraw-Hill, New York City, N.Y., 1962.

Harvey, C. A., and G. Stein, "Quadratic Weights for Asymptotic Regulator Properties," *IEEE Trans. on Aut. Contr.,* **AC-23,** 3, 1978.

Hnatek, E. R., *A User's Handbook of D/A and A/D Converters,* J. Wiley, New York City, N.Y., 1976.

Householder, A. S., *The Theory of Matrices in Numerical Analysis,* Blaisdell, Waltham, Mass., 1964.

Hurewicz, W., *Theory of Servomechanisms,* Chapter 5, Radiation Laboratory Series, Vol. 25, McGraw-Hill, New York City, N.Y., 1947.

James, H. M., N. B. Nichols, and R. S. Phillips, *Theory of Servomechanisms,* Radiation Laboratory Series, Vol. 25, McGraw-Hill, New York City, N.Y., 1947.

Jury, E. I., *Theory and Application of the z-transform Method,* J. Wiley, New York City, N.Y., 1964.

Kailath, T., *A Course in Linear System Theory,* Prentice-Hall, Englewood Cliffs, N.J., 1979.

Källström, C., "Computing EXP(A) and $\int$EXP(As)ds," *Report 7309,* Lund Institute of Technology, Division of Automatic Control, March, 1973.

Kalman, R. E. "On the General Theory of Control Systems," *Proc. First International Congress of Automatic Control,* Moscow, USSR, 1960.

Kalman, R. E., and T. S. Englar, "A User's Manual for the Automatic Synthesis Program," *NASA Contractor Report NASA CR-475,* June 1966.

Kalman, R. E., Y. C. Ho, and K. S. Narendra, "Controllability of Linear Dynamical

Systems,'' *Contributions to Differential Equations,* Vol. 1, No. 2, J. Wiley, New York City, N.Y., 1961.

Katz, P., and J. D. Powell, "Sample Rate Selection for Aircraft Digital Control," *AIAA Journal* **13** (8), March 1973.

Kendal, M. C., and A. Stuart, *The Advanced Theory of Statistics,* Vol. 2, Griffin, London, 1961.

Kuo, B. C., *Automatic Control Systems,* 3rd ed., Prentice-Hall, Englewood Cliffs, N.J., 1975.

Kwackernack, H., and R. Sivan, *Linear Optimal Control Systems,* J. Wiley, New York City, N.Y., 1972.

Luenberger, D. G., *Introduction to Linear and Nonlinear Programming,* Addison-Wesley, Reading, Mass., 1973.

Luenberger, D. G. "Observing the State of a Linear System," *IEEE Trans. Military Electr.* **MIL-8,** 1964.

Mantey, P., "Eigenvalue Sensitivity and State Variable Selection," *IEEE Trans. on Aut. Contr.* **AC-13,** 3, 1968.

Mantey, P. E., and G. F. Franklin, "Comment on 'Digital Filter Design Techniques in the Frequency Domain'," *Proc. IEEE,* **55,** 12, 2196–2197, 1967.

Mason, S. J., "Feedback Theory: Further Properties of Signal Flowgraphs," *Proc. IRE,* **44,** 970–6, 1956.

Mayr, O., *The Origins of Feedback Control,* M.I.T. Press, Cambridge, Mass., 1970.

Mendel, J. M., *Discrete Techniques of Parameter Estimation,* Marcel Dekker, New York City, N.Y., 1973.

Moler, C., and C. van Loan, "Nineteen Dubious Ways to Compute the Exponential of a Matrix," *SIAM Review,* **20,** 4, 1978.

Morrison, R. L., "Microcomputers Invade the Linear World," *IEEE Spectrum,* **15,** 7, 38–41, 1978.

Nyquist, H. "Regeneration Theory," *Bell System Technical Journal,* **11,** 126–147, 1932.

Ogata, K., *Modern Control Engineering,* Prentice-Hall, Englewood Cliffs, N.J., 1970.

Parzen, E., *Stochastic Processes,* Holden-Day, San Francisco, 1962.

Peled, U., and J. D. Powell, "The Effect of Prefilter Design on Sample Rate Selection in Digital Control Systems," *Proceedings of AIAA G&C Conference, Paper no. 78–1308,* August 1978.

Rabiner, L. R., and C. M. Rader, eds., *Digital Signal Processing,* IEEE Press, New York City, N.Y., 1972.

Ragazzini, J. R., and G. F. Franklin, *Sampled-Data Control Systems,* McGraw-Hill, New York City, N.Y., 1958.

Rosenbrock, H. H., *State Space and Multivariable Theory,* John Wiley, New York City, N.Y., 1970.

Rosenbrock, H. H., and C. Storey, *Mathematics of Dynamical Systems,* J. Wiley, New York City, N.Y., 1970.

Saucedo, R., and E. E. Shiring, *Introduction to Continuous and Digital Control Systems,* Macmillan, New York City, N.Y., 1968.

Schmidt, L. A., "Designing Programmable Digital Filters for LSI Implementation," *Hewlett-Packard Journal,* **29,** 13, 1978.

Scientific American, *Automatic Control,* Simon and Schuster, New York City, N.Y., 1955.

Slaughter, J., "Quantization Errors in Digital Control Systems," *IEEE Trans. Aut. Contr.,* **AC-9,** 70–74, 1964.

Söderström, T., L. Ljung, and I. Gustavsson, "A Comparative Study of Recursive Identification Methods," *Lund Report 7427,* Dec. 1974.

Strang, G., *Linear Algebra and Its Applications,* Academic Press, New York City, N.Y., 1976.

Terman, F. E., *Radio Engineering,* McGraw-Hill, New York City, N.Y., 1932.

Truxal, J. G., *Automatic Feedback Control System Synthesis,* McGraw-Hill, New York City, N.Y., 1955.

Tustin, A., "A Method of Analyzing the Behavior of Linear Systems in Terms of Time Series," *JIEE* (London), **94,** pt. IIA, 130–142, 1947.

Vaughn, D. R., "A Nonrecursive Algebraic Solution for the Discrete Riccati Equation," *IEEE Trans. Aut. Contr.* **AC-15,** 5, Oct. 1970.

Widrow, B., "A Study of Rough Amplitude Quantization by Means of Nyquist Sampling Theory," *IRE Trans. on Circuit Theory,* **CT-3,** 4, 266–276, 1956.

Wiener, N., *Interpolation, Extrapolation and Smoothing of Stationary Time Series,* J. Wiley, New York City, N.Y., 1948.

Wilde, D. J., and C. S. Beightler, *Foundations of Optimization,* Prentice-Hall, Englewood Cliffs, N.J., 1967.

Woodson, H. H., and J. R. Melcher, *Electromechanical Dynamics: Part I, Discrete Systems,* J. Wiley, New York City, N.Y., 1968.

Index